Structured Light and Its Applications: An Introduction to Phase-Structured Beams and Nanoscale Optical Forces

Structured Light and Its Applications: An Introduction to Phase-Structured Beams and Nanoscale Optical Forces

DAVID L. ANDREWS

University of East Anglia, UK

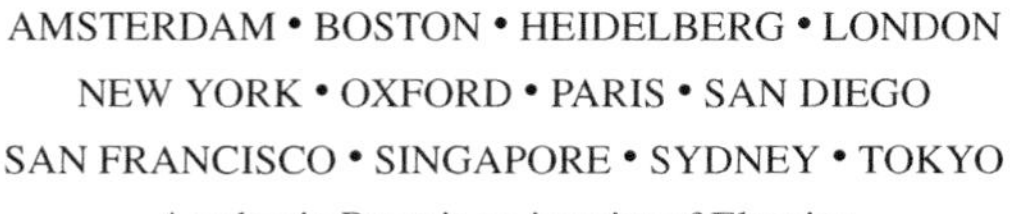

Academic Press is an imprint of Elsevier
30 Corporate Drive, Suite 400, Burlington, MA 01803, USA
525 B Street, Suite 1900, San Diego, California 92101-4495, USA
84 Theobald's Road, London WC1X 8RR, UK

This book is printed on acid-free paper. ♾

Design Direction: Eric De Cicco

Cover Design: Dutton & Sherman Design

Library of Congress Cataloging-in-Publication Data
Application submitted

British Library Cataloguing-in-Publication Data
A catalogue record for this book is available from the British Library.

ISBN: 978-0-12-374027-4

For information on all Academic Press publications
visit our Web site at www.books.elsevier.com

Printed and bound by CPI Group (UK) Ltd, Croydon, CR0 4YY

Transferred to Digital Print 2011

Contents

Author Affiliations xi
Preface xiii

1 Introduction to Phase-Structured Electromagnetic Waves 1
Les Allen and Miles Padgett

1.1 Introduction 1
1.2 Laguerre–Gaussian Beams and Orbital Angular Momentum 2
1.3 Bessel and Mathieu Beams 7
1.4 General Solution of the Wave Equation 8
1.5 Classical or Quantum? 8
1.6 Creating Laguerre–Gaussian Beams with Lenses and Holograms 9
1.7 Coherence: Spatial and Temporal 11
1.8 Transformations Between Basis Sets 12
1.9 Conclusion 14
References 15

2 Angular Momentum and Vortices in Optics 19
Gerard Nienhuis

2.1 Introduction 19
2.2 Classical Angular Momentum of Fields and Particles 22
2.2.1 Angular Momentum of Particles and Radiation 22
2.2.2 Rate of Change of Contributions to Angular Momentum 24
2.3 Separation of Radiative Angular Momentum in **L** and **S** 24
2.3.1 Classical Description 24
2.3.2 Quantum Operators 25
2.4 Multipole Fields and Their Vortex Structure 27
2.4.1 Spherical Multipole Fields 27
2.4.2 Cylindrical Multipole Fields 30
2.5 Angular Momentum of Monochromatic Paraxial Beams 33
2.5.1 Paraxial Approximation 33
2.5.2 Angular Momentum of a Monochromatic Beam 34
2.5.3 Uniform Orbital and Spin Angular Momentum 36
2.5.4 Nonuniform Polarization 38

2.6 Quantum Description of Paraxial Beams 40
2.6.1 Quantum Operators for Paraxial Fields 40
2.6.2 Quantum Operators for Spin and Orbital Angular Momentum 41
2.7 Nonmonochromatic Paraxial Beam 42
2.7.1 Angular Momentum of Nonmonochromatic Beam 42
2.7.2 Spin of Rotating Polarization 43
2.7.3 Orbital Angular Momentum of Rotating Mode Pattern 44
2.7.4 Angular Momentum of Rotating Nonuniform Polarization 46
2.8 Operator Description of Classical Paraxial Beams 48
2.8.1 Dirac Notation of Paraxial Beams 48
2.8.2 Paraxial Beams and Quantum Harmonic Oscillators 49
2.8.3 Raising and Lowering Operators for Modes 51
2.8.4 Orbital Angular Momentum and the Hermite–Laguerre Sphere 53
2.9 Dynamics of Optical Vortices 55
2.9.1 Invariant Mode Patterns 55
2.9.2 Rotating Patterns of Vortices with Same Orientation 57
2.9.3 Vortex Creation and Annihilation 57
2.10 Conclusion 59
References 60

3 Singular Optics and Phase Properties 63
Enrique J. Galvez

3.1 Fundamental Phase Singularities 64
3.2 Beams with Composite Vortices 69
3.3 Noninteger Vortex Beams 72
3.4 Propagation Dynamics 74
3.5 Conclusions 74
Acknowledgments 75
References 75

4 Nanoscale Optics: Interparticle Forces 79
Luciana C. Dávila Romero and David L. Andrews

4.1 Introduction 79
4.2 QED Description of Optically Induced Pair Forces 82
4.2.1 Quantum Foundations 82
4.2.2 Defining the Geometry 85
4.2.3 Tumbling Cylindrical Pair 87
4.2.4 Collinear Pair 90
4.2.5 Cylindrical Parallel Pair 92

4.2.6 Spherical Particles 94
4.2.7 Spherical Particles in a Laguerre–Gaussian Beam 96
4.3 Overview of Applications 98
4.4 Discussion 101
Acknowledgments 102
References 102

5 Near-Field Optical Micromanipulation 107
Kishan Dholakia and Peter J. Reece

5.1 Introduction 107
5.1.1 What Is the Near Field? 108
5.1.2 Optical Geometries for the Near Field and Initial Guiding Studies 109
5.2 Theoretical Considerations for Near-Field Trapping 111
5.3 Experimental Guiding and Trapping of Particles in the Near Field 114
5.3.1 Near-Field Surface Guiding and Trapping 114
5.3.2 Trapping Using TIR Objectives 122
5.3.3 Micromanipulation Using Optical Waveguides 126
5.4 Emergent Themes in the Near Field 129
5.4.1 Optical Force Induced Self-Organization of Particles in the Near Field 129
5.4.2 Near-Field Trapping with Advanced Photonic Architectures 132
5.5 Conclusions 134
Acknowledgments 134
References 134

6 Holographic Optical Tweezers 139
Gabriel C. Spalding, Johannes Courtial, and Roberto Di Leonardo

6.1 Background 139
6.2 *Example* Rationale for Constructing Extended Arrays of Traps 140
6.3 Experimental Details 142
6.3.1 The Standard Optical Train 142
6.4 Algorithms for Holographic Optical Traps 149
6.4.1 Random Mask Encoding 151
6.4.2 Superposition Algorithms 152
6.4.3 Gerchberg–Saxton Algorithms 153
6.4.4 Direct-Search Algorithm and Simulated Annealing 156
6.4.5 Summary 156
6.4.6 Alternative Means of Creating Extended Optical Potential Energy Landscapes 157

6.5 The Future of Holographic Optical Tweezers 162
Acknowledgments 162
References 162

7 Atomic and Molecular Manipulation Using Structured Light 169
Mohamed Babiker and David L. Andrews

7.1 Introduction 169
7.2 A Brief Overview 170
7.3 Transfer of OAM to Atoms and Molecules 171
7.4 Doppler Forces and Torques 172
7.4.1 Essential Formalism 173
7.4.2 Transient Dynamics 175
7.4.3 Steady State Dynamics 178
7.4.4 Dipole Potential 179
7.5 The Doppler Shift 180
7.5.1 Trajectories 181
7.5.2 Multiple Beams 181
7.5.3 Two- and Three-Dimensional Molasses 184
7.6 Rotational Effects on Liquid Crystals 185
7.7 Comments and Conclusions 187
Acknowledgments 191
References 191

8 Optical Vortex Trapping and the Dynamics of Particle Rotation 195
Timo A. Nieminen, Simon Parkin, Theodor Asavei, Vincent L.Y. Loke, Norman R. Heckenberg, and Halina Rubinsztein-Dunlop

8.1 Introduction 195
8.2 Computational Electromagnetic Modeling of Optical Trapping 196
8.3 Electromagnetic Angular Momentum 199
8.4 Electromagnetic Angular Momentum of Paraxial and Nonparaxial Optical Vortices 202
8.5 Nonparaxial Optical Vortices 205
8.6 Trapping in Vortex Beams 211
8.7 Symmetry and Optical Torque 218
8.8 Zero Angular Momentum Optical Vortices 226
8.9 Gaussian "Longitudinal" Optical Vortex 228
8.10 Conclusion 231
References 231

9 Rotation of Particles in Optical Tweezers **237**
Miles Padgett and Jonathan Leach

9.1 Introduction 237
9.2 Using Intensity Shaped Beams to Orient and Rotate Trapped Objects 238
9.3 Angular Momentum Transfer to Particles Held in Optical Tweezers 240
9.4 Out of Plane Rotation in Optical Tweezers 242
9.5 Rotation of Helically Shaped Particles in Optical Tweezers 243
9.6 Applications of Rotational Control in Optical Tweezers 244
References 247

10 Rheological and Viscometric Methods **249**
Simon J.W. Parkin, Gregor Knöner, Timo A. Nieminen, Norman R. Heckenberg, and Halina Rubinsztein-Dunlop

10.1 Introduction 249
10.2 Optical Torque Measurement 251
10.2.1 Measuring Spin Angular Momentum 251
10.2.2 Measuring Orbital Angular Momentum 253
10.3 A Rotating Optical Tweezers Based Microviscometer 254
10.3.1 Experimental Setup for a Spin Based Microviscometer 255
10.3.2 Results and Analysis 256
10.3.3 Orbital Angular Momentum Used for Microviscometry 261
10.4 Applications 264
10.4.1 Picolitre Viscometry 264
10.4.2 Medical Samples 265
10.4.3 Flow Field Measurements 266
10.5 Conclusion 268
References 268

11 Orbital Angular Momentum in Quantum Communication and Information **271**
Sonja Franke-Arnold and John Jeffers

11.1 Sending and Receiving Quantum Information 273
11.1.1 Generation of Entangled OAM States 275
11.1.2 Detection of OAM States at the Single Photon Level 277
11.1.3 Intrinsic Security 279
11.2 Exploring the OAM State Space 280
11.2.1 Superpositions of OAM States 280
11.2.2 Generating Entangled Superposition States 283

11.2.3 Storing OAM Information 284
11.3 Quantum Protocols 286
11.3.1 Advantages of Higher Dimensions 286
11.3.2 Communication Schemes 287
11.4 Conclusions and Outlook 290
Acknowledgments 291
References 291

12 Optical Manipulation of Ultracold Atoms **295**
G. Juzeliūnas and P. Öhberg

12.1 Background 295
12.2 Optical Forces and Atom Traps 296
12.3 The Quantum Gas: Bose–Einstein Condensates 299
12.3.1 Bose–Einstein Condensation in a Cloud of Atoms 300
12.3.2 The Condensate and Its Description 301
12.3.3 Phase Imprinting the Quantum Gas 303
12.4 Light-Induced Gauge Potentials for Cold Atoms 308
12.4.1 Background 308
12.4.2 General Formalism for the Adiabatic Motion of Atoms in Light Fields 309
12.5 Light-Induced Gauge Potentials for the Λ Scheme 311
12.5.1 General 311
12.5.2 Adiabatic Condition 313
12.5.3 Effective Vector and Trapping Potentials 314
12.5.4 Co-Propagating Beams with Orbital Angular Momentum 315
12.5.5 Counterpropagating Beams with Shifted Transverse Profiles 317
12.6 Light-Induced Gauge Fields for a Tripod Scheme 320
12.6.1 General 320
12.6.2 The Case where $S_{12} = 0$ 322
12.7 Ultra-Relativistic Behavior of Cold Atoms in Light-Induced Gauge Potentials 323
12.7.1 Introduction 323
12.7.2 Formulation 324
12.7.3 Quasi-Relativistic Behavior of Cold Atoms 325
12.7.4 Proposed Experiment 327
12.8 Final Remarks 329
References 330

Index **335**

Author Affiliations

Les Allen	Universities of Glasgow and Strathclyde, UK
David L. Andrews	University of East Anglia, UK
Theodor Asavei	University of Queensland, Australia
Mohamed Babiker	University of York, UK
Johannes Courtial	University of Glasgow, UK
Luciana C. Dávila Romero	University of East Anglia, UK
Kishan Dholakia	University of St. Andrews, UK
Sonja Franke-Arnold	University of Glasgow, UK
Roberto Di Leonardo	Università di Roma, Italy
Enrique J. Galvez	Colgate University, USA
Norman R. Heckenberg	University of Queensland, Australia
John Jeffers	University of Strathclyde, UK
Gediminas Juzeliūnas	Institute of Theoretical Physics and Astronomy of Vilnius University, Lithuania
Gregor Knöner	University of Queensland, Australia
Jonathan Leach	University of Glasgow, UK
Vincent L.Y. Loke	University of Queensland, Australia
Timo A. Nieminen	University of Queensland, Australia
Gerard Nienhuis	Universiteit Leiden, The Netherlands
Patrick Öhberg	Heriot-Watt University, UK
Miles Padgett	University of Glasgow, UK
Simon Parkin	University of Queensland, Australia
Peter J. Reece	University of St. Andrews, UK
Halina Rubinsztein-Dunlop	University of Queensland, Australia
Gabriel C. Spalding	Illinois Wesleyan University, USA

Preface

It is difficult to over state the profoundly unexpected character of the properties that have recently emerged in connection with optical beams having complex structures, or indeed the astonishingly wide range of prospective applications for the noncontact optical manipulation of matter. Extending well beyond the established methods of focused laser trapping and tweezers, many of these methods offer new opportunities for subwavelength resolution nanooptics, with potential applications extending from biological cell handling, through microfluidics, nanofabrication and photonics, to laser cooling, atom trapping, quantum informatics and the control of Bose–Einstein condensates. Many of these new methods exploit the distinctive angular momentum, nodal architecture, and phase properties of beams with a complex wavefront structure.

The interest in these topics is, however, by no means limited to technical applications. This is an area in which theory and experiment have a particularly vigorous dynamic: theory is constantly informing and suggesting new experiments, while experimental results challenge and invite new theory. Just as the technology first advanced through the application of fundamental theory, so too have the results necessitated some significant shifts and developments in our understanding of photon properties. The shock of discovering that photon angular momentum is not limited to intrinsic spin has not yet entirely abated. To quote Rodney Loudon, writing in another context in the millennium edition of his *Quantum Theory of Light*, "It is no longer so straightforward to explain what is meant by a 'photon'."

Research on structured light and optical forces has cultivated a new, vigorous and distinctive interdisciplinary area, with strong interest and activity worldwide. It is the general objective of this volume to introduce and elucidate the key developments, and to exhibit their growth and interplay. From its very inception, it was my delight to receive enthusiastic agreements to contribute from many of the leading experts in the field. I believe that the comprehensive and definitive work they have produced not only serves as a benchmark of the subject matter—it also conveys in ringing tones the excitement of those involved in the field. My indebtedness to them all will be evident, and they have my very sincere thanks.

David L. Andrews
Norwich, October 2007

Chapter 1

Introduction to Phase-Structured Electromagnetic Waves

Les Allen[1] *and Miles Padgett*[2]

[1]*Universities of Glasgow and Strathclyde, UK*
[2]*University of Glasgow, UK*

1.1 INTRODUCTION

A cursory inspection of the indices of the classic books on electromagnetic theory or optics shows that, in free space, propagation is generally investigated only for spherical waves and the mythological plane wave. The fields of the propagating wave discussed in such books have, in general, no detailed phase or amplitude structure. In the early days of lasers, a beam was specified solely in terms of its polarization and intensity profiles. Usually the profile of the laser was described in terms of the rectangular symmetric Hermite–Gaussian modes of amplitude given by Allen and colleagues [1]

$$\begin{aligned} u_{nm}^{\mathrm{HG}}(x,y,z) = {} & \left(\frac{2}{\pi n!m!}\right)^{1/2} 2^{-N/2}(1/w)\exp\left[-ik\left(x^2+y^2\right)/2R\right] \\ & \times \exp\left[-\left(x^2+y^2\right)/w^2\right]\exp\left[-i(n+m+1)\psi\right] \\ & \times H_n\left(x\sqrt{2}/w\right)H_m\left(y\sqrt{2}/w\right), \end{aligned} \tag{1}$$

where $H_n(x)$ is the Hermite polynomial of order n. Inspection reveals that each maximum of equation (1) is π out of phase with its nearest neighbor, and it is due to this phase discontinuity that the lobes remain separated upon propagation of the beam. To

STRUCTURED LIGHT AND ITS APPLICATIONS

this phase structure is added that of the nearly spherical phase-fronts associated with beam divergence or focusing.

In 1990, Tamm and Weiss [43] investigated the TEM^*_{01} hybrid mode created by stable oscillation between frequency degenerate TEM_{10} and TEM_{01} modes and found a helical structure specified by an azimuthal phase. Subsequently, Harris and colleagues [24] produced similar laser beams with distributions having a continuous cophasal surface of helical form. The structures of these beams were displayed by characteristic spiral fringes, when interfered with their mirror image or a plane wave. The exact form of these fringes and how they change with propagation may be readily calculated (Soskin and colleagues [41]). The phase singularity on the beam axis is an example of an optical vortex, present throughout natural light fields and studied extensively from the 1970s onwards by Nye and Berry [36].

1.2 LAGUERRE–GAUSSIAN BEAMS AND ORBITAL ANGULAR MOMENTUM

The study of azimuthally phased beams and optical vortices took on an entirely new form after the realization by Allen and colleagues [1] that such beams carried an orbital angular momentum. It had been known for many decades that photons could carry an orbital angular momentum. It had been well known in nuclear multipole radiation, where it accounted for the conservation of angular momentum in decay processes (Blatt and Weisskopf [47]). However, it had not been appreciated that light beams readily created in a laboratory could also carry a well-defined orbital angular momentum. It transpires that one can produce Laguerre–Gaussian modes as well as Hermite–Gaussian, as shown in Figures 1.1 and 1.2. The helically phased beams generated by Tamm and Weiss and by Harris are related to Laguerre–Gaussian modes, which comprise concentric rings and, more importantly, have an associated azimuthal phase term.

$$\begin{aligned} u_{nm}^{\mathrm{LG}}(r,\varphi,z) = {} & \left(\frac{2}{\pi n! m!}\right)^{1/2} \min(n,m)!(1/w)\exp\left[-ikr^2/2R\right] \\ & \times \exp\left[-\left(r^2/w^2\right)\right]\exp\left[-i(n+m+1)\psi\right] \\ & \times \exp[-il\varphi](-1)^{\min(n,m)}\left(r\sqrt{2}/w\right)L_{\min(n,m)}^{n-m}\left(2r^2/w^2\right), \end{aligned} \tag{2}$$

where $L_p^l(x)$ is a generalized Laguerre polynomial. This amplitude distribution, and that of the Hermite–Gaussian modes given previously, are solutions to the paraxial form of the wave equation. The latter is derived from the scalar wave equation, which for a monochromatic wave, $\psi(x,y,z,t)\chi(x,y,z)\exp(-iwt)$, reduces to the

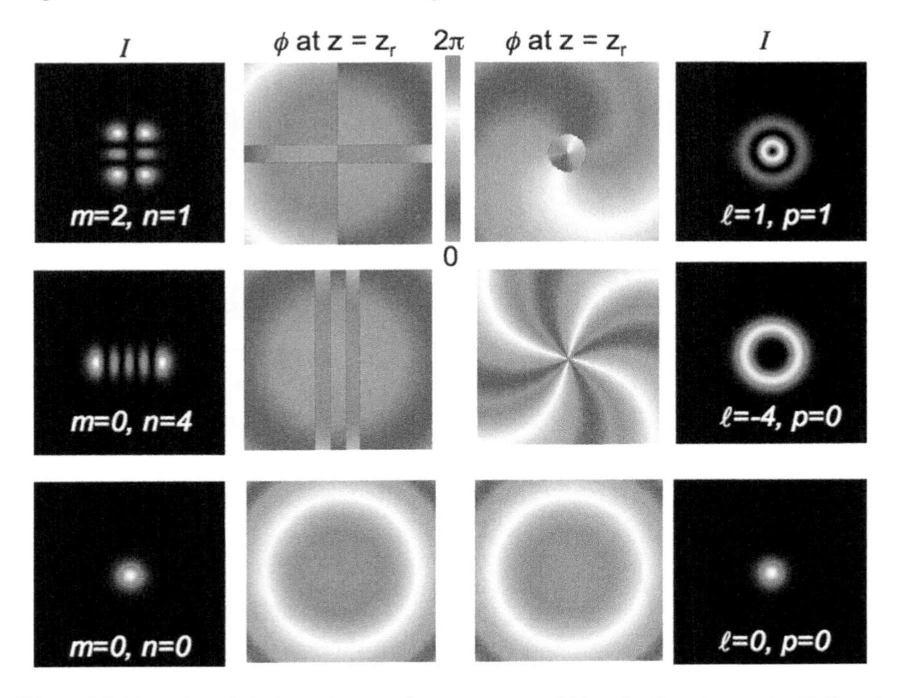

Figure 1.1 Examples of the intensity and phase structures of Hermite–Gaussian modes (*left*) and Laguerre–Gaussian modes (*right*), plotted at a distance from the beam waist equal to the Rayleigh range. See color insert.

Helmholtz equation for the space dependence

$$(\nabla^2 + k^2)\chi(x, y, z) = 0. \tag{3}$$

For a beam where $\chi(x, y, z) = u(x, y, z)\exp(ikz)$, which spreads only very slowly, this gives

$$i\frac{\partial u}{\partial z} = -\frac{1}{2k}\left(\frac{\partial^2}{\partial x^2} + \frac{\partial^2}{\partial y^2}\right)u. \tag{4}$$

This paraxial form of the wave equation is identical to the Schrödinger equation with t replaced by z. As the operator of the z-component of orbital angular momentum can be represented by $L_z = -i\hbar\partial/\partial\varphi$, the identification of orbital angular momentum is strongly suggested by analogy between paraxial optics and quantum mechanics (Allen and colleagues [1], Marcuse [48]).

It is easy to understand why orbital angular momentum is present, although why it is quantized in integer multiples of $\hbar$ is not so obvious. In any light beam, the wave-

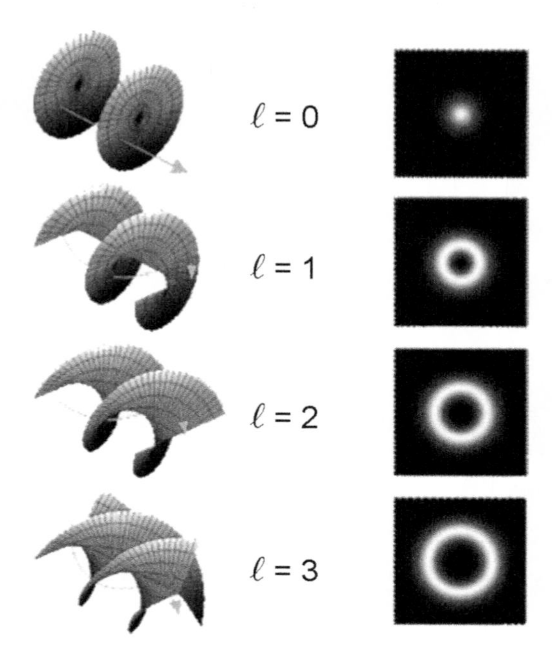

Figure 1.2 Helical phase fronts corresponding to the $\exp(i\ell\phi)$ phase structure, and corresponding intensity profiles of Laguerre–Gaussian modes. See color insert.

vector can be defined at every transverse position of the beam as being perpendicular to the phase fronts and represents the direction of the linear momentum flow within the beam of $\hbar k$ per photon. For a helically phased wave, the skew angle of the wave vector with respect to the optical axis may be shown to be simply ℓ/kr (Padgett and Allen [38]). This gives an azimuthal component to the linear momentum flow of $\ell\hbar/r$, and hence an angular momentum of $\ell\hbar$ per photon. It follows that for a beam of maximum radius R, the largest value of ℓ possible is $\ell \leqslant kR$.

Unlike spin angular momentum that has only two independent states, spin up or spin down, orbital angular momentum has an unlimited number of orthogonal states, corresponding to the integer values of ℓ. Although the link between spin angular momentum and circular polarization is well established, the link between orbital angular momentum and certain aspects of the structure of the beam is not so obvious. It is tempting, for instance, to directly associate the orbital angular momentum to the ℓ-value of any optical vortex, but this is not correct. The center of the vortex is a position of zero intensity, and consequently carries no momentum or energy. The angular momentum is rather, associated with regions of high intensity, a fact best illustrated in [34] by Molina-Terriza and colleagues, who showed that after the beam passes through the focus of a cylindrical lens, the azimuthal component of the linear momentum near the vortex center is reversed, while the total orbital angular momentum of the beam remains the same (Padgett and Allen [39]). The reversal of the vortex is simply image inversion in geometrical optics and has no angular momentum implications.

As already mentioned, orbital angular momentum arises whenever a beam has phase fronts inclined with respect to the propagation direction, and it is possible to create vortex-free beams that contain a well-defined, although unquantized, orbital angular momentum [15]. Within a geometric optics approximation, this is equivalent to the light rays that make up the beam being skewed with respect to the axis. Although seemingly simplistic, this skew-ray model predicts the correct result in most experimental situations [16,13,44]. It is worth re-emphasizing that the light's orbital angular momentum simply arises from the azimuthal components of its linear momentum acting at a radius from the beam axis, although this is very often integrated over the whole beam.

More formally, on the cycle-averaged linear momentum density, **p**, and the angular momentum density, **j**, of a light beam may be calculated from the electric field, **E**, and the magnetic field, **B** (Jackson, [49])

$$\mathbf{p} = \varepsilon_0 \langle \mathbf{E} \times \mathbf{B} \rangle, \tag{5}$$

$$\mathbf{j} = \varepsilon_0 \big(\mathbf{r} \times \langle \mathbf{E} \times \mathbf{B} \rangle \big) = (\mathbf{r} \times \mathbf{p}). \tag{6}$$

The total linear and angular momentum per unit length may be found by integrating these expressions over the area of the beam. In this classical treatment, equation (6) encompasses both the spin and orbital angular momentum of a light beam. The spin angular momentum of light is, of course, very often expressed as resulting from the spin of individual photons, but the approaches are equivalent. Barnett [9] has shown that the angular momentum flux for a light beam can be separated into spin and orbital parts. Unlike the angular momentum density approach, the separation is gauge invariant and does not rely on the paraxial approximation.

It is clear from equation (2), that for there to be angular momentum j_z in propagation direction z, the light field must have linear momentum in the azimuthal direction. This requires a z-component to the electromagnetic field. As emphasized by Simmons and Guttman [50], an idealized plane wave of infinite extent, which has only transverse fields, does not carry angular momentum, irrespective of its degree of polarization. However, for beams of finite extent, a z-component of the electromagnetic field can arise in two distinct ways. Because the electric field is always perpendicular to the inclined phase-front, a beam with helical phase-fronts has an oscillating component in the propagation direction. Less obviously, it transpires that a circularly polarized beam, where $\mathbf{E}_x$ is in quadrature with $\mathbf{E}_y$, has an electric field in the propagation direction proportional to the radial intensity gradient. This is, of course, consistent with the argument with respect to the notional plane wave, as its gradient is then zero.

Within the paraxial approximation, the local value of the linear momentum density of a light beam is given by

$$\begin{aligned}\mathbf{p} &= \frac{\varepsilon_0}{2}(\mathbf{E}^* \times \mathbf{B} + \mathbf{E} \times \mathbf{B}^*) \\ &= i\omega\frac{\varepsilon_0}{2}(u^*\nabla u - u\nabla u^*) + \omega k \varepsilon_0 |u|^2 \mathbf{z} + \omega\sigma\frac{\varepsilon_0}{2}\frac{\partial |u|^2}{\partial r}\mathbf{\Phi},\end{aligned} \tag{7}$$

where $u \equiv u(r, \phi, z)$ is the complex scalar function describing the distribution of the field amplitude. Here σ describes the degree of polarization of the light: $\sigma = \pm 1$ for right- and left-hand circularly polarized light, and $\sigma = 0$ for linearly polarized light. The spin angular momentum density component arising from polarization is encompassed in the final term and depends upon σ and on the intensity gradient. The cross-product of this momentum density with the radius vector $\mathbf{r} \equiv (r, 0, z)$ yields an angular momentum density. The angular momentum density in the z-direction depends upon the $\mathbf{\Phi}$ component of $\mathbf{p}$, such that

$$j_z = r p_\phi. \tag{8}$$

For many mode functions, u, such as for circularly polarized Laguerre–Gaussian modes, equations (7) and (8) can be evaluated analytically and integrated over the whole beam. The angular momentum in the propagation direction is readily shown to be equivalent to $\sigma\hbar$ per photon for the spin and $l\hbar$ per photon for the orbital angular momentum (Allen and colleagues [5]). A theoretical discussion of the behavior of local momentum density has been published elsewhere (Allen and Padgett [3]), and it should be noted that the local spin and orbital angular momentum by no means behave in the same way.

The orbital angular momentum arises for all light beams from the ϕ-component of linear momentum about a radius vector. If this value changes as the axis about which it is calculated shifts, the angular momentum is said to be *extrinsic*. For a Laguerre–Gaussian beam the local value of $\mathbf{p}$ is found to change when equations (7) and (8) are evaluated about a displaced axis. However, the total spin angular momentum integrated over the whole beam remains constant. As is well known, the spin angular momentum never depends upon a choice of axis, and so is intrinsic. Berry [51] showed that for certain circumstances the orbital angular momentum does not depend upon the axis of the beam and so may also be called intrinsic. If the radius vector $\mathbf{r}$ in the expression for the angular momentum for the whole beam

$$\mathbf{J}_z = \varepsilon_0 \iint dx\, dy\, \mathbf{r} \times \langle \mathbf{E} \times \mathbf{B} \rangle \tag{9}$$

is shifted by a constant amount $\mathbf{r}_0 \equiv (r_{x0}, r_{y0}, r_{z0})$, it is easy to show that the change in the z-component of angular momentum is given by

$$\Delta J_z = (r_{x0} \times P_y) + (r_{y0} \times P_x)$$
$$= r_{x0}\varepsilon_0 \iint dx\,dy\,\langle \mathbf{E} \times \mathbf{B} \rangle_y + r_{y0}\varepsilon_0 \iint dx\,dy\,\langle \mathbf{E} \times \mathbf{B} \rangle_x. \tag{10}$$

The component r_{z0} plays no role because it has no cross-product in the z-direction. The change in angular momentum, ΔJ_z, trivially equals zero if $r_{0x} = r_{0y} = 0$, and the evaluation is with respect to the beam axis. More importantly, there is also no change in the angular momentum if z is stipulated as the direction for which the transverse momentum currents $\varepsilon_0 \iint dx\,dy\,\langle \mathbf{E} \times \mathbf{B} \rangle_x$ and $\varepsilon_0 \iint dx\,dy\,\langle \mathbf{E} \times \mathbf{B} \rangle_y$ are exactly zero. We may therefore choose to call the orbital angular momentum *quasi-intrinsic* (O'Neil and colleagues [37]).

1.3 BESSEL AND MATHIEU BEAMS

Although our emphasis has been on Laguerre–Gaussian beams, they are not the only beams possessing a phase structure that carries orbital angular momentum. Any beam with a helical phase structure $\exp(i\ell\phi)$ carries orbital angular momentum $\ell\hbar$ per photon. Durnin's description of Bessel beams as diffraction-free [17,18], has led to a topic of considerable research interest (McGloin and Dholakia [31]). Durnin realized that in the Helmoltz equation solution,

$$u(x, y, z) = \exp(ik_z z) \int_0^{2\pi} A(\varphi) \exp\bigl(ik_r(x\cos\varphi + y\sin\varphi)\bigr)\, d\varphi, \tag{11}$$

the intensity $|u(x, y, z)|^2$ is independent of z. For the case of $A(\varphi)$ constant, he found that

$$u(r, z) \propto \exp(ik_z z) J_0(k_r r), \tag{12}$$

where J_0 is a Bessel function of the first kind. A beam with this distribution of amplitude and phase is known as a *zero-order Bessel beam*. Phase functions of the form $A(\phi) \propto \exp(il\varphi)$ lead to higher-order Bessel beams with amplitude

$$u(r, \varphi, z) \propto \exp(ik_z z) \exp(il\varphi) J_m(k_r r). \tag{13}$$

More recently, closed forms for $u(x, y, z)$ and $A(\varphi)$ have been found for families of elliptical Bessel generalizations known as Mathieu beams. Bessel beams are circularly symmetric, while Mathieu beams are characterized by an ellipticity parameter. Both types can possess orbital angular momentum [23,14,30].

1.4 GENERAL SOLUTION OF THE WAVE EQUATION

In this chapter, the treatment of electromagnetic waves refers only to solutions of the Helmholtz equation. This assumes that the laser fields of interest are compatible with the paraxial approximation. In an attempt to see what aspects, if any, of optical angular momentum might be artefacts of this approximation, Barnett and Allen [10] found the exact solution of the full Maxwell theory for the general form for a monochromatic beam propagating in the z-direction for which the x- and y-components of the electric field have azimuthal dependence $\exp(i\ell\phi)$. The electric field may be written

$$\mathbf{E} = (\alpha\mathbf{x} + \beta\mathbf{y})E(z,\rho)\exp(il\varphi) + \mathbf{z}E_x, \tag{14}$$

where each component satisfies the Helmholtz equation, and E_z has to be chosen to ensure the transversality of the electric field. The general solution for $E(z,\rho)$ has the form

$$E(z,\rho) = \int_0^k d\kappa\, E(\kappa)J_l(\kappa\rho)\exp\bigl(i\sqrt{k^2-\kappa^2}z\bigr). \tag{15}$$

It is found that the electric field of the beam has the form

$$\begin{aligned}\mathbf{E} = {} & \int_0^k d\kappa\, E(\kappa)\exp(il\varphi)\exp\bigl(i\sqrt{k^2-\kappa^2}z\bigr) \\ & \times \left((\alpha\mathbf{x}+\beta\mathbf{y})J_l(\kappa\rho) + \mathbf{z}\frac{\kappa}{\sqrt{k^2-\kappa^2}}\bigl[(i\alpha-\beta)\exp(-i\varphi)J_{l-1}(\kappa\rho)\right. \\ & \left. -(i\alpha+\beta)\exp(i\varphi)J_{l+1}(\kappa\rho)\bigr]\right),\end{aligned} \tag{16}$$

where we note that the Bessel function $J_l(\kappa\rho)$ of order ℓ appears naturally. It is perhaps, therefore, not surprising that the Bessel beam is an allowed solution in the paraxial approximation. In principle, it should be possible to derive all other field distributions arising from a variety of different approximations from the general solution, although in practice it is not easy to do so.

1.5 CLASSICAL OR QUANTUM?

The implication of an integer orbital angular momentum per photon is that it is a quantum property. However, we have so far considered the angular momentum properties of the helically phased beam purely in classical terms. The angular momentum per photon arises from the ratio of the orbital angular momentum to energy being equal to ℓ/ω. Multiplication of the top and bottom of this ratio by $\hbar$ gives the orbital angular momentum per photon. However, the orbital angular momentum is a direct consequence of the mode structure of the beam, and hence it is reasonable to expect that the orbital angular momentum properties of the beam are also present at the single photon level. It has been shown that for three-wave interactions, such as frequency-doubling, the orbital angular momentum is conserved within the optical fields (Dholakia and colleagues [16]). However, from a quantum perspective, it is parametric down-conversion that holds greater interest. Down-converted photon pairs possessing orbital angular momentum are known to be entangled (Mair and colleagues [32]). This entanglement arises from the complex overlap of the three spatial modes within the nonlinear crystal (Franke-Arnold and colleagues [19]). These experiments were analogous to those of Aspect and colleagues [8] for polarized light and, hence, entanglement of spin angular momentum.

A method of describing the propagation of light beams by means of operators has been developed by van Enk and Nienhuis [45]. They showed that the Gouy phase of a Gaussian beam is equal to the dynamic phase of a quantum mechanical harmonic oscillator with time-dependent energy. It follows that l is the eigenvalue of the angular momentum operator and so represents the quantized orbital angular momentum.

1.6 CREATING LAGUERRE–GAUSSIAN BEAMS WITH LENSES AND HOLOGRAMS

Polarized beams carrying spin angular momentum are readily produced by means of a wave-plate to convert linear light to circularly polarized light. Beijersbergen and colleagues [12] realized that a Hermite–Gaussian beam with no angular momentum could be similarly transformed with cylindrical lenses into a Laguerre–Gaussian beam carrying orbital angular momentum, as shown in Figure 1.3.

Although this conversion process is highly efficient, each Laguerre–Gaussian mode requires a particular Hermite–Gaussian mode as the input. The prerequisite of a specific Hermite–Gaussian mode limits the range of Laguerre–Gaussian modes that can be produced. Consequently, numerically computed holograms have been the most common method of creating helical beams, as they can generate any Laguerre–Gaussian mode from the any initial beam (Figure 1.4).

The interference pattern between a plane wave and the beam one desires to produce is recorded as a hologram on photographic film. Once developed, the film can be illuminated by a plane wave to produce a first-order diffracted beam that has both the intensity and phase of the desired beam. The big advantage of this approach lies in the current availability of high-quality spatial light modulators (SLMs). These

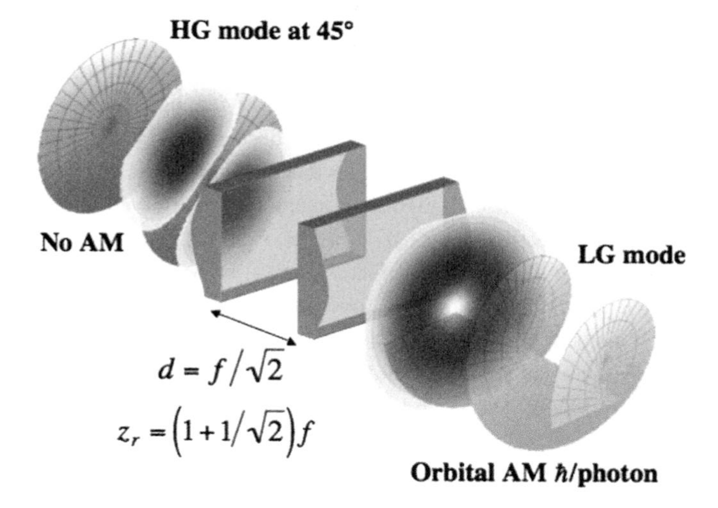

Figure 1.3 Cylindrical lens mode converter used to transform Hermite–Gaussian modes into Laguerre–Gaussian modes. See color insert.

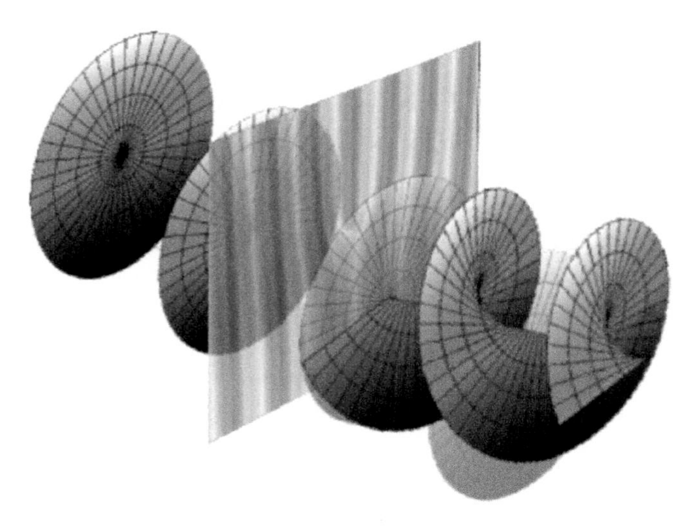

Figure 1.4 The "classic" forked hologram for the production of helically phased beams. See color insert.

pixellated liquid crystal devices take the place of the holographic film, and the interference pattern need not be experimentally derived, but can be simply calculated and displayed on the device. Used in this fashion, such devices are reconfigurable computer-calculated holograms that allow a simple laser beam to be converted into any beam with an exotic phase and amplitude structure. A further advantage is that the pattern can be changed many times per second, so the transformed beam can be adjusted to meet the experimental requirements. Figure 1.4 shows how a comparatively simple forked pattern, pioneered by Soskin, transforms the plane wave output from that of a conventional laser mode into a Laguerre–Gaussian mode carrying orbital angular momentum (Bazhenov and colleagues [11]). In recent years, spatial light modulators have been used in applications as diverse as optical tweezers (Grier [22]) and adaptive optics (Wright and colleagues [46]). The holograms can also be designed to create elaborate superpositions of modes, such as the combination of Laguerre–Gaussian modes required to form optical vortex links and knots (Leach and colleagues [26]).

1.7 COHERENCE: SPATIAL AND TEMPORAL

Circularly polarized light can, at least in principle, be produced from light sources that are both spatially and temporally incoherent. The situation for orbital angular momentum is not so clear. Pure-valued orbital angular momentum states are described by an azimuthal phase term, $\exp(i\ell\phi)$. This implicitly assumes that the beam is spatially coherent. However, providing that the beam is spatially coherent, temporal coherence is not required as each individual spectral component may have an $\exp(i\ell\phi)$ phase.

White-light beams have been recently produced from spatially filtered thermal sources and supercontinuum lasers. Although in principle the SLM could be programmed directly with the required phase profile $u(x, y)$, imperfections in the linearity and pixilation of the device mean that the resulting beam would in effect be the superposition of several different diffraction orders. When used with a white-light source, this problem is further exacerbated by the fact that the phase change is wavelength dependent, so that the weighting between different diffraction orders changes with wavelength. These limitations in the performance of the SLM can be overcome by adopting the forked hologram design, which introduces an angular deviation between the diffraction orders and allows a spatial filter positioned in the Fourier-plane of the SLM to select the first-order diffracted beam. Each wavelength component of this first-order beam has the required phase structure irrespective of any nonlinearity in the SLM. Unfortunately, the angular separation of this off-axis design makes the SLM inherently dispersive. Recently, this angular dispersion has been compensated by imaging the plane of the SLM to a secondary dispersive component such as a prism (Leach and Padgett [27]), or grating (Mariyenko and colleagues [33]), as shown in Figure 1.5.

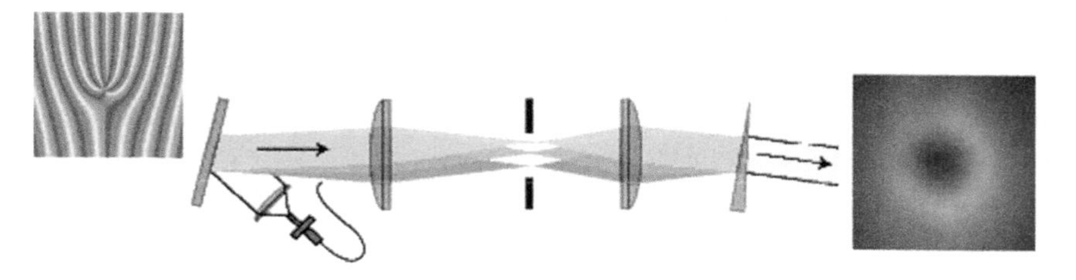

Figure 1.5 Use of a prism to compensate for the chromatic dispersion of the forked hologram allows the creation of white-light optical vortices. See color insert.

1.8 TRANSFORMATIONS BETWEEN BASIS SETS

Transformations of the polarization state on passage of light through an optical component fall into two distinct groups. The first is that of a birefringent wave-plate, which introduces a phase shift between linear polarization states that have orientations defined with respect to the plate (Hecht and Zajac [52]). The second is a rotation of linear states, independent of their orientation, which is equivalent to a phase shift between circular polarization states. The latter rotation is associated with optical activity (Hecht and Zajac [52]).

Wave-plates introducing a phase delay ϕ may be described in terms of the Jones Matrices, by a column vector

$$\begin{pmatrix} 1 \\ e^{i\phi} \end{pmatrix}$$

acting on the orthogonal polarization states

$$\begin{pmatrix} 0 \\ 1 \end{pmatrix} \quad \text{and} \quad \begin{pmatrix} 1 \\ 0 \end{pmatrix}.$$

Rotation of the wave-plate through θ with respect to the polarization is represented by the multiplication of the wave-plate vector by a rotation matrix

$$\begin{pmatrix} \cos\theta & \sin\theta \\ -\sin\theta & \cos\theta \end{pmatrix}.$$

For Hermite–Gaussian modes and Laguerre–Gaussian modes of order $\theta = 1$, a similar set of matrices may be written. The HG_{01} and HG_{10} modes are described by the vectors

$$\begin{pmatrix} 0 \\ 1 \end{pmatrix} \quad \text{and} \quad \begin{pmatrix} 1 \\ 0 \end{pmatrix},$$

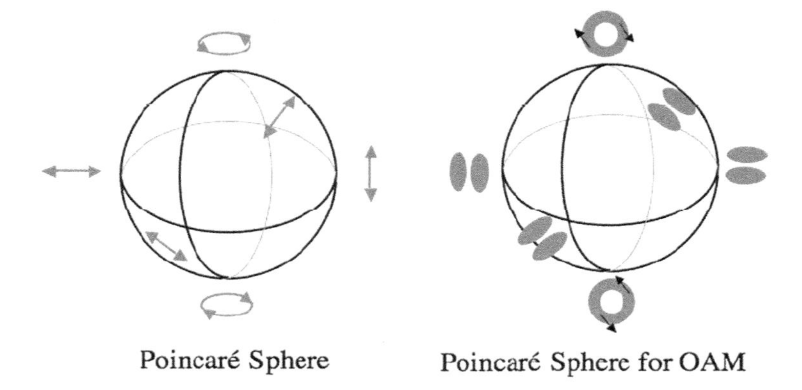

Figure 1.6 The Poincaré sphere and its equivalent geometric construction for modes of $N = 1$.

which are the same as for the linear polarization states, while the Laguerre–Gaussian modes are represented by

$$\begin{pmatrix} \sqrt{2} \\ \sqrt{2}i \end{pmatrix} \quad \text{and} \quad \begin{pmatrix} \sqrt{2} \\ -\sqrt{2}i \end{pmatrix},$$

which are the same as for the circular polarization states. For polarization, these Jones matrices and their associated transformations may be represented by the Poincaré sphere, and an equivalent geometrical construction can be made for Hermite–Gaussian modes and Laguerre–Gaussian modes (Padgett and Courtial [40]), as shown in Figure 1.6.

As stated earlier, Beijersbergen and colleagues [12] recognized that the equivalent transformation to that of a wave-plate may be performed on the orbital angular momentum states by a cylindrical lens mode converter. Consequently, the appropriate Jones Matrix for a cylindrical lens mode converter is that of a wave-plate. A special case is the π cylindrical lens mode converter, which is effectively an image inverter and therefore is identical in its optical properties to a Dove prism (Gonzalez and colleagues [21]). The analogous transformation of wave-plates and cylindrical lens mode converters extends beyond $N = 1$ to modes of any order (Allen and colleagues [2]), where the phase shift of an HG mode is proportional to $(m - n)$, and a $\pi/2$ cylindrical lens mode converter transforms every HG_{mn} mode into a $\mathrm{LG}^{m-n}_{\min(m,n)}$ mode (Beijersbergen and colleagues [12]).

It is not as easy to understand what constitutes the equivalent transformation to the rotation of linear polarization. For polarization, a rotation of the linear state through an angle α corresponds to a relative phase shift of the circular polarization states, $\sigma = \pm 1$, of $\Delta\phi = \sigma\alpha$. For HG modes, the equivalent rotation is simply a rotation of the mode around its own axis. Inspection of higher-order rotation matrices (Allen and colleagues [2]) confirms that rotation by α of a high-order HG mode is also exactly equivalent to a phase shift of the constituent LG modes of $\Delta\phi = l\alpha$. It has been shown that transmission of light through two Dove prisms in succession rotates the image by twice the angle between them (Leach and colleagues [28]). This is precisely the same geometrical transformation as that introduced to the polarization state by the application of two half-wave plates.

Hermite–Gaussian modes and Laguerre–Gaussian modes are complete orthonormal sets, and any amplitude distribution can be described by an appropriate complex superposition of modes from either set. Consequently, the rotation of an arbitrary image can be completely described in terms of the rotation of the constituent HG modes or by the phase delay of the corresponding LG modes. In general, a beam rotation can either be applied locally to each point in the beam, or globally about a specific axis (Allen and Padgett [4]). For polarization, either mapping results in the same transformation in the beam. However, for the mode structure, a rotation must be a global rotation around the beam axis. Similarly, whereas optical activity rotates the polarization at every point within the beam, an image rotation is a global mapping around the beam axis. It has been both predicted (Andrews and colleagues [6]) and observed (Araoka and colleagues [7]) that optically active materials do not alter the orbital angular momentum state.

1.9 CONCLUSION

The exploration of phase-structured light and its implications to optical angular momentum are perhaps only just beginning. The study of the interaction of light and atoms has much scope for development. Emerging and exciting applications of orbital angular momentum exist for optical micro-machines. The increased dimensionality (Molina-Terriza and colleagues [35]) of the Hilbert space will undoubtedly play a role in new approaches to encryption and for higher data densities (Gibson and colleagues [20]) for application in the areas of quantum communication and processing. It has also been proposed that orbital angular momentum might provide new elements in both laboratory- and observatory-based astronomy (Swartzlander [42], Lee and colleagues [29], Harwit [25]), including improved telescope design or the hunt for extraterrestrial signaling.

REFERENCES

[1] L. Allen, M.W. Beijersbergen, R.J.C. Spreeuw, J.P. Woerdman, Orbital angular momentum of light and the transformation of Laguerre–Gaussian laser modes, *Phys. Rev. A 45* (1992) 8185–8189.

[2] L. Allen, J. Courtial, M.J. Padgett, Matrix formulation for the propagation of light beams with orbital and spin angular momenta, *Phys. Rev. E 60* (1999) 7497–7503.

[3] L. Allen, M.J. Padgett, The Poynting vector in Laguerre–Gaussian beams and the interpretation of their angular momentum density, *Opt. Commun. 184* (2000) 67–71.

[4] L. Allen, M. Padgett, Equivalent geometric transformations for spin and orbital angular momentum of light, *J. Mod. Opt. 54* (2007) 487–491.

[5] L. Allen, M.J. Padgett, M. Babiker, Progress in Optics, vol. XXXIX, Elsevier, Amsterdam, 1999, pp. 291–372.

[6] D.L. Andrews, L.C.D. Romero, M. Babiker, On optical vortex interactions with chiral matter, *Opt. Commun. 237* (2004) 133–139.

[7] F. Araoka, T. Verbiest, K. Clays, A. Persoons, Interactions of twisted light with chiral molecules: An experimental investigation, *Phys. Rev. A 71* (2005) 3.

[8] A. Aspect, J. Dalibard, G. Roger, Experimental test of Bell inequalities using time-varying analysers, *Phys. Rev. Lett. 49* (1982) 1804–1807.

[9] S.M. Barnett, Optical angular momentum flux, *J. Opt. B: Quantum Semiclass. Opt. 4* (2002) S7–S16.

[10] S.M. Barnett, L. Allen, Orbital angular momentum and nonparaxial light-beams, *Opt. Commun. 110* (1994) 670–678.

[11] V.Y. Bazhenov, M.V. Vasnetsov, M.S. Soskin, Laser beams with screw dislocations in their wavefronts, *JETP Lett. 52* (1990) 429–431.

[12] M.W. Beijersbergen, L. Allen, H. Vanderveen, J.P. Woerdman, Astigmatic laser mode converters and transfer of orbital angular momentum, *Opt. Commun. 96* (1993) 123–132.

[13] V. Chavez, D. McGloin, M.J. Padgett, W. Dultz, H. Schmitzer, K. Dholakia, Observation of the transfer of the local angular momentum density of a multiringed light beam to an optically trapped particle, *Phys. Rev. Lett. 91* (2003) 4.

[14] S. Chavez-Cerda, M.J. Padgett, I. Allison, G.H.C. New, J.C. Gutierrez-Vega, A.T. O'Neil, I. MacVicar, J. Courtial, Holographic generation and orbital angular momentum of high-order Mathieu beams, *J. Opt. B: Quantum Semiclass. Opt. 4* (2002) S52–S57.

[15] J. Courtial, K. Dholakia, L. Allen, M.J. Padgett, Gaussian beams with very high orbital angular momentum, *Opt. Commun. 144* (1997) 210–213.

[16] K. Dholakia, N.B. Simpson, M.J. Padgett, L. Allen, Second-harmonic generation and the orbital angular momentum of light, *Phys. Rev. A 54* (1996) R3742–R3745.

[17] J. Durnin, Exact solutions for non-diffracting beams I. The scalar theory, *J. Opt. Soc. Am. A 4* (1987) 651–654.

[18] J. Durnin, J.J. Miceli, J.H. Eberly, Diffraction-free beams, *Phys. Rev. Lett. 58* (1987) 1499–1501.

[19] S. Franke-Arnold, S.M. Barnett, M.J. Padgett, L. Allen, Two-photon entanglement of orbital angular momentum states, *Phys. Rev. A 65* (2002) 033823.

[20] G. Gibson, J. Courtial, M.J. Padgett, M. Vasnetsov, S.M. Pas'ko, V. Barnett, S. Franke-Arnold, Free-space information transfer using light beams carrying orbital angular momentum, *Opt. Exp. 12* (2004) 5448–5456.

[21] N. Gonzalez, G. Molina-Terriza, J.P. Torres, How a Dove prism transforms the orbital angular momentum of a light beam, *Opt. Exp. 14* (2006) 9093–9102.

[22] D.G. Grier, A revolution in optical manipulation, *Nature 424* (2003) 810–816.

[23] J.C. Gutierrez-Vega, M.D. Iturbe-Castillo, S. Chavez-Cerda, Alternative formulation for invariant optical fields: Mathieu beams, *Opt. Lett. 25* (2000) 1493–1495.

[24] M. Harris, C.A. Hill, P.R. Tapster, J.M. Vaughan, Laser modes with helical wave-fronts, *Phys. Rev. A 49* (1994) 3119–3122.

[25] M. Harwit, Photon orbital angular momentum in astrophysics, *Astrophys. J. 597* (2003) 1266–1270.

[26] J. Leach, M. Dennis, J. Courtial, M. Padgett, Laser beams—Knotted threads of darkness, *Nature 432* (2004) 165.

[27] J. Leach, M.J. Padgett, Observation of chromatic effects near a white-light vortex, *New J. Phys. 5* (2003) 154.1–154.7.

[28] J. Leach, M.J. Padgett, S.M. Barnett, S. Franke-Arnold, J. Courtial, Measuring the orbital angular momentum of a single photon, *Phys. Rev. Lett. 88* (2002) 257901.

[29] J.H. Lee, G. Foo, E.G. Johnson, G.A. Swartzlander, Experimental verification of an optical vortex coronagraph, *Phys. Rev. Lett. 97* (2006) 053901.

[30] C. Lopez-Mariscal, M.A. Bandres, J.C. Gutierrez-Vega, S. Chavez-Cerda, Observation of parabolic nondiffracting optical fields, *Opt. Exp. 13* (2005) 2364–2369.

[31] D. McGloin, K. Dholakia, Bessel beams: Diffraction in a new light, *Contemp. Phys. 46* (2005) 15–28.

[32] A. Mair, A. Vaziri, G. Weihs, A. Zeilinger, Entanglement of the orbital angular momentum states of photons, *Nature 412* (2001) 313–316.

[33] I. Mariyenko, J. Strohaber, C. Uiterwaal, Creation of optical vortices in femtosecond pulses, *Opt. Exp. 13* (2005) 7599–7608.

[34] G. Molina-Terriza, J. Recolons, J.P. Torres, L. Torner, E.M. Wright, Observation of the dynamical inversion of the topological charge of an optical vortex, *Phys. Rev. Lett. 87* (2001) 023902.

[35] G. Molina-Terriza, J.P. Torres, L. Torner, Management of the angular momentum of light: Preparation of photons in multidimensional vector states of angular momentum, *Phys. Rev. Lett. 88* (2002) 4.

[36] J.F. Nye, M.V. Berry, Dislocations in wave trains, *Proc. R. Soc. London, Series A: Math. Phys. Eng. Sci. 336* (1974) 165–190.

[37] A.T. O'Neil, I. MacVicar, L. Allen, M.J. Padgett, Intrinsic and extrinsic nature of the orbital angular momentum of a light beam, *Phys. Rev. Lett. 88* (2002) 053601.

[38] M.J. Padgett, L. Allen, The Poynting vector in Laguerre–Gaussian laser modes, *Opt. Commun. 121* (1995) 36–40.

[39] M.J. Padgett, L. Allen, Orbital angular momentum exchange in cylindrical-lens mode converters, *J. Opt. B: Quantum Semiclass. Opt. 4* (2002) S17–S19.

[40] M.J. Padgett, J. Courtial, Poincaré-sphere equivalent for light beams containing orbital angular momentum, *Opt. Lett. 24* (1999) 430–432.

[41] M.S. Soskin, V.N. Gorshkov, M.V. Vasnetsov, J.T. Malos, N.R. Heckenberg, Topological charge and angular momentum of light beams carrying optical vortices, *Phys. Rev. A 56* (1997) 4064–4075.

[42] G.A. Swartzlander, Peering into darkness with a vortex spatial filter, *Opt. Lett. 26* (2001) 497–499.

[43] C. Tamm, C.O. Weiss, Bistability and optical switching of spatial patterns in a laser, *J. Opt. Soc. Am. B: Opt. Phys. 7* (1990) 1034–1038.

[44] G.A. Turnbull, D.A. Robertson, G.M. Smith, L. Allen, M.J. Padgett, Generation of free-space Laguerre–Gaussian modes at millimetre-wave frequencies by use of a spiral phaseplate, *Opt. Commun. 127* (1996) 183–188.

[45] S.J. Van Enk, G. Nienhuis, Eigenfunction description of laser beams and orbital angular momentum of light, *Opt. Commun. 94* (1992) 147–158.

[46] A.J. Wright, B.A. Patterson, S.P. Poland, J.M. Girkin, G.M. Gibson, M.J. Padgett, Dynamic closed-loop system for focus tracking using a spatial light modulator and a deformable membrane mirror, *Opt. Exp. 14* (2006) 222–228.

[47] J.M. Blatt, V.F. Weisskopf, Theoretical Nuclear Physics, Wiley, New York, 1952.

[48] D. Marcuse, Light Transmission Optics, Van Nostrand, New York, 1972.

[49] J.D. Jackson, Classical Electrodynamics, Wiley, New York, 1962.

[50] J.W. Simmons, M.J. Guttman, States, Wawes and Photons, Addison–Wesley, Reading, MA, 1970.

[51] M.V. Berry, Paraxialbeams of spinning light, in: M.S. Soskin, M.V. Vasnetson (Eds.), *Singulat Optics*, in: *SPIE*, vol. 3487, 1998, pp. 6–11.

[52] H. Hecht, A. Zajac, Optics, Addison–Wesley, Reading, MA, 1974.

Chapter 2

Angular Momentum and Vortices in Optics

Gerard Nienhuis

Universiteit Leiden, The Netherlands

In this chapter, we will review the concept of angular momentum of the classical and the quantum radiation field. We will discuss the separation in an orbital part and a spin part, both for an arbitrary field and for a light beam in the paraxial limit. We will also cover the properties of the corresponding quantum operators. Light fields where the density of angular momentum is simply related to the energy (or photon) density are of special interest. Such states tend to contain vortices in the form of phase singularities. Examples are spherical and cylindrical multipole fields, which are exact solutions of Maxwell's equations. Laguerre–Gaussian beams are well-known analogs in the paraxial limit. Angular momentum also arises in light fields that are physically rotating in time. Paraxial beams of light can be mapped onto the wave functions of a two-dimensional harmonic oscillator during a half cycle. This map is applied to analyze the propagation properties of optical vortices.

2.1 INTRODUCTION

Angular momentum (AM) is one of the fundamental conserved quantities in physics, along with energy and linear momentum. Whereas energy conservation arises from invariance for time translation, and momentum conservation reflects the homogeneity of space, it is the isotropy of space that gives rise to conservation of AM. As a consequence of this isotropy, a rotated version of a possible history of the state of a closed physical system is again a possible history. The relation between angular-momentum conservation and space isotropy is most obvious in a quantum description,

STRUCTURED LIGHT AND ITS APPLICATIONS

where a quantity is conserved when the corresponding operator commutes with the Hamiltonian. This explains why a rotated version of a possible time-dependent state is still a solution of the Schrödinger equation, since rotation of a state is generated by the vector operator of AM.

For any classical system, the density of AM is defined by

$$\mathbf{j}(\mathbf{r}) = \mathbf{r} \times \mathbf{p}(\mathbf{r}) \tag{1}$$

in terms of the momentum density $\mathbf{p}$. For a material system, the total AM $\mathbf{J} = \int d\mathbf{r}\,\mathbf{j}$ is conveniently separated into an orbital part and a spin part. In the case of the Earth orbiting around the Sun, the spin arises from the daily rotation around its axis, while the motion in its orbit gives rise to an orbital AM. For an arbitrary material system, it is sufficient to separate the arm as $\mathbf{r} = \mathbf{R} + \mathbf{r}'$, with $\mathbf{R}$ the position of the center of mass, and $\mathbf{r}'$ the position in the center-of-mass system. This gives a corresponding separation of $\mathbf{j}$. After integration over position, the total AM is accordingly separated as

$$\mathbf{J} = \mathbf{L} + \mathbf{S}. \tag{2}$$

Here $\mathbf{L} = \mathbf{R} \times \mathbf{P}$ is the AM associated with the center-of-mass motion, with $\mathbf{P} = \int d\mathbf{r}\,\mathbf{p}$ the total momentum. This contribution $\mathbf{L}$ varies with the choice of the origin, so that it has an extrinsic nature. By a proper choice of the origin, the contribution $\mathbf{L}$ can always be made to vanish. In contrast, $\mathbf{S}$ is the AM in the center-of-mass system. For a rigid body, $\mathbf{S}$ is the AM corresponding to rotation about its center of mass, which gives it the significance of a spin. In general, the contribution $\mathbf{S}$ has an intrinsic nature in that it does not depend on the choice of the origin. For a closed material system, $\mathbf{L}$ and $\mathbf{S}$ are separately conserved.

The classical notion of spin should not be confused with the intrinsic spin of a quantum particle. A quantum spin cannot be understood as arising from a mechanical rotation of a mass distribution, and it cannot be expressed as the integration over position of a density as in equation (1). A mechanical rotation of a rigid material body could only lead to integer values of the eigenvalues of any spin component (in units of $\hbar$). Another difference between classical and quantum AM is that the orbital AM of a quantum system cannot always be made to vanish by a proper choice of the origin. In fact, an electron in an atom can very well have a nonzero orbital AM, although it is commonly evaluated in the center-of-mass system. This results from the fact that the wave function of an electron has a finite extension.

In optics, the properties of AM and its decomposition are less straightforward. It is common knowledge that when an atom absorbs a circularly polarized photon, it picks up an AM $\pm\hbar$ in the propagation direction. This follows from the selection rules $\Delta m = \pm 1$ for the magnetic quantum number of the atom. More generally, according to

Maxwell's theory, an electromagnetic field has a density of energy w and momentum $\mathbf{p}$ as given by the expressions

$$w = \frac{1}{2}\left(\epsilon_0 \mathbf{E}^2 + \frac{1}{\mu_0}\mathbf{B}^2\right), \qquad \mathbf{p} = \epsilon_0 \mathbf{E} \times \mathbf{B}. \tag{3}$$

In a vacuum, the momentum density is identical to energy flow (the Poynting vector) $\mathbf{E} \times \mathbf{H}$ divided by c^2. This momentum density also gives rise to a density of AM [1], according to equation (1). From this optical expression, it is obvious that circular polarization is not required for a light field to have AM. One might expect that the AM of light can be separated into a contribution arising from polarization, and a contribution arising from rotary phase gradients, analogous to the separation in spin and orbital AM in quantum fields. This separation is known to be problematic for light [2–4]. On the other hand, for a light beam in the paraxial limit, the separation of the AM in the propagation direction in a spin and an orbital part is well understood [5,6].

In analogy to a quantum particle, orbital AM of light arises when the phase increases by a multiple of 2π along a closed contour that encircles a dislocation line, where the phase is undetermined, and the intensity vanishes [7]. The structure of the corresponding phase gradient around a phase singularity resembles a vortex in a circulating fluid. The resulting optical vortices are a prime example of singularities in optics [8]. They are mathematically identical to vortices that can arise in wave phenomena in general, including quantum-mechanical probability waves [9]. Also, the light polarization can have singular points or lines [10].

In this contribution we discuss the concept and the various manifestations of the AM of light, both for arbitrary electromagnetic fields and for paraxial beams. In the classical case, we allow for nonmonochromatic light. We also consider the quantum operators for AM, which are the generators of rotation. In all cases, we discuss the possibility and the significance of the separation of AM in a spin and an orbital contribution. We then compare the various contributions to the AM of a closed classical system of charges and fields. In the next section, we will demonstrate the general possibility of a separation of optical AM into an intrinsic part, which is independent of the choice of the origin, and an extrinsic part. This separation is independent of gauge, and for a free field these parts are separately conserved. It is tempting to view these parts as spin and orbital AM. This is misleading in the sense that in the quantum description these separate parts do not produce AM operators with the correct commutation rules. This is related to the fact that it is not possible in general to separately rotate the polarization of a Maxwell field while leaving the amplitude pattern unchanged. Multipole fields are exact solutions of Maxwell's equations, which are also reminiscent of eigenstates of AM in quantum mechanics. Accordingly, they tend to display singularities in phase and polarization at the origin or at the axis. Later, we will analyze the angular-momentum properties as well as the phase and polarization singularities of spherical and cylindrical multipole fields. We will then discuss the AM of paraxial light beams and its separation into an orbital and a polarization part. In this case, the description simplifies, and the orbital and polarization AM along the optical axis has all the expected properties. In the special case of beams with uniform orbital AM, a phase singularity arises on the axis. Superposition of such beams with different polarization gives light beams with a polarization that varies across the transverse plane, with a polarization vortex on the axis. In a quantum description, angular-momentum operators are generators of rotation. In order to illustrate this for paraxial beams, we need a quantum version of the paraxial approximation which we will cover in the following section. One would expect that AM can also be generated when a light beam is physically rotating in time around its axis [11]. A beam with a time-dependent transverse field pattern is not stationary. This raises the question of AM of nonmonochromatic light beams, which we discuss in the next section. Finally, we address the topic of the behavior of vortices under propagation of a paraxial beam.

2.2 CLASSICAL ANGULAR MOMENTUM OF FIELDS AND PARTICLES

When we consider the electromagnetic field together with its sources, the separation of the total AM in a radiation part and a matter part is delicate. The main reason is that a part of the field (the Coulomb field) is inseparably linked to the charges, and it moves along with them. We will clarify this in the present section.

2.2.1 Angular Momentum of Particles and Radiation

We consider a classical system consisting of pointlike particles (with position $\mathbf{r}_\alpha$, mass m_α, and charge q_α) and the electromagnetic field, which is described by the magnetic field $\mathbf{B}$, and the electric field $\mathbf{E}$. The dynamics of the fields is described by Maxwell's (microscopic) equations, and the motion of the particles obeys Newton's Second Law. The momentum density of the electromagnetic field is given in equation (3), while the kinetic (not the canonical) momentum of the particles is $m_\alpha \dot{\mathbf{r}}_\alpha$. For the total AM of the field and the particles, this gives the expression

$$\mathbf{J} = \mathbf{J}_{\text{kin}} + \mathbf{J}_{\text{field}}, \tag{4}$$

with

$$\mathbf{J}_{\text{kin}} = \sum_\alpha m_\alpha \mathbf{r}_\alpha \times \dot{\mathbf{r}}_\alpha, \qquad \mathbf{J}_{\text{field}} = \epsilon_0 \int d\mathbf{r}\, \mathbf{r} \times (\mathbf{E} \times \mathbf{B}). \tag{5}$$

The various contributions to the AM are conveniently analyzed when we apply Helmholtz's theorem to make the standard separation of the electric field $\mathbf{E} = \mathbf{E}^{\parallel} + \mathbf{E}^{\perp}$ into a longitudinal and a transverse part, where the longitudinal part has vanishing curl ($\nabla \times \mathbf{E}^{\parallel} = 0$) and the transverse part has vanishing divergence ($\nabla \cdot \mathbf{E}^{\perp} = 0$) [1,4]. As usual, we express the magnetic field as well as the transverse electric field in terms of the purely transverse vector potential $\mathbf{A}$, such that $\mathbf{B} = \nabla \times \mathbf{A}$, $\mathbf{E}^{\perp} = -\dot{\mathbf{A}}$. The transverse vector potential $\mathbf{A}$ is uniquely determined by the instantaneous magnetic field, and vice versa. The choice of a transverse vector potential, so that $\nabla \cdot \mathbf{A} = 0$, is equivalent to using the Coulomb gauge. This transverse vector potential is identical to the transverse part of the vector potential in any other gauge. From Maxwell's scalar equation $\rho = \epsilon_0 \nabla \cdot \mathbf{E}^{\parallel}$, it follows that the longitudinal electric field $\mathbf{E}^{\parallel}$ is identical to the Coulomb field that corresponds to the instantaneous positions of the charges. This field may be viewed as traveling along with the charged particles, so that it is not an independent degree of freedom. On the other hand, the transverse field $\mathbf{E}^{\perp}$ together with the magnetic field $\mathbf{B}$ represents the radiation field. This is an independent degree of freedom, which is fully described by the transverse vector potential [4].

The separation of the electric field into a longitudinal (Coulomb) part and a transverse (radiative) part leads to a corresponding separation of the field AM as

$$\mathbf{J}_{\text{field}} = \mathbf{J}_{\text{Coul}} + \mathbf{J}_{\text{rad}}, \tag{6}$$

with

$$\mathbf{J}_{\text{Coul}} = \epsilon_0 \int d\mathbf{r}\, \mathbf{r} \times \left(\mathbf{E}^{\parallel} \times \mathbf{B}\right) = \sum_{\alpha} q_{\alpha} \mathbf{r}_{\alpha} \times \mathbf{A}(\mathbf{r}_{\alpha}). \tag{7}$$

The last expression follows after substituting the identity $\mathbf{B} = \nabla \times \mathbf{A}$, and applying partial integration. In the next section, we will further analyze the radiative AM

$$\mathbf{J}_{\text{rad}} = \epsilon_0 \int d\mathbf{r}\, \mathbf{r} \times \left(\mathbf{E}^{\perp} \times \mathbf{B}\right). \tag{8}$$

It is natural to view the sum of the kinetic and the Coulomb contribution as the contribution of the particle motion, which leads to the expression as [4]

$$\mathbf{J}_{\text{kin}} + \mathbf{J}_{\text{Coul}} \equiv \mathbf{J}_{\text{part}} = \sum_{\alpha} \mathbf{r}_{\alpha} \times \mathbf{p}_{\alpha}, \tag{9}$$

where $\mathbf{p}_{\alpha} = m_{\alpha} \dot{\mathbf{r}}_{\alpha} + q_{\alpha} \mathbf{A}(\mathbf{r}_{\alpha})$ is the canonical momentum of particle α.

The total AM from equation (4) takes the form

$$\mathbf{J} = \mathbf{J}_{\text{part}} + \mathbf{J}_{\text{rad}}, \tag{10}$$

where the two contributions correspond to the particles and the radiation field, as independent degrees of freedom. When the arm $\mathbf{r} = \mathbf{R} + \mathbf{r}'$ as in the previous section, the AM $\mathbf{J}_{\text{part}} = \mathbf{L}_{\text{part}} + \mathbf{S}_{\text{part}}$ of the particles is separated in an intrinsic and an extrinsic contribution.

2.2.2 Rate of Change of Contributions to Angular Momentum

The rate of change of $\mathbf{J}_{\text{kin}}$ follows from the expression for the Lorentz force. It is convenient to express the particle positions and velocities in terms of the charge density $\rho(\mathbf{r})$ and current density $\mathbf{u}(\mathbf{r})$, which are sums of delta functions $q_\alpha \delta(\mathbf{r} - \mathbf{r}_\alpha)$ and $q_\alpha \dot{\mathbf{r}}_\alpha \delta(\mathbf{r} - \mathbf{r}_\alpha)$. Then we find that [12]

$$\frac{d}{dt}\mathbf{J}_{\text{kin}} = \int d\mathbf{r}\, \mathbf{r} \times \left(\rho \mathbf{E}^{\perp} + \mathbf{u} \times \mathbf{B}\right). \tag{11}$$

The contribution of the longitudinal electric field $\mathbf{E}^{\parallel}$ can be omitted here, since ρ is proportional to $\nabla \cdot \mathbf{E}^{\parallel}$. By using Maxwell's equations, one can express the rate of change of $\mathbf{J}_{\text{Coul}}$ in the form [12]

$$\frac{d}{dt}\mathbf{J}_{\text{Coul}} = -\int d\mathbf{r}\, \mathbf{r} \times \left(\rho \mathbf{E}^{\perp} + \mathbf{j}^{\parallel} \times \mathbf{B}\right). \tag{12}$$

The longitudinal part $\mathbf{j}^{\parallel}$ of the current density is determined by the time derivative of the charge density—by continuity. The rate of change of $\mathbf{J}_{\text{rad}}$ can also be evaluated, with the result [12]

$$\frac{d}{dt}\mathbf{J}_{\text{rad}} = -\int d\mathbf{r}\, \mathbf{r} \times \left(\mathbf{j}^{\perp} \times \mathbf{B}\right) = -\frac{d}{dt}\mathbf{J}_{\text{part}}. \tag{13}$$

The last equality is obvious when adding equations (11) and (12). This confirms that the total AM from equation (10) of particles and radiation is conserved, as it should.

2.3 SEPARATION OF RADIATIVE ANGULAR MOMENTUM IN L AND S

In this section, we will discuss the possibility of expressing the AM $\mathbf{J}_{\text{rad}}$ of the radiation field as a sum $\mathbf{L}_{\text{rad}} + \mathbf{S}_{\text{rad}}$ of an intrinsic and an extrinsic part.

2.3.1 Classical Description

The expression in equation (8) for the AM of the radiation field can also be separated after expressing the magnetic field in the vector potential and applying partial integration [4]. This leads to the result

$$\mathbf{J}_{\text{rad}} = \mathbf{L}_{\text{rad}} + \mathbf{S}_{\text{rad}}, \tag{14}$$

with

$$\mathbf{L}_{\text{rad}} = \epsilon_0 \sum_i \int d\mathbf{r}\, E_i^{\perp}(\mathbf{r} \times \nabla) A_i, \qquad \mathbf{S}_{\text{rad}} = \epsilon_0 \int d\mathbf{r}\, \mathbf{E}^{\perp} \times \mathbf{A}. \tag{15}$$

Since $\mathbf{A}$ is the transverse vector potential, these quantities are independent of gauge. The contribution $\mathbf{L}_{\text{rad}}$ varies with the choice of the origin, just as an orbital AM, so that it has an extrinsic nature. Moreover, it is determined by the phase gradient of the field. On the other hand, the contribution $\mathbf{S}_{\text{rad}}$ does not change for a different choice of the origin, and it is determined by the polarization of the field. This gives it the flavor of a spin AM. It is also interesting to consider the expressions for the rates of change of these quantities, which are found as [12]

$$\frac{d}{dt}\mathbf{L}_{\text{rad}} = -\sum_i \int d\mathbf{r}\, j_i^{\perp}(\mathbf{r} \times \nabla) A_i, \qquad \frac{d}{dt}\mathbf{S}_{\text{rad}} = -\int d\mathbf{r}\, \mathbf{j}^{\perp} \times \mathbf{A}. \tag{16}$$

This demonstrates that the change of $\mathbf{L}_{\text{rad}}$ and $\mathbf{S}_{\text{rad}}$ is entirely due to the interactions with the transverse part of the current. For a free field, in the absence of charges, $\mathbf{L}_{\text{rad}}$ and $\mathbf{S}_{\text{rad}}$ are separately conserved.

2.3.2 Quantum Operators

The expressions in equation (15) for the contributions $\mathbf{L}_{\text{rad}}$ and $\mathbf{S}_{\text{rad}}$ to the radiative AM are quite suggestive for their interpretation as orbital and spin parts. However, this interpretation is problematic. This is clear when we consider the quantized version of the same system of particles and fields that we discussed in Section 2.2. Quantization is straightforward once we have expressed the system in terms of a set of generalized canonical pairs of coordinates and momenta [4], and it leads to the well-known form of the operators for the vector potential $\hat{\mathbf{A}}$, the magnetic field $\hat{\mathbf{B}} = \nabla \times \hat{\mathbf{A}}$, and the transverse electric field $\hat{\mathbf{E}}^{\perp}$, in terms of creation and annihilation operators. For our purpose, it is convenient to choose a continuum of plane-wave modes, each characterized by a wave vector $\mathbf{k}$, and one out of two possible normalized circular polarization vectors $\mathbf{e}_{\pm}$ in the plane normal to $\mathbf{k}$. The helicity of the vector $\mathbf{e}_{+}$ is parallel to $\mathbf{k}$, whereas $\mathbf{e}_{-}$ has opposite helicity. Then the expression for the operators $\hat{\mathbf{E}}^{\perp}$ for the transverse electric field and $\hat{\mathbf{A}}$ for the vector potential have the form [13]

$$\hat{\mathbf{E}}^{\perp} = i \int d\mathbf{k} \sqrt{\frac{\hbar\omega}{2\epsilon_0(2\pi)^3}} e^{i\mathbf{k}\cdot\mathbf{r}} \big(\mathbf{e}_+(\mathbf{k})\hat{a}_+(\mathbf{k}) + \mathbf{e}_-(\mathbf{k})\hat{a}_-(\mathbf{k})\big) + \text{H.c.},$$

$$\hat{\mathbf{A}}(\mathbf{r}) = \int d\mathbf{k} \sqrt{\frac{\hbar}{2\epsilon_0\omega(2\pi)^3}} e^{i\mathbf{k}\cdot\mathbf{r}} \big(\mathbf{e}_+(\mathbf{k})\hat{a}_+(\mathbf{k}) + \mathbf{e}_-(\mathbf{k})\hat{a}_-(\mathbf{k})\big) + \text{H.c.}, \quad (17)$$

where the mode operators obey the commutation rule $[\hat{a}_+(\mathbf{k}'), \hat{a}_+^\dagger(\mathbf{k})] = \delta(\mathbf{k} - \mathbf{k}')$, and so on, and where $\omega = ck$ is the mode frequency.

The quantum operator for the quantity $\mathbf{S}_{\text{rad}}$ is obtained by substituting the quantum operators $\hat{\mathbf{A}}$ and $\hat{\mathbf{E}}^{\perp}$ in the expression in equation (15). The result can be put in the intuitive form

$$\hat{\mathbf{S}}_{\text{rad}} = \int d\mathbf{k}\, \hbar \frac{\mathbf{k}}{k} \big(\hat{a}_+^\dagger(\mathbf{k})\hat{a}_+(\mathbf{k}) - \hat{a}_-^\dagger(\mathbf{k})\hat{a}_-(\mathbf{k})\big). \quad (18)$$

This expression is the continuum version of an expression for a finite quantization volume [12,14]. It simply illustrates that each photon with wave vector $\mathbf{k}$ and polarization vector $\mathbf{e}_+(\mathbf{k})$ contributes to $\mathbf{S}_{\text{rad}}$ a unit $\hbar$ in the direction of $\mathbf{k}$. A photon with the opposite circular polarization $\mathbf{e}_-$ gives the opposite contribution.

An obvious property of the quantum operator $\mathbf{S}_{\text{rad}}$ is that its three components ($\hat{S}_x$, $\hat{S}_y$, and $\hat{S}_z$) commute, simply because the creation and annihilation operators for different modes commute. In fact, number states of all modes with circular polarization are common eigenstates of all three components. This implies that $\hat{\mathbf{S}}_{\text{rad}}$ cannot be viewed as a proper AM operator, which generates rotations. On the other hand, the operator $\hat{\mathbf{J}}_{\text{rad}}$ is an AM momentum operator, as is exemplified by the commutation rule $[\hat{J}_{x,\text{rad}}, \hat{J}_{y,\text{rad}}] = i\hbar\hat{J}_{z,\text{rad}}$, and so on. As a result, the commutation rules for the components of the quantum operator $\hat{\mathbf{L}}_{\text{rad}}$ take the form [12,15]

$$[\hat{L}_{x,\text{rad}}, \hat{L}_{y,\text{rad}}] = i\hbar(\hat{L}_{z,\text{rad}} - \hat{S}_{z,\text{rad}}), \quad (19)$$

and so on. These remarkable commutation properties can be traced back to the fact that a rotation of the polarization of a radiation field without rotating the field pattern itself would violate the transversality of the field. The vector operator $\hat{\mathbf{S}}_{\text{rad}}$ is a proper quantum operator within the space of physical states. Its transformation properties resemble a rotation of the polarization pattern only insofar as it is allowed within the constraint of transversality [15].

The Hamiltonian of the radiation field also has a familiar form. After substitution of the expressions for the field operators, one may easily verify that

$$\hat{\mathbf{H}}_{\text{rad}} = \frac{1}{2}\int d\mathbf{r}\left(\epsilon_0 \hat{\mathbf{E}}^2(\mathbf{r}) + \frac{1}{\mu_0}\hat{\mathbf{B}}^2(\mathbf{r})\right)$$
$$= \sum_s \int d\mathbf{k}\, \hbar\omega\big(\hat{a}_s^\dagger(\mathbf{k})\hat{a}_s(\mathbf{k}) + \hat{a}_s(\mathbf{k})\hat{a}_s^\dagger(\mathbf{k})\big), \tag{20}$$

with $s = \pm$ indicating the helicity of the circular polarization. Again, this Hamiltonian obviously commutes with the components of the spin operator (equation (18)), which proves that $\hat{\mathbf{S}}_{\text{rad}}$ (as well as $\hat{\mathbf{L}}_{\text{rad}}$) is conserved for a free field.

2.4 MULTIPOLE FIELDS AND THEIR VORTEX STRUCTURE

The quantum field operators in equation (17) are represented as an expansion over plane-wave modes $\propto \exp(i\mathbf{k}\cdot\mathbf{r})$. These modes are solutions of Maxwell's equations with a well-defined value of the momentum. This momentum amounts to $\hbar\mathbf{k}$ for each photon energy $\hbar\omega$. This statement is somewhat misleading, since for any other solution of Maxwell's equations the total momentum (the volume integral of the momentum density in equation (3)) is equally well-defined. What is characteristic for plane waves is that the momentum density is proportional to the energy density, with the ratio $\mathbf{k}/\omega$. In analogy, one might expect that the solutions of Maxwell's equations suitable for the discussion of AM are multipole fields. We shall discuss both the spherical and cylindrical versions of multipole fields. We are interested in their angular-momentum properties and their vortex structure.

2.4.1 Spherical Multipole Fields

The fields emitted by electric or magnetic multipoles are called *multipole fields* [1], and they are best represented in spherical coordinates r, θ, and ϕ in terms of the spherical harmonics $Y_{lm}(\theta,\phi)$. They are derived from scalar solutions of the Helmholtz wave equation $\nabla^2\psi = -k^2\psi$ in the separated form

$$\psi_{lm}(\mathbf{r}) = g_l(r)Y_{lm}(\theta,\phi), \tag{21}$$

where the function g_l is a solution of the radial wave equation

$$\left(\frac{d^2}{dr^2} + \frac{2}{r}\frac{d}{dr} + k^2 - \frac{l(l+1)}{r^2}\right)g_l(r) = 0. \tag{22}$$

The function g_l is a linear combination of the spherical Bessel function $j_l(kr)$, and the spherical Hankel function $n_l(kr)$. The Hankel function has a singularity in the origin, and the Bessel functions are regular.

It is customary to distinguish two types of multipole fields. In the transverse electric (TE) fields, the electric field has a vanishing radial component, whereas the fields with a vanishing radial magnetic field are termed *transverse magnetic* (TM). Both are fully specified by a monochromatic vector potential $\mathbf{A}$. A TE field with frequency $\omega = ck$ is given in spherical coordinates by the expression

$$\mathbf{A}_{lm}^{\mathrm{TE}}(\mathbf{r}, t) = (\mathbf{r} \times \nabla) g_l(r) Y_{lm}(\theta, \phi) e^{-i\omega t} + \text{c.c.} \tag{23}$$

It is noteworthy that the operator $\mathbf{r} \times \nabla$ in spherical coordinates only contains derivatives with respect to the angles θ and ϕ, so that it does not act on g_l. The corresponding electric and magnetic fields are then

$$\begin{aligned} \mathbf{E}_{lm}^{\mathrm{TE}}(\mathbf{r}, t) &= i\omega(\mathbf{r} \times \nabla) g_l(r) Y_{lm}(\theta, \phi) e^{-i\omega t} + \text{c.c.}, \\ \mathbf{B}_{lm}^{\mathrm{TE}}(\mathbf{r}, t) &= \nabla \times (\mathbf{r} \times \nabla) g_l(r) Y_{lm}(\theta, \phi) e^{-i\omega t} + \text{c.c.} \end{aligned} \tag{24}$$

A TM field is specified by the vector potential

$$\mathbf{A}_{lm}^{\mathrm{TM}}(\mathbf{r}, t) = -i\frac{c}{\omega} \nabla \times (\mathbf{r} \times \nabla) g_l(r) Y_{lm}(\theta, \phi) e^{-i\omega t} + \text{c.c.}, \tag{25}$$

which gives the expressions for the fields

$$\begin{aligned} \mathbf{E}_{lm}^{\mathrm{TM}}(\mathbf{r}, t) &= c\nabla \times (\mathbf{r} \times \nabla) g_l(r) Y_{lm}(\theta, \phi) e^{-i\omega t} + \text{c.c.}, \\ \mathbf{B}_{lm}^{\mathrm{TM}}(\mathbf{r}, t) &= -i\frac{\omega}{c}(\mathbf{r} \times \nabla) g_l(r) Y_{lm}(\theta, \phi) e^{-i\omega t} + \text{c.c.} \end{aligned} \tag{26}$$

Both the TE and the TM fields are defined for all integer values of $l \geqslant 1$, and for all integer values of m with $-l \leqslant m \leqslant l$. The isotropic spherical harmonic Y_{00} obviously does not produce a nonvanishing field. All fields have a vanishing divergence and they obey the wave equation, so that they represent solutions of Maxwell's equations. Expressions (24) and (26) reflect the dual character of the free electric and magnetic field, which means that a solution of Maxwell's equations in a vacuum gives another solution after the substitution $\mathbf{E} \to c\mathbf{B}$ and $c\mathbf{B} \to -\mathbf{E}$. It is this substitution that transforms TE and TM fields into each other.

The rotational transformation properties of the set of fields $\mathbf{A}_{lm}^{\mathrm{TE}}$ for a fixed value of l is the same as for the spherical harmonics Y_{lm}, and the same statement holds for the TM fields. Since rotation operators are generated by the quantum operator $\hat{\mathbf{J}}$ for AM, this implies that each photon in a TE mode or a TM mode with mode numbers l and m carries an AM $\hbar m$ in the z-direction. Also, the contribution of each photon to the quantity $\hat{\mathbf{J}}^2$ is equal to $\hbar^2 l(l+1)$.

These expressions determine the density of energy and of AM of the multipole field. After time averaging the general expression (3) over an optical cycle, the energy density for the fields (24) or (26) is found as

$$w(\mathbf{r}) = \epsilon_0\big(\omega^2\big|(\mathbf{r}\times\nabla)g_l(r)Y_{lm}(\theta,\phi)\big|^2 + c^2\big|\nabla\times(\mathbf{r}\times\nabla)g_l(r)Y_{lm}(\theta,\phi)\big|^2\big). \quad (27)$$

This expression holds both for the TE and the TM field. For the TE field, the first term in equation (27) is the electric contribution to the energy, whereas this term represents the magnetic contribution of the TM field. The density of AM is found by substituting the results in equations (24) and (26) in the general expression (3). After time averaging, we find

$$\mathbf{j}(\mathbf{r}) = -i\omega\epsilon_0 l(l+1)|g_l|^2 Y_{lm}^*(\mathbf{r}\times\nabla)Y_{lm} + \text{c.c.}, \quad (28)$$

again both for TE and TM fields. The z-component of the angular-momentum density takes the simple expression

$$j_z(\mathbf{r}) = 2\omega\epsilon_0 l(l+1)m|g_l|^2|Y_{lm}|^2. \quad (29)$$

One would expect that the ratio between j_z/w is identical to m/ω, which means that for each photon the AM in the z-direction is $\hbar m$. Although the expression's densities of energy and AM bear some similarity, this local ratio does not hold exactly. On the other hand, in the far field, where $kr \gg 1$, this ratio between AM and energy is obtained for each spherical shell, for the field emitted by a radiating multipole in the origin [1]. When we integrate the AM density (equation (28)) over the surface of a sphere with radius r, we obtain

$$\frac{d\mathbf{J}}{dr} \equiv \int d\Omega\,\mathbf{j}(\mathbf{r}) = 2\omega\epsilon_0 l(l+1)m|g_l|^2\mathbf{e}_z. \quad (30)$$

Likewise, we can obtain an expression for the energy density integrated over a sphere of radius r when we substitute in equation (27) the identity in spherical coordinates and spherical unit vectors

$$\nabla\times(\mathbf{r}\times\nabla)g_l Y_{lm} = \left[\left(\frac{1}{r}+\frac{d}{dr}\right)g_l\right]\big[\mathbf{e}_r\times(\mathbf{r}\times\nabla)Y_{lm}\big] + \frac{i}{r}\mathbf{e}_r l(l+1)g_l Y_{lm}, \quad (31)$$

and integrate the result over the spherical angles. The result is simple to obtain, since the two vector contributions in equation (31) are orthogonal. The result demonstrates that the ratio between dJ_z/dr and $dW/dr \equiv \int d\Omega\, w(\mathbf{r})$ is not independent of r in general.

This same relation (equation (31)) can also be used to study the singularity in phase or polarization of the multipole fields at the origin. When the multipole field is emitted by an electric multipole at the origin, it has the form of the TM field, with g_l proportional to the outgoing spherical Hankel function $h_l^+(kr)$. A magnetic multipole at the origin emits a TE field, specified by this same function h_l^+ [1]. For a source-free field, the function g_l is proportional to the spherical Bessel function $j_l(kr)$. This function is regular at the origin, and proportional to r^l. This determines the phase and the polarization of the field at the surface of a small sphere around the origin. At the origin the fields vanish, and both the phase and the polarization are undetermined. Therefore, the origin of the spherical multipole fields has a three-dimensional vortex structure, both in phase and in polarization. The behavior of the TE electric field vector in equation (24) near the origin is given by

$$\mathbf{E}_{lm}^{\mathrm{TE}}(\mathbf{r},t) \propto r^l(\mathbf{r}\times\nabla)Y_{lm}e^{-i\omega t} + \text{c.c.} \tag{32}$$

Its polarization is everywhere orthogonal to $\mathbf{r}$, so that it is tangential to the sphere. The polarization properties of the TM electric field in equation (26) can be analyzed by using the identity in equation (31). This gives the electric field proportional to

$$\mathbf{E}_{lm}^{\mathrm{TM}}(\mathbf{r},t) \propto r^{l-1}\big(\mathbf{e}_r\times(\mathbf{r}\times\nabla)Y_{lm} + i\mathbf{e}_r lY_{lm}\big)e^{-i\omega t} + \text{c.c.} \tag{33}$$

The first term on the right side is tangential, and the last term has a radial orientation. (Recall that the operator $\mathbf{r}\times\nabla$ in spherical coordinates only depends on the spherical angles θ and ϕ.) These three-dimensional vortical structures are relatively simple for low values of l, and they become quite complicated with increasing l-values. They consist of combinations of phase singularities. The situation is more transparent for vortices in two dimensions. These will be considered in the following sections.

2.4.2 Cylindrical Multipole Fields

Most common light beams have a cylindrical rather than a spherical nature. Cylindrical multipole fields arise from scalar solutions of the Helmholtz wave equation of the cylindrically separated type

$$\psi_{m\kappa}(\mathbf{r}) = G_{mK}(R)e^{i\kappa z}e^{im\phi}, \tag{34}$$

where R is the cylindrically radial parameter $\sqrt{x^2+y^2}$, and κ is the z-component of the wave vector. The integer m is the azimuthal index. The radial wave equation is

$$\left(\frac{d^2}{dR^2} + \frac{1}{r}\frac{d}{dR} + K^2 - \frac{m^2}{R^2}\right)G_{mK}(R) = 0. \tag{35}$$

The frequency ω of the wave is determined by $\omega^2 = c^2(\kappa^2 + K^2)$, and K is the transverse (xy) component of the wave vector.

In analogy to the expressions (23) and (25) for the vector potential of the spherical multipole fields, the cylindrical fields are given by

$$\mathbf{A}_{m\kappa}^{\mathrm{TE}}(\mathbf{r},t) = (\mathbf{e}_z \times \nabla)\psi_{m\kappa}e^{-i\omega t} + \text{c.c.},$$
$$\mathbf{A}_{m\kappa}^{\mathrm{TM}}(\mathbf{r},t) = -i\frac{c}{\omega}\nabla \times (\mathbf{e}_z \times \nabla)\psi_{m\kappa}e^{-i\omega t} + \text{c.c.} \tag{36}$$

This also determines the electric and the magnetic fields in these two cases. Expressed in cylindrical coordinates and unit vectors $\mathbf{e}_R$, $\mathbf{e}_\phi$ and $\mathbf{e}_z$, we find the expressions for the TE fields in the form

$$\mathbf{E}_{m\kappa}^{\mathrm{TE}} = \left(\mathbf{e}_R\frac{\omega m}{R}G_{mK} + i\omega\mathbf{e}_\phi\frac{dG_{mK}}{dR}\right)e^{i\kappa z}e^{im\phi}e^{-i\omega t} + \text{c.c.},$$
$$\mathbf{B}_{m\kappa}^{\mathrm{TE}} = \left(-i\kappa\mathbf{e}_R\frac{dG_{mK}}{dR} + \mathbf{e}_\phi\frac{\kappa m}{R}G_{mK} - \mathbf{e}_z K^2 G_{mK}\right)e^{i\kappa z}e^{im\phi}e^{-i\omega t} + \text{c.c.} \tag{37}$$

Notice that the electric field $\mathbf{E}_{m\kappa}^{\mathrm{TE}}$ is restricted to the xy-plane, which justifies the name of these fields. The TM fields follow from these expressions by the duality transformation $\mathbf{E}_{m\kappa}^{\mathrm{TM}} = c\mathbf{B}_{m\kappa}^{\mathrm{TE}}$, $\mathbf{B}_{m\kappa}^{\mathrm{TM}} = -\mathbf{E}_{m\kappa}^{\mathrm{TE}}/c$.

When substituting these results in equation (3), it is now easy to find expressions for the time-averaged density of energy and momentum for these fields. The results are the same for TE fields and TM fields. As a result of the cylindrical symmetry, the energy density depends exclusively upon the distance R from the axis. This is also true for the components of the momentum density along the cylindrical unit vectors. By using equation (1), one finds the expression for the z-component of AM, in the form

$$j_z(\mathbf{r}) = R\mathbf{p}\cdot\mathbf{e}_\phi = 2\omega\epsilon_0 K^2 m\left|G_{mK}(R)\right|^2. \tag{38}$$

The expressions in equations (37) for the fields can display singularities on the axis, where $R = 0$. For a source-free field, the radial function $G_{m\kappa}(R)$ has to be regular at the origin, which makes it proportional to the Bessel function $J_m(KR)$. The corresponding Bessel beams have the peculiar property that they are diffraction-free [16,17]. The price one has to pay, however, is that the power of the beam passing a transverse plane is infinite. Near the axis, the Bessel function $J_m(KR)$ is proportional to $R^{|m|}$. From equations (37), we can find the form of the TE and TM electric fields near the axis, and they can be put in Cartesian form by using the identities $(\mathbf{e}_R + i\mathbf{e}_\phi)e^{i\phi} = \mathbf{e}_x + i\mathbf{e}_y$ and $R\exp(i\phi) = x + iy$. For values of $m \geqslant 1$, we find

$$\mathbf{E}_{m\kappa}^{\mathrm{TE}}(\mathbf{r},t) \propto (\mathbf{e}_x + i\mathbf{e}_y)(x+iy)^{m-1}e^{i\kappa z}e^{-i\omega t} + \text{c.c.},$$
$$\mathbf{E}_{m\kappa}^{\mathrm{TM}}(\mathbf{r},t) \propto \left(i\kappa m(\mathbf{e}_x + i\mathbf{e}_y)(x+iy)^{m-1} + K^2\mathbf{e}_z(x+iy)^m\right)e^{i\kappa z}e^{-i\omega t} + \text{c.c.} \tag{39}$$

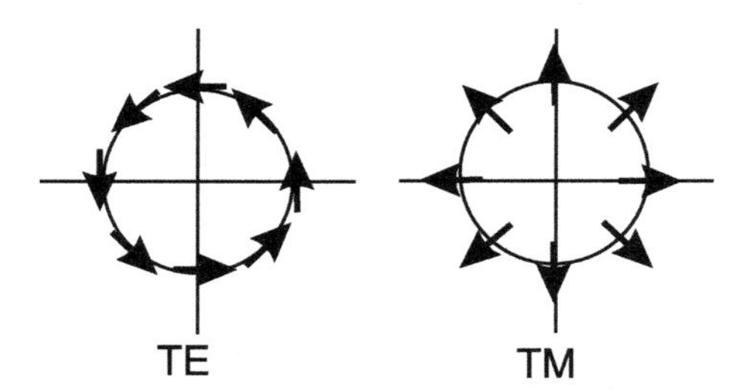

Figure 2.1 Sketch of the linear polarization in the transverse plane for the cylindrical TE and TM modes with $m = 0$. Arrows indicate the direction of linear polarization.

These fields have an m-fold rotational symmetry. For the corresponding negative values of $m \leqslant -1$, the electric fields have the form

$$\mathbf{E}_{m\kappa}^{\mathrm{TE}}(\mathbf{r}, t) \propto (\mathbf{e}_x - i\mathbf{e}_y)(x - iy)^{|m|-1} e^{i\kappa z} e^{-i\omega t} + \text{c.c.},$$
$$\mathbf{E}_{m\kappa}^{\mathrm{TM}}(\mathbf{r}, t) \propto \big(-i\kappa|m|(\mathbf{e}_x - i\mathbf{e}_y)(x - iy)^{|m|-1} + K^2\mathbf{e}_z(x - iy)^{|m|}\big) e^{i\kappa z} e^{-i\omega t} + \text{c.c.} \tag{40}$$

The TE fields are circularly polarized at the origin, while the phase has a vortex with charge $m - 1$ for positive m-values, and charge $-|m| + 1 = m + 1$ for negative m-values. (For a phase vortex with charge m, the phase increases with the value $2\pi m$ along a closed contour around the origin.) This means that the TE fields with $m = \pm 1$ have no phase vortex at the origin. The TM fields have the same structure, with an additional linearly polarized component in the z-direction, and a phase vortex with charge m. The isotropic case $m = 0$ requires special attention. In that case, also the second term in the power expansion of the Bessel function J_0 is needed to obtain the lowest nonvanishing order of the field. When we substitute the approximation $J_0(KR) \approx 1 - (KR)^2/4$, we obtain as expressions for the electric fields with $m = 0$ in Cartesian notation

$$\mathbf{E}_{0\kappa}^{\mathrm{TE}}(\mathbf{r}, t) \propto (-\mathbf{e}_x y + \mathbf{e}_y x) e^{i\kappa z} e^{-i\omega t} + \text{c.c.},$$
$$\mathbf{E}_{0\kappa}^{\mathrm{TM}}(\mathbf{r}, t) \propto \big(i\kappa(\mathbf{e}_x x + \mathbf{e}_y y) - \mathbf{e}_z\big) e^{i\kappa z} e^{-i\omega t} + \text{c.c.} \tag{41}$$

The TE field near the origin has a linear polarization in the azimuthal direction $\mathbf{e}_\phi$, so that it has a polarization vortex. At the origin, the polarization is undetermined, and the field vanishes. The TM field has a finite z-component at the origin, while the transverse component has a polarization vortex, with radial polarization around the origin. The polarization around the axis in the transverse plane is shown in Figure 2.1.

Both the spherical and the cylindrical multipole fields are exact solutions of Maxwell's equations. On the other hand, their energy content is infinite, and the intensity of the Bessel beams is infinite, just as for plane-wave fields. This makes these Bessel beams unrealistic as a representation of light beams. In the next sections, we will discuss the AM and vortex properties of light beams that do not suffer from this drawback.

2.5 ANGULAR MOMENTUM OF MONOCHROMATIC PARAXIAL BEAMS

A light field in a vacuum is commonly described by a complex field $\mathbf{f}(\mathbf{r}, t)$, so that the electric field is given by $\mathbf{E} = \mathbf{f} + \mathbf{f}^*$. From Maxwell's equations, it follows that $\mathbf{f}$ has a vanishing divergence and that it obeys the wave equation so that

$$c^2\nabla^2\mathbf{f} = \frac{\partial^2\mathbf{f}}{\partial t^2}, \qquad \nabla \cdot \mathbf{f} = 0. \tag{42}$$

2.5.1 Paraxial Approximation

The paraxial approximation for the description of a radiation field applies when the wave vectors of the field fall within a narrow cone with a small opening angle. This is the case for light beams, as they are produced by lasers. Light fields with wide angles also produce paraxial fields after passing through small apertures. The electric field of a monochromatic light beam with frequency ω that propagates in a vacuum in the positive z-direction is written in the standard complex notation as the product of a plane wave and a slowly varying envelope as

$$\mathbf{f}(\mathbf{r}, t) = \mathbf{u}(\mathbf{R}, z)e^{i(kz-\omega t)}, \tag{43}$$

with $\omega = ck$. Here $\mathbf{R} = (x, y)$ is the 2D transverse position vector, and $\mathbf{r} = (\mathbf{R}, \mathbf{z})$ is the position vector in three dimensions.

The propagation equation for $\mathbf{u}$ follows from the conditions in equation (42) for $\mathbf{f}$. The paraxial approximation is justified when $|\partial u/\partial \mathbf{R}|/(ku) \ll 1$. In this case, the transverse profile of $\mathbf{u}$ varies only slowly with z, so that the second derivative with respect to z can be ignored. Then the propagation of the light beam is well described by the paraxial wave equation [18,19]

$$\left(\frac{\partial^2}{\partial \mathbf{R}^2} + 2ik\frac{\partial}{\partial z}\right)\mathbf{u}(\mathbf{R}, z) = 0, \tag{44}$$

where the vector field $\mathbf{u}$ lies in the transverse (xy) plane. The paraxial approximation can be viewed as a lowest-order term of an expansion in the small paraxial parameter

$\delta = 1/(k\gamma_0)$, with γ_0 the beam waist [18]. For a monochromatic field, the complex vector potential and the complex magnetic field are simply determined by the electric field (equation (43)). The lowest-order terms are

$$\mathbf{a}(\mathbf{r}, \mathbf{t}) = \frac{\mathbf{1}}{\mathbf{i}\omega}\mathbf{f}(\mathbf{R}, \mathbf{z}), \qquad \mathbf{b}(\mathbf{r}, \mathbf{t}) = \frac{\mathbf{1}}{\mathbf{c}}\mathbf{e_z} \times \mathbf{f}(\mathbf{R}, \mathbf{z}). \tag{45}$$

Note that to the zeroth order, these fields obey the relations $\mathbf{f} = -\dot{\mathbf{a}}$ and $\dot{\mathbf{b}} + \nabla \times \mathbf{f} = 0$, in accordance with Maxwell's equations. Equations (45) show that the components of the complex vector potential **a** and the complex magnetic field **b** in the transverse plane have the same pattern as the electric field. The vector potential is a quarter phase ahead of the electric field, while the magnetic-field pattern has a polarization vector that is equal to the electric polarization vector rotated over $\pi/2$ in the positive (anticlockwise) direction.

The components of the real physical transverse electric field **E**, the vector potential **A**, and the magnetic field **B** in the transverse xy-plane are denoted as $\mathbf{E}_t$, $\mathbf{A}_t$ and $\mathbf{B}_t$. These components are determined by the identities

$$\mathbf{E}_t(\mathbf{r}, \mathbf{t}) = \mathbf{f}(\mathbf{r}, \mathbf{t}) + \text{c.c.}, \qquad \mathbf{A_t}(\mathbf{r}, \mathbf{t}) = \mathbf{a}(\mathbf{r}, \mathbf{t}) + \text{c.c.},$$
$$\mathbf{B}_t(\mathbf{r}, \mathbf{t}) = \mathbf{b}(\mathbf{r}, \mathbf{t}) + \text{c.c.}, \tag{46}$$

in terms of the corresponding complex fields **f**, **a**, and **b**. The z-components of the fields **E**, **A**, and **B** are nonvanishing in higher order. Since the fields are all divergence-free, their first-order terms are proportional to the transverse divergence of **f**, **a**, and **b**, and we find

$$E_z = \frac{i}{k}\frac{\partial}{\partial \mathbf{R}} \cdot \mathbf{f} + \text{c.c.}, \qquad A_z = \frac{i}{k}\frac{\partial}{\partial \mathbf{R}} \cdot \mathbf{a} + \text{c.c.},$$
$$B_z = \frac{i}{k}\frac{\partial}{\partial \mathbf{R}} \cdot \mathbf{b} + \text{c.c.} \tag{47}$$

2.5.2 Angular Momentum of a Monochromatic Beam

The momentum density has a leading term $\epsilon_0 \mathbf{E}_t \times \mathbf{B}_t$, which points in the z-direction. After using the expressions (46) and eliminating the rapidly oscillating terms by averaging over a few optical cycles, the zeroth-order contribution to the momentum density is found as

$$p_z(\mathbf{R}, \mathbf{z}) = \frac{2\epsilon_0}{c}\mathbf{u}^* \cdot \mathbf{u}. \tag{48}$$

It is easy to verify that the leading term in the Poynting vector $\mathbf{S} = \mathbf{E} \times \mathbf{H}$ is equal to its z-component $S_z = c^2 p_z = cw$, with

$$w(\mathbf{R}, \mathbf{z}) = \epsilon_0 \big(\mathbf{E}_t^2 + c^2 \mathbf{B}_t^2\big)/2 = 2\epsilon_0 \mathbf{u}^* \cdot \mathbf{u}, \tag{49}$$

the energy density of the beam. The energy per unit length is denoted as

$$\frac{dW}{dz} = \int d\mathbf{R}\, w(\mathbf{R}, z) = 2\epsilon_0 \int d\mathbf{R}\, \mathbf{u}^* \cdot \mathbf{u}. \tag{50}$$

When we use the photon energy $\hbar\omega$ as an energy quantum, the photon density is $n = w/(\hbar\omega)$, and the momentum density (equation (48)) amounts to $n\hbar k$, which corresponds to a unit $\hbar k$ per photon.

However, we are not interested in the AM arising from this photon momentum along the axis, but in the component j_z of the angular-momentum density in the propagation direction. Since

$$j_z = \mathbf{R} \times \mathbf{p}_t, \tag{51}$$

this z-component arises from the components of the momentum density in the transverse (xy) plane, which according to equation (3) arises from combinations of $\mathbf{E}$ and $\mathbf{B}$ in the xy-plane and in the z-direction. (Since both $\mathbf{R}$ and $\mathbf{p}_t$ are vectors in the xy-plane, their cross-product points in the z-direction, and we can treat it as a scalar.) To first order in δ, the transverse component of the momentum density is

$$\mathbf{p}_t = \epsilon_0 \big[E_z(\mathbf{e}_z \times \mathbf{B}_t) + B_z(\mathbf{E}_t \times \mathbf{e}_z)\big]. \tag{52}$$

After substituting the expressions (46) and (47) and averaging over an optical cycle, one finds after some rewriting that j_z can be separated into the sum $j_z = l + s$, where l and s are given by the expressions in cylindrical coordinates

$$l(\mathbf{R}, \mathbf{z}, \mathbf{t}) = \epsilon_0 \mathbf{f}^* \cdot \frac{\partial}{\partial \phi}\mathbf{a} + \text{c.c.}, \qquad s(\mathbf{R}, \mathbf{z}, \mathbf{t}) = -\epsilon_0 \mathbf{f}^* \times R\frac{\partial}{\partial R}\mathbf{a} + \text{c.c.} \tag{53}$$

These expressions for the densities of orbital and spin AM have striking similarities and differences with the integrands in the expressions (15) for an arbitrary radiation field. Both contain a product of the electric field and (a spatial derivative of) the vector potential. A major difference is that in the case of a paraxial beam, the separation in l and s can be made for the local densities. This is not true for an arbitrary radiation field.

In the special case of a monochromatic field, where $\mathbf{f}$ and $\mathbf{a}$ are given by equations (43) and (45), the separation (equation (53)) takes the form

$$l(\mathbf{R}, \mathbf{z}) = \frac{\epsilon_0}{i\omega}\mathbf{u}^* \cdot \frac{\partial}{\partial \phi}\mathbf{u} + \text{c.c.}, \qquad s(\mathbf{R}, \mathbf{z}) = -\frac{\epsilon_0}{i\omega} R\frac{\partial}{\partial R}\big(\mathbf{u}^* \times \mathbf{u}\big). \tag{54}$$

The contribution l is determined by the phase gradient of the two components of $\mathbf{u}$ in the azimuthal direction. This expression has the flavor of a density of orbital AM, as is obvious when we compare it to the expression for the z-component of the orbital AM of a particle in elementary quantum mechanics. On the other hand, the separation of j_z in l and s holds exactly for the contributions to the density of AM. The quantity s arises from the gradient in the radial direction of the cross-product $(\mathbf{u}^* \times \mathbf{u})/i$ of the transverse mode amplitude. We recall that for an arbitrary radiation field, the separation (equation (14)) of $\mathbf{J}$ into $\mathbf{L}$ and $\mathbf{S}$ could only be made for the total angular momenta, integrated over the entire space. It is remarkable that for a paraxial beam, the separation of j_z as $l + s$ arises for the densities in each point of space separately. Even so, expression (54) for l is identical to the integrand in expression (15) for $\mathbf{L}_{\text{rad}}$, when $\mathbf{A}$ and $\mathbf{E}^{\perp}$ are represented by their paraxial expressions (46).

The spin per unit beam length is given by the integral $\Sigma \equiv \int d\mathbf{R}\, s(\mathbf{R}, \mathbf{z})$, and the orbital AM per unit length is equal to $\Lambda \equiv \int d\mathbf{R}\, l(\mathbf{R}, \mathbf{z})$. We use partial integration with respect to ϕ for Λ, and with respect to R for Σ, and we obtain

$$\Lambda = \frac{2\epsilon_0}{\omega} \int d\mathbf{R}\, \mathbf{u}^* \cdot \frac{1}{i} \frac{\partial}{\partial \phi} \mathbf{u}, \qquad \Sigma = \frac{2\epsilon_0}{\omega} \int d\mathbf{R}\, (\mathbf{u}^* \times \mathbf{u})/i. \tag{55}$$

It is easy to show that both Σ and Λ are invariant under free propagation [20]. One also easily verifies that the integrand of this expression for Σ coincides with the integrand in equation (15) for the z-component of $\mathbf{S}_{\text{rad}}$. Again, this is remarkable, since the integration in equation (55) runs only over the transverse plane, not over the entire volume.

When we separate the complex vector field $\mathbf{u}(\mathbf{R}, z)$ as $\mathbf{u} = u\mathbf{e}$, with $\mathbf{e}$ the complex normalized local polarization vector, and $u = |\mathbf{u}|$ the local field strength, we arrive at the identity $2\epsilon_0(\mathbf{u}^* \times \mathbf{u})/i = \sigma w$, where the cross-product $\sigma = (\mathbf{e}^* \times \mathbf{e})/i$ is the local helicity of the beam. The helicity σ is a real number that is zero for linear polarization, and it takes the value ± 1 for circular polarization $\mathbf{e}_{\pm} = (\mathbf{e}_x \pm i\mathbf{e}_y)/\sqrt{2}$. The spin density in equation (54) is found to be localized in the region of the radial gradient of the product σw. However, equation (55) for Σ may be read as an integration of $n\hbar\sigma$, which is the product of the photon density $n = w/(\hbar\omega)$ and the spin $\hbar\sigma$ per photon, where both the photon density and the helicity may depend on the transverse position $\mathbf{R}$.

2.5.3 Uniform Orbital and Spin Angular Momentum

The expressions (54) and (55) generalize the results for a monochromatic beam with uniform polarization. In that case, we can write $\mathbf{u}(\mathbf{R}, z) = \mathbf{e}u(\mathbf{R}, z)$, where the polarization vector $\mathbf{e}$ is independent of position. Then the helicity σ is uniform over the cross-section of the beam, and we recover from equations (54) the known expressions [5]

$$l(\mathbf{R}, \mathbf{z}) = \frac{\epsilon_0}{i\omega} u^* \frac{\partial}{\partial \phi} u + \text{c.c.}, \qquad s(\mathbf{R}, z) = -\frac{\sigma}{2\omega} R \frac{\partial}{\partial R} w(\mathbf{R}, z). \tag{56}$$

This shows that the spin density is determined by the radial derivative of the energy density. The integrated spin momentum obeys the relation $\Sigma = \sigma W/\omega$, which corresponds to $\hbar\sigma$ per photon, as expected [3]. However, the spin is localized in the region of the gradient of energy density, so that it vanishes in the region of uniform intensity. On the other hand, when a fraction of the light is absorbed by a particle, or when it is cut out by an aperture, the relation $\Sigma = \sigma W/\omega$ also applies for this fraction. In this sense, it is justified to say that light with a uniform helicity σ carries a spin $\hbar\sigma$ per photon [2].

Of special interest are mode profiles of the form

$$u(R, \phi, z) = F_m(R, z) \exp(im\phi), \tag{57}$$

where the ϕ-dependence is given by the factor $\exp(im\phi)$. In order that the mode be continuous, m must be an integer. Then the density of orbital AM is equal to $l = mw/\omega = n\hbar m$, and the orbital AM per photon is $\hbar m$. These modes have a phase singularity in the origin, which corresponds to a vortex of charge m. They are eigenmodes of the differential operator $\partial/\partial\phi$. However, it would be confusing to state that they are eigenmodes of orbital AM. In the classical context we are discussing here, orbital AM is just a classical quantity, not an operator. For any classical beam, the amount of orbital AM has a well-defined specific value, and the same is true for the spin. What is special about these modes is that the density of orbital AM is proportional to the energy density. In this sense, the orbital AM can be said to be uniform over the beam profile. The modes in equation (57) have an orbital AM that can be quantified as $\hbar m$ per photon. Since the paraxial wave equation (44) is isotropic, this ϕ-dependence is conserved during free propagation. The radial mode function F_m obeys the radial paraxial wave equation

$$\left(\frac{\partial^2}{\partial R^2} + \frac{1}{R}\frac{\partial}{\partial R} - \frac{m^2}{R^2} + 2ik\frac{\partial}{\partial z} \right) F_m(R, z) = 0. \tag{58}$$

A well-known example is provided by the Laguerre–Gaussian modes [5,21]. For these modes, the radial mode functions are denoted as $F_{mp}(R, z)$, where p is the radial mode number. The function F_{mp} is the product of a Gaussian function, a factor $R^{|m|}$, and an associated Laguerre polynomial that depends only upon the absolute value $|m|$ [21]. These mode functions have the special property that their radial shape is invariant during free propagation, apart from a scaling factor. The z-dependent scaling factor is the width of the radial profile. Around the beam axis, the profiles of these beams are proportional to $R^{|m|}\exp(im\phi) = (x \pm iy)^{|m|}$, depending upon the sign of m. This shows that the beams have a phase singularity of charge m.

2.5.4 Nonuniform Polarization

When a beam with nonuniform polarization passes a polarizer, the mode profile of the outgoing beam depends upon the setting of the polarizer. This means that the mode function $\mathbf{u}$ does not factorize into the form $\mathbf{e}u(\mathbf{R}, z)$ with a fixed polarization vector. On the quantum level, this means that for each photon in the beam, its polarization and its translational degrees of freedom are entangled. At present, light beams with a nonuniform linear polarization and axial symmetry are widely studied. They can be generated by using liquid–crystal converters [22]. Another technique is based on spatially varying dielectric gratings [23,24].

As an example, we consider the superposition of two Laguerre–Gaussian light beams with opposite azimuthal mode number $\pm m$, and with opposite circular polarizations. We consider a monochromatic beam characterized by the mode pattern

$$\mathbf{u}(R, \phi, z) = F_{mp}(R, z)\big[\mathbf{e}_{+}e^{-im\phi} + \mathbf{e}_{-}e^{im\phi}\big]/\sqrt{2}. \tag{59}$$

The mode function in equation (59) is real everywhere, and it is the superposition of two components with orbital AM per photon $\mp m\hbar$, and spin AM $\pm\hbar$ per photon. The vector multiplying F_{mp} in equation (59) is a ϕ-dependent linear polarization vector $\mathbf{e}(\phi) = \mathbf{e}_x\cos(m\phi) + \mathbf{e}_y\sin(m\phi)$. This polarization vector is in the x-direction for $\phi = 0$, and along a circle around the beam axis the polarization direction makes m full rotations in the positive direction. The directions of linear polarization as a function of ϕ are indicated by the black arrows in Figure 2.2. The linearly polarized field oscillates in phase everywhere along such a circle. For negative values of m, the polarization direction rotates in the negative direction along the circle.

In the special case when $m = 1$, the number of rotations is 1, and the pattern is rotationally invariant. Then the density of AM $j_z = l + s$ is zero for each one of the separate contributions in equation (59), and the beam is invariant for rotation around the axis. The polarization direction is always in the radial direction. When we replace ϕ by $\phi - \phi_0$ in the right side of equation (59), the pattern is still isotropic, and the polarization direction makes an angle ϕ_0 with the radial direction.

The density of orbital and spin AM of the mode in equation (59) can be evaluated with equation (54), and both are found to be zero. In fact, this mode is a superposition of two terms with orbital AM equal to $\mp\hbar m$, and spin $\pm\hbar$ per photon. The energy density is

$$w(R, z) = 2\epsilon_0\big|F_{mp}(R, z)\big|^2. \tag{60}$$

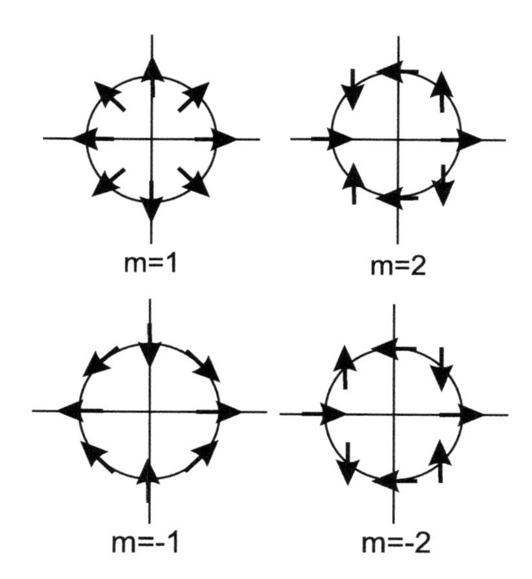

Figure 2.2 Sketch of the position-dependent linear polarization for a mode as described in equation (59). Arrows indicate the direction of linear polarization.

Accordingly, near the axis, the pattern of phase and polarization is described by the expression

$$\mathbf{u}(x, y) \propto (\mathbf{e}_x + i\mathbf{e}_y)(x - iy)^m + (\mathbf{e}_x + i\mathbf{e}_y)(x + iy)^m. \qquad (61)$$

This describes a singularity in phase and polarization with a mixed charge.

An interesting generalization is the case of a similar superposition of modes with opposite circular polarization, and ϕ-dependent phase terms with two arbitrary m-values. This gives a transverse mode function

$$\mathbf{u}(R, \phi) = F(R)\big[\mathbf{e}_+ e^{im'\phi} + \mathbf{e}_- e^{im\phi}\big]/\sqrt{2}, \qquad (62)$$

prepared in a single transverse plane, where the azimuthal mode numbers m and m' are now arbitrary integer numbers. We omitted the z-dependence of the mode, since in the general case the two terms will undergo different diffraction, so for different transverse planes the radial mode functions will no longer be identical. When we extract a phase factor $\exp(i(m + m')\phi/2)$, the remaining polarization vector is $\mathbf{e}(\phi) = \mathbf{e}_x \cos((m - m')\phi/2) + \mathbf{e}_y \sin((m - m')\phi/2)$. The number of rotations of the polarization vector along a circle around the beam axis is then $(m - m')/2$. This is a half-integer value when $m - m'$ is odd. The polarization pattern is illustrated in Figure 2.3 for the cases that $m - m' = \pm 1$. The overall phase factor $\exp(i(m + m')\phi/2)$ indicates that the phase of the polarized field varies along the circle.

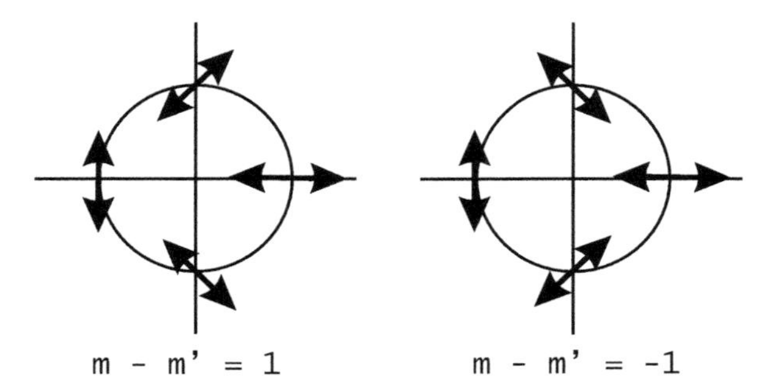

Figure 2.3 Sketch of the position-dependent linear polarization for a mode as described in equation (62). Arrows indicate the direction of linear polarization.

2.6 QUANTUM DESCRIPTION OF PARAXIAL BEAMS

2.6.1 Quantum Operators for Paraxial Fields

The expressions (17) for the quantum operators of the transverse electric field and the vector potential can also be given in the special case of the paraxial approximation. The results are useful in cases that only photons in paraxial beams are considered. It is possible to apply a formal canonical quantization procedure to the paraxial formalism by starting with a Lagrangian density [25]. It is simpler to apply the paraxial approximation to the general expressions (17) for the transverse electric field and the vector potential, and to restrict the values for the wave vectors **k** to the paraxial regime around the z-axis. This means that the component of **k** in the xy-plane is small compared to the z-component $k_z \approx k = \omega/c$. We are free to choose any orthonormal basis of paraxial modes. Since we are also interested in the paraxial operators for AM, it is convenient to use the Laguerre–Gaussian modes with circular polarization. These modes are denoted as

$$\mathbf{u}_{mps}(\mathbf{R}, z|\omega) = \mathbf{e}_s e^{im\phi} F_{mp}(R, z|\omega), \tag{63}$$

with $s = \pm$ for the two circular polarizations, and F_{mp} the radial part of the Laguerre–Gaussian modes at frequency ω, with m and p the azimuthal and the radial mode index. The modes are labeled by the discrete parameters m, p, s, and the continuous variable ω. The normalization is ensured by the condition

$$2\pi \int R\, dR\, F^*_{mp}(R, z|\omega) F_{mp'}(R, z|\omega) = \delta_{pp'}. \tag{64}$$

The annihilation operator for photons in these modes are indicated as $\hat{a}_{mps}(\omega)$, with the commutation rules $[\hat{a}_{m'p's'}(\omega'), \hat{a}^{\dagger}_{mps}(\omega)] = \delta_{mm'}\delta_{pp'}\delta_{ss'}\delta(\omega - \omega')$.

The operator expressions for the transverse electric field and the vector potential are now obtained from equation (17), after taking ω as a variable replacing k_z, while expanding the plane-wave terms in the xy-direction in the transverse modes (equation (63)). The operators for the components of the electric field and the vector potential in the xy-plane are now $\hat{\mathbf{E}}_t = \hat{\mathbf{f}} + \hat{\mathbf{f}}^\dagger$, $\hat{\mathbf{A}}_t = \hat{\mathbf{a}} + \hat{\mathbf{a}}^\dagger$, where the operators $\hat{\mathbf{f}}$ and $\hat{\mathbf{a}}$ are found as

$$\hat{\mathbf{f}}(\mathbf{r}) = i \sum_{mps} \int d\omega \sqrt{\frac{\hbar\omega}{4\pi\epsilon_0 c}} \mathbf{u}_{mps}(\mathbf{R}, z|\omega) e^{i\omega z/c} \hat{a}_{mps}(\omega),$$

$$\hat{\mathbf{a}}(\mathbf{r}) = \sum_{mps} \int d\omega \sqrt{\frac{\hbar}{4\pi\epsilon_0 \omega c}} \mathbf{u}_{mps}(\mathbf{R}, z|\omega) e^{i\omega z/c} \hat{a}_{mps}(\omega). \tag{65}$$

The Hamiltonian is found by integrating the expression (73) over space, which gives the expected result

$$\hat{H} = \sum_{mps} \int d\omega \, \hbar\omega \hat{a}^\dagger_{mps}(\omega) \hat{a}_{mps}(\omega). \tag{66}$$

It generates the correct time evolution for the field operators in the Heisenberg picture

$$\hat{a}_{mps}(\omega, t) = \exp(i\hat{H}t/\hbar)\hat{a}_{mps}(\omega)\exp(-i\hat{H}t/\hbar) = e^{-i\omega t}\hat{a}_{mps}(\omega, 0). \tag{67}$$

The Heisenberg operators for the fields $\hat{\mathbf{f}}(\mathbf{r}, t)$ and $\hat{\mathbf{a}}(\mathbf{r}, t)$ are given by the expansions in equation (65), with the exponentials replaced by $\exp(-i\omega(t - z/c))$.

2.6.2 Quantum Operators for Spin and Orbital Angular Momentum

As indicated in equation (53), the densities of orbital and spin AM in the paraxial limit can be expressed in terms of the complex fields **f** and **a**. The quantum operators for the components of **L** and **S** in the propagation direction are found by substituting equations (65) into equation (53) and integrating over space. The result can be worked out completely, with the result

$$\hat{S}_z = \sum_{mps} \int d\omega \, \hbar s \hat{a}^\dagger_{mps}(\omega) \hat{a}_{mps}(\omega),$$

$$\hat{L}_z = \sum_{mps} \int d\omega \, \hbar m \hat{a}^\dagger_{mps}(\omega) \hat{a}_{mps}(\omega). \tag{68}$$

These operators commute with each other and with the Hamiltonian equation (66). The operator for the total AM is denoted as $\hat{J}_z = \hat{S}_z + \hat{L}_z$.

The operator $\hat{S}_z$ generates rotations of the polarization, while $\hat{L}_z$ generates rotations of the mode pattern. This is demonstrated by the transformations

$$e^{i\alpha\hat{S}_z/\hbar}\hat{a}_{mps}(\omega)e^{-i\alpha\hat{S}_z/\hbar} = e^{-i\alpha s}\hat{a}_{mps}(\omega),$$
$$e^{i\alpha\hat{L}_z/\hbar}\hat{a}_{mps}(\omega)e^{-i\alpha\hat{L}_z/\hbar} = e^{-i\alpha m}\hat{a}_{mps}(\omega). \tag{69}$$

In a rotated state $\exp(-i\alpha\hat{S}_z/\hbar)|\Psi\rangle$ of the paraxial field, the expectation value of the electric field operator $\hat{\mathbf{f}}$ has a polarization that is rotated over an angle α compared with the polarization in the state $|\Psi\rangle$. In the rotated state $\exp(-i\alpha\hat{L}_z/\hbar)|\Psi\rangle$, it is the mode pattern that is rotated over angle α, while the polarization is left unchanged.

2.7 NONMONOCHROMATIC PARAXIAL BEAM

2.7.1 Angular Momentum of Nonmonochromatic Beam

We now consider a paraxial beam of nonmonochromatic light, which contains a discrete set of frequencies ω_n. As a generalization of equation (43), the transverse electric field for such a polychromatic beam can be written as in equation (46), where

$$\mathbf{f}(\mathbf{R}, z, t) = \sum_n \mathbf{u}_n(\mathbf{R}, z)e^{-i\omega_n(t-z/c)}. \tag{70}$$

The mode function $\mathbf{u}_n(\mathbf{R}, z)$ is a solution of the paraxial wave equation (44), with the frequency ω replaced by ω_n. This accounts for the frequency-dependent diffraction of the mode. The electric field also determines the complex vector potential $\mathbf{a}$ and the magnetic field $\mathbf{b}$, in the form

$$\mathbf{a}(\mathbf{R}, z, t) = \sum_n \frac{1}{i\omega_n}\mathbf{u}_n(\mathbf{R}, z)e^{-i\omega_n(t-z/c)},$$
$$\mathbf{b}(\mathbf{r}, \mathbf{t}) = \frac{1}{c}\mathbf{e}_z \times \mathbf{f}(\mathbf{R}, z). \tag{71}$$

The components of the real physical fields in the transverse plane are determined by equation (46).

It is trivial to generalize these results to the case of pulsed beams of light. It is sufficient to replace the summations in equations (70) and (71) by an integration over a continuous band of frequency values, so that

$$\mathbf{f}(\mathbf{R}, z, t) = \int d\omega\, \mathbf{u}(\mathbf{R}, z|\omega)e^{-i\omega(t-z/c)},$$
$$\mathbf{a}(\mathbf{R}, z, t) = \int d\omega \frac{1}{i\omega}\mathbf{u}(\mathbf{R}, z|\omega)e^{-i\omega(t-z/c)}. \tag{72}$$

The expression for the complex magnetic field **b** in equations (71) remains valid.

The electric and magnetic fields contribute equally to the energy density w of a polychromatic beam to zeroth order in δ. This energy density can be expressed in the complex field **f**. We assume that the differences in frequency are relatively small, and we evaluate the energy while averaging over rapidly oscillating terms at frequencies $\omega_n + \omega_{n'}$. On the other hand, we account for terms oscillating at the beat frequencies $\omega_n - \omega_{n'}$. Then the energy density is found in the form

$$w(\mathbf{R}, \mathbf{z}, \mathbf{t}) = 2\epsilon_0 \mathbf{f}^* \cdot \mathbf{f}. \tag{73}$$

This expression is a product of two summations over n and n'. The result depends explicitly on time, due to the presence of the beat terms with $n \neq n'$.

The density of AM in the z-direction to first order in δ can be obtained by using equation (51), with the transverse momentum given by equation (52). The z-components of the fields are related to the transverse parts by equation (47). Then the expression (53) for the separation of the density of AM into the sum $j_z = l + s$ is still valid, where in the present nonmonochromatic case, the complex electric field and the complex transverse vector potential in equations (70) and (71). For the general case of nonmonochromatic light, the transverse beam profile must be expected to depend upon time, and the same will be true for the energy density w and the density of AM j. This time dependence arises from the terms with $n \neq n'$ in the double summation over frequency values.

2.7.2 Spin of Rotating Polarization

A linear polarization vector that performs a rotation with angular frequency Ω can be described as [26]

$$\mathbf{e}(z, t) = \frac{1}{\sqrt{2}}\left(\mathbf{e}_+ e^{-i\Omega(t-z/c)} + \mathbf{e}_- e^{i\Omega(t-z/c)}\right). \tag{74}$$

As a function of time, the polarization rotates in the positive direction with angular frequency Ω in the transverse plane. As a function of the propagation coordinate z, the polarization rotates in the negative direction, with a pitch $2\pi c/\Omega$. The polarization vector can be viewed as indicating the direction of the steps of a winding staircase that is rotating about its axis. A light beam with a uniform rotating polarization is described by the complex electric field

$$\mathbf{f}(\mathbf{r}, t) = \mathbf{e}(z, t) u(\mathbf{R}, z) e^{-i\omega(t-z/c)}, \tag{75}$$

with $u(\mathbf{R}, z)$ a solution of the paraxial wave equation (44). This field describes the superposition of two light fields with frequencies $\omega \pm \Omega$. Strictly speaking, the mode profiles for the two frequency components will obey the paraxial wave equation (44) with different wave numbers so that they undergo slightly different diffraction. As a result, when the mode profiles are the same in one transverse plane, they will be slightly different in different planes. In practical cases where $\Omega \ll \omega$, this difference can be neglected for distances of the order of the diffraction length of the beam.

In order to evaluate the density of AM, we also need an expression for the complex vector potential $\mathbf{a}$ for the present case of a bichromatic field. This vector potential takes the form

$$\mathbf{a}(\mathbf{R}, z, t) = \frac{1}{\sqrt{2}} u(\mathbf{R}, z) e^{-i\omega(t-z/c)} \times \left(\mathbf{e}_+ \frac{1}{i(\omega + \Omega)} e^{-i\Omega(t-z/c)} + \mathbf{e}_- \frac{1}{i(\omega - \Omega)} e^{i\Omega(t-z/c)} \right). \tag{76}$$

The effect of the rotation is that the contribution of positive circular polarization is slightly diminished, while the contribution of negative circular polarization is enhanced. As a result, the contributions from the two circular polarizations to the spin density s do not exactly cancel each other. We find from equation (53), that

$$s(\mathbf{R}, \mathbf{z}) = \epsilon_0 \frac{\Omega}{\omega^2 - \Omega^2} R \frac{\partial}{\partial R} (u^* u). \tag{77}$$

The spin Σ per unit length takes the form

$$\Sigma = -W \frac{\Omega}{\omega^2 - \Omega^2}, \tag{78}$$

with $W = 2\epsilon_0 \int d\mathbf{R}\, u^* u$ the energy per unit length.

It is remarkable that rotation in the positive direction leads to a negative value of the spin AM. This may be a bit counterintuitive. It can be understood, however, by noting that the positive and negative circular polarization contribute equally to the beam intensity in this case. Since the positive circular polarization has the higher frequency, there are less photons with this polarization than with the opposite polarization. Since photons with circular polarization carry a spin ± 1, this explains equation (78).

2.7.3 Orbital Angular Momentum of Rotating Mode Pattern

A rotating mode pattern arises when we expand the pattern in terms $\propto \exp(im\phi)$, and replace ϕ by $\phi - \Omega t$. Each term with azimuthal mode number m will then pick up a frequency shift $m\Omega$. The simplest example of a rotating mode pattern is a linear combination of Laguerre–Gaussian modes with opposite mode numbers $\pm m$, with the

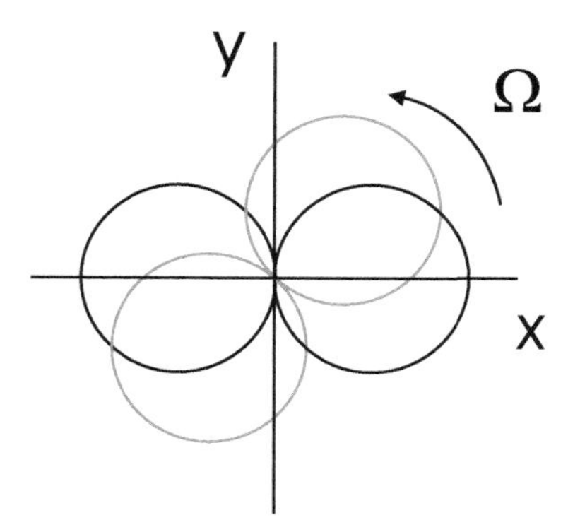

Figure 2.4 Sketch of a Hermite–Gaussian beam profile, rotating with angular velocity Ω in a fixed transverse plane. This corresponds to the energy density (equation (81)) in the case that $m = 1$. Black circles indicate the orientation of the profile at one instance of time; gray circles indicate the orientation direction at a time $\pi/4\Omega$ later.

complex electric field

$$\mathbf{f}(R,\phi,z,t) = \sqrt{2}\mathbf{e}\,F_{mp}(R,z)\cos\big[m\big(\phi - \Omega(t - z/c)\big)\big]e^{-i\omega(t-z/c)}. \tag{79}$$

As a function of the azimuthal angle ϕ, it takes the form of a circular standing wave. The expression for the vector potential in equation (71) is

$$\mathbf{a}(R,\phi,z,t) = \frac{1}{\sqrt{2}}\mathbf{e}\,F_{mp}(R,z)\bigg[\frac{1}{i(\omega + m\Omega)}e^{im\phi}e^{-im\Omega(t-z/c)} + \frac{1}{i(\omega - m\Omega)}e^{-im\phi}e^{im\Omega(t-z/c)}\bigg]e^{-i\omega(t-z/c)}, \tag{80}$$

and the energy density for this rotating mode pattern is

$$w(R,\phi,z,t) = 4\epsilon_0\big|F_{mp}(R,z)\big|^2\cos^2\big[m\big(\phi - \Omega(t - z/c)\big)\big]. \tag{81}$$

For the density of orbital AM we find that

$$l(\rho,\phi,z,t) = -\frac{m^2\Omega}{\omega^2 - m^2\Omega^2}w(\rho,\phi,z,t). \tag{82}$$

So in this special case of a circular standing wave, the orbital AM density is proportional to the energy density. This generalizes a result obtained earlier by a different argument for the special case $m = 1$ [11], where the superposition of two Laguerre–Gaussian modes just produces a rotating first-order Hermite–Gaussian beam. This case, where the energy pattern is proportional to $\cos^2(\phi - \Omega t)$, is shown in Figure 2.4. The gray curve refers to a situation a little later than the black curve in the rotation cycle. For each value of the azimuthal angle ϕ, the curves cut out a radius that measures the energy density of the beam as a function of ϕ.

The relation between the orbital AM Λ and the energy W, both per unit length, is simply

$$\Lambda = -\frac{m^2 \Omega}{\omega^2 - m^2 \Omega^2} W. \tag{83}$$

Notice that for an intensity pattern rotating in a positive direction, the orbital AM is negative. It would seem that a beam of light has a negative moment of inertia. However, as discussed in [11], a beam of light cannot be considered as a rigid body. The result in equation (83) can be understood by conceiving the rotating beam as the superposition of two beams with frequency $\omega \pm m\Omega$, with the same power, where the photons with the higher frequency carry an orbital AM $m\hbar$ and an energy $\hbar(\omega + m\Omega)$, and the lower-frequency photons carry an orbital AM $-m\hbar$ and an energy $\hbar(\omega - m\Omega)$. Since the powers are the same, there are more photons with the lower than with the higher frequency, and the sign of the net orbital AM is determined by the more abundant lower-frequency photons. This argument gives precisely the correct negative ratio (equation (83)) between the orbital AM and the energy.

2.7.4 Angular Momentum of Rotating Nonuniform Polarization

As an example of rotating nonuniform polarization, we start from the patterns sketched in Figure 2.4. The position-dependent linear polarization is set into rotation with angular velocity Ω when the two components of the mode function in equation (59) are given opposite frequency shifts $\pm\Omega$. The electric field is then given by

$$\begin{aligned} \mathbf{f}(R, \phi, z, t) = \frac{1}{\sqrt{2}} F_m(R, z)\big[&\mathbf{e}_+ e^{-im\phi} e^{-i\Omega(t-z/c)} \\ &+ \mathbf{e}_- e^{im\phi} e^{i\Omega(t-z/c)}\big] e^{-i\omega(t-z/c)}. \end{aligned} \tag{84}$$

The expression in square brackets indicates the polarization as a function of time t, azimuthal angle ϕ, and propagation coordinate z. A rotation of the polarization with angular velocity Ω is equivalent to a rotation of the mode pattern with the angular velocity $-\Omega/m$. This is also shown in Figure 2.5.

The corresponding vector potential is

$$\begin{aligned} \mathbf{a}(R, \phi, z, t) = \sqrt{2} F_m(R, z)\bigg(&\mathbf{e}_+ \frac{1}{i(\omega + \Omega)} e^{-im\phi} e^{-i\Omega(t-z/c)} \\ &+ \mathbf{e}_- \frac{1}{i(\omega - \Omega)} e^{im\phi} e^{i\Omega(t-z/c)}\bigg) e^{-i\omega(t-z/c)}. \end{aligned} \tag{85}$$

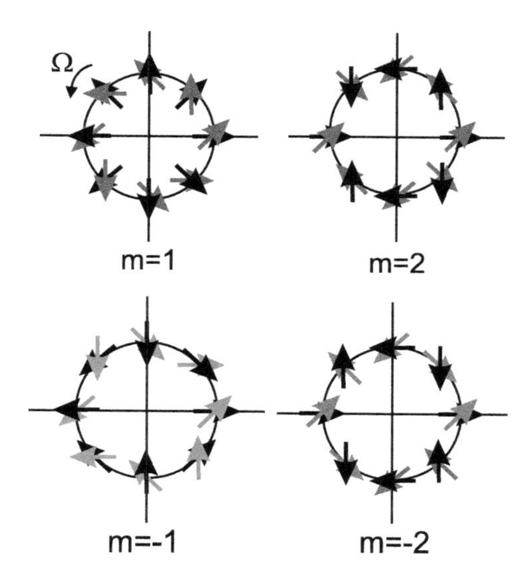

Figure 2.5 Sketch of the position-dependent pattern of rotating linear polarizations as described by equation (84). Dark arrows indicate the direction of the linear polarization at one instance of time; gray arrows indicate the linear polarization at a time $\pi/4\Omega$ later.

The energy density in this case of a rotating beam is still given by equation (60), which is both stationary and rotationally invariant.

The density of AM of the beam is remarkably simple. This is due to the fact that the mode is a biorthogonal sum of two polarizations ($\mathbf{e}_\pm$) and two amplitude modes ($\exp(\mp im\phi)$). In quantum-mechanical terms, the polarization and the transverse degrees of freedom of a photon in this mode are maximally entangled. From equation (53), we find as expressions for the angular-momentum densities

$$l(R,z) = \frac{m\Omega}{\omega^2 - \Omega^2} w(R,z), \qquad s(R,z) = \frac{1}{2}\frac{\Omega}{\omega^2 - \Omega^2} R \frac{\partial}{\partial R} w(R,z). \tag{86}$$

Both l and s are proportional to the angular velocity Ω. Just as the energy density is symmetric, the densities of orbital AM and spin are axially symmetric in this case. Integration over a transverse plane gives the relations between Λ, Σ and W as

$$\Lambda = \frac{m\Omega}{\omega^2 - \Omega^2} W, \qquad \Sigma = -\frac{\Omega}{\omega^2 - \Omega^2} W. \tag{87}$$

The expression (87) for Σ is identical to equation (78). This is understandable, since both situations refer to linear polarization rotating at the angular velocity Ω. The expression (87) for Λ follows from equation (83) when Ω is replaced by $-\Omega/m$, which is the effective angular velocity of the mode pattern. According to equation

(87), the total AM per unit length is $\Lambda + \Sigma = (m-1)\Omega W/(\omega^2 - \Omega^2)$. In the special case that $m = 1$, the total AM disappears, which is in accordance with the fact that in that case the beam (equation (59)) is invariant for rotation around the axis.

The situation of the rotating position-dependent linear polarizations is confusingly similar to the case of a circularly polarized Laguerre–Gaussian light field. In this case, the electric-field vector has the same distribution as shown in Figure 2.5, and it is rotating in the positive direction, as indicated. In this case, the field is monochromatic, the polarization is uniform, and the electric-field vector is rotating at the optical frequency ω. The rotational frequency shift in this situation of a well-determined AM has been analyzed in [27]. In contrast, in the situation discussed in the present section, the light is bichromatic while the polarization is linear at all times. The direction of this linear polarization is nonuniform and rotating at the much smaller frequency Ω.

2.8 OPERATOR DESCRIPTION OF CLASSICAL PARAXIAL BEAMS

2.8.1 Dirac Notation of Paraxial Beams

As is well known, the paraxial wave equation (44) for a monochromatic beam is mathematically equivalent to the Schrödinger equation of a free quantum particle in two dimensions, where the propagation coordinate z replaces time. This analogy suggests denoting the mode function $\mathbf{u}$ as a function of $\mathbf{R}$ in a single transverse plane as a state vector $|\mathbf{u}(z)\rangle$ in Dirac notation [28]. The scalar product of two state vectors

$$\langle \mathbf{v}|\mathbf{u}\rangle = \int d\mathbf{R}\, \mathbf{v}^*(\mathbf{R}) \cdot \mathbf{u}(\mathbf{R}) \tag{88}$$

involves an integration over the transverse coordinates as well as the scalar product of the two vector functions. When we introduce the transverse derivative in terms of a momentum operator $\hat{\mathbf{P}} = -i\partial/\partial\mathbf{R}$, the paraxial wave equation (44) takes the form

$$\frac{\partial}{\partial z}\big|\mathbf{u}(z)\big\rangle = -\frac{i}{2k}\hat{\mathbf{P}}^2\big|\mathbf{u}(z)\big\rangle, \tag{89}$$

where $k = \omega/c$. In this notation, the orbital AM Λ per unit length of a monochromatic beam as given in equation (55) is

$$\Lambda = \frac{2\epsilon_0}{\omega}\langle \mathbf{u}|\hat{\mathbf{R}} \times \hat{\mathbf{P}}|\mathbf{u}\rangle, \tag{90}$$

which is reminiscent of the expression for the orbital AM of a quantum particle in two dimensions. The orbital AM per photon is $\Lambda\hbar\omega/W$.

From the paraxial wave equation in operator form (89), it is obvious that the effect on a state vector of free propagation over a distance z is described by the operator

$$\hat{U}(z) = \exp\left(-\frac{i}{2k}\hat{\mathbf{P}}^2 z\right). \tag{91}$$

Likewise, the effect of an ideal thin lens with focal length f on the state vector of a mode is given by the operator $\exp(-ik\hat{\mathbf{R}}^2/(2f))$. In this way, any optical lens system can be described by a product of operators for lenses and free propagation [20]. Astigmatic lenses, with different focal length for two orthogonal axes, can be directly incorporated in the description. It is sufficient to replace $\hat{\mathbf{R}}^2/f$ by $\hat{\mathbf{R}} \cdot \mathsf{f}^{-1} \cdot \hat{\mathbf{R}}$, with f a real symmetric matrix with the two focal lengths of the lens as eigenvalues.

We emphasize that although the operator notation is borrowed from quantum mechanics, the description is applied here to classical paraxial optics.

2.8.2 Paraxial Beams and Quantum Harmonic Oscillators

It is well known that the paraxial wave equation has a complete set of solutions in the form of a Gaussian function multiplied by a Hermite polynomial [21,29]. These Hermite–Gaussian mode functions have a pattern that is invariant under propagation, apart from a scaling factor. They closely resemble the eigenfunctions of the two-dimensional quantum harmonic oscillator (HO). In fact, the Hermite–Gaussian mode functions can be written as [30]

$$u_{n_x n_y}(\mathbf{r}) = \frac{1}{\gamma}\psi_{n_x}\left(\frac{x}{\gamma}\right)\psi_{n_y}\left(\frac{y}{\gamma}\right)\exp\left(\frac{ikR^2}{2\rho} - i\chi(n_x + n_y + 1)\right). \tag{92}$$

Here the functions $\psi_n(\xi)$ for $n = 0, 1, \ldots$ are the real normalized energy eigenfunctions of the one-dimensional quantum HO in dimensionless form. Hence they are eigenfunctions of the Hamiltonian

$$\hat{H}_\xi = \frac{1}{2}\left(-\frac{\partial^2}{\partial\xi^2} + \xi^2\right) \tag{93}$$

with eigenvalue $n + 1/2$. The explicit form of the normalized eigenfunctions is

$$\psi_n(\xi) = \frac{1}{\sqrt{2^n n! \sqrt{\pi}}} e^{-\xi^2/2} H_n(\xi), \tag{94}$$

in terms of the Hermite polynomials H_n, and the eigenvalues are $n_x + 1/2$. The mode functions $u_{n_x n_y}(\mathbf{r})$ are determined by three mode parameters that depend on the propagation coordinate z. The width γ determines the spot size, ρ is the radius of curvature of the wave fronts, and χ is the Gouy phase, which determines the phase delay over the beam focus. When the transverse plane $z = 0$ coincides with the focal plane, the z-dependence of these parameters is determined by the equalities

$$\frac{1}{\gamma^2} - \frac{ik}{\rho} = \frac{k}{b + iz}, \qquad \tan\chi = \frac{z}{b}. \tag{95}$$

Here b is the diffraction length (or Rayleigh range) of the beam. The Gouy phase increases by an amount π from $z = -\infty$ to ∞. The mode functions in equation (92) are exact normalized solutions of the paraxial wave equation (44). The spot size at focus is $\gamma_0 = \gamma(0) = \sqrt{b/k}$.

It is noteworthy that the Gouy phase term in the Hermite–Gaussian modes (equation (92)) is proportional to the eigenvalue $n_x + n_y + 1$ of the two-dimensional quantum HO. This allows us to express an arbitrary solution of the paraxial wave equation in terms of an arbitrary time-dependent solution of the Schrödinger equation of the HO, provided that we replace time by the Gouy phase χ. In dimensionless notation, this equation takes the form

$$\frac{\partial}{\partial\chi}\Psi(\xi, \eta, \chi) = -i(\hat{H}_\xi + \hat{H}_\eta)\Psi(\xi, \eta, \chi). \tag{96}$$

Obviously, wave functions $\psi_{n_x}(\xi)\psi_{n_y}(\eta)\exp(-i(n_x + n_y + 1)\chi)$ are solutions of this Schrödinger equation, and by taking linear combination of these one can obtain the most general solution $\Psi(\xi, \eta, \chi)$. On the other hand, an arbitrary solution of the paraxial wave equation is a linear combination of the HG modes (equation (92)). We conclude that an arbitrary solution $\Psi(\xi, \eta, \chi)$ of equation (96) gives an arbitrary solution $u(\mathbf{r})$ of equation (44), by the identification [31]

$$u(\mathbf{r}) = \frac{1}{\gamma}\Psi(\xi, \eta, \chi)\exp\left(\frac{ikR^2}{2\rho}\right), \tag{97}$$

with $\xi = x/\gamma$, $\eta = y/\gamma$, and where the parameters γ, ρ and χ are specified by equation (95) as functions of z. The overlap of two modes is the same as the overlap of the two corresponding wave functions, or in Dirac notation $\langle v(z)|u(z)\rangle = \langle\Phi(\chi)|\Psi(\chi)\rangle$, when Ψ represents u, and Φ represents v. In particular, a normalized mode corresponds to a normalized wave function.

This identification (equation (97)) is exact, and it works both ways: there is a one-to-one correspondence between a time-dependent state of the two-dimensional HO and a monochromatic paraxial beam of light. For a given HO wave function, we find a mode function after choosing as free parameter the diffraction length b, which is a measure of the size of the focal region. Moreover, a solution of the quantum HO remains a solution under a shift of time. If we substitute $\Psi(\xi, \eta, \chi - \chi_0)$ for $\Psi(\xi, \eta, \chi)$ in equation (97), we find a different paraxial beam in general. When Ψ is a stationary state of the HO, such a shift makes no difference, and the corresponding mode pattern retains its shape during propagation, apart from scaling by the width γ. For an arbitrary mode, a phase shift $\chi_0 = \pi/2$ leads to an interchange of the mode pattern in focus and in the far field. Since the Gouy phase increases by an amount π, the mode function u from $z = -\infty$ to ∞ corresponds to a half cycle of the oscillator.

In this identification, the polarization of the paraxial beam has been ignored. The polarization is accounted for by decomposing the vector solution $\mathbf{u}(\mathbf{r})$ of the paraxial wave equation into two scalar solutions $u_\pm(\mathbf{r})$, one for both circular polarizations $\mathbf{e}_\pm$. They can be represented by two corresponding HO wave functions $\Psi_\pm(\xi, \eta, \chi)$.

There is a simple relation between the orbital AM of the paraxial beam and the orbital AM of the HO. It is convenient to work with mode functions that are normalized to 1, so that

$$\langle u|u\rangle = \langle \Psi|\Psi\rangle = 1. \tag{98}$$

Then the orbital AM per photon is given by

$$\frac{\hbar\omega\Lambda}{W} = \langle u|\frac{\hbar}{i}\frac{\partial}{\partial\phi}|u\rangle = \langle \Psi|\frac{\hbar}{i}\frac{\partial}{\partial\phi}|\Psi\rangle. \tag{99}$$

Here we used that the azimuthal angle ϕ in the xy-plane is the same as in the $\xi\eta$-plane.

In conclusion, there is a one-to-one correspondence between the description of a freely propagating paraxial beam, and the time-dependent wave function of a two-dimensional quantum HO during a half cycle.

2.8.3 Raising and Lowering Operators for Modes

The correspondence between HO wave functions and paraxial modes implies the existence of ladder operators that raise or lower the mode indices n_x and n_y, in analogy to the ladder operators coupling the energy eigenstates of an HO. For the functions $\psi_{n_x}(\xi)$, these operators take the well-known form

$$\hat{a}_\xi = \frac{1}{\sqrt{2}}\left(\xi + \frac{\partial}{\partial\xi}\right), \qquad \hat{a}_\xi^\dagger = \frac{1}{\sqrt{2}}\left(\xi - \frac{\partial}{\partial\xi}\right), \tag{100}$$

and the ladder operators $\hat{a}_\eta$ and $\hat{a}_\eta^\dagger$ have a similar form. They obey the bosonic commutation rules $[\hat{a}_\xi, \hat{a}_\xi^\dagger] = [\hat{a}_\eta, \hat{a}_\eta^\dagger] = 1$, and so on. The correspondence in equation (97) between HO states and modes then leads to z-dependent raising operators for HG modes [30]. Likewise, the LG modes correspond to circular eigenstates of the HO, which are the eigenstates of the Hamiltonian and of the angular-momentum operator.

The HG modes $|u_{n_x n_y}\rangle$ correspond to the HO eigenstates

$$|\Psi_{n_x n_y}\rangle = \frac{1}{\sqrt{n_x! n_y!}} \left(\hat{a}_\xi^\dagger\right)^{n_x} \left(\hat{a}_\eta^\dagger\right)^{n_y} |\Psi_{00}\rangle. \tag{101}$$

The circular eigenstates are obtained by applying to the ground state the circular raising operators

$$\hat{a}_\pm^\dagger = \frac{1}{\sqrt{2}} \left(\hat{a}_\xi^\dagger \pm i \hat{a}_\eta^\dagger\right). \tag{102}$$

The LG modes u_{mp} then corresponds to the HO eigenstates

$$|\Psi_{mp}\rangle = \frac{1}{\sqrt{n_+! n_-!}} \left(\hat{a}_+^\dagger\right)^{n_+} \left(\hat{a}_-^\dagger\right)^{n_-} |\Psi_{00}\rangle. \tag{103}$$

The azimuthal quantum number m and the radial mode number p are related to the circular raising numbers n_+ and n_- by the relations $m = n_+ - n_-$ and $p = \min(n_+, n_-)$. These states are eigenstates of the Hamiltonian $\hat{H}_\xi + \hat{H}_\eta$ with eigenvalue $n_+ + n_- + 1 = 2p + |m| + 1$, and of the angular-momentum operator

$$\frac{1}{i} \frac{\partial}{\partial \phi} = \hat{a}_+^\dagger \hat{a}_+ - \hat{a}_-^\dagger \hat{a}_-, \tag{104}$$

with eigenvalue $n_+ - n_- = m$. This confirms that the Laguerre–Gaussian modes are eigenmodes of the operator in equation (104) with eigenvalue m, so that their azimuthal dependence is given by the factor $\exp(im\phi)$. The orbital AM per photon (equation (99)) in these modes is therefore $\hbar m$.

The notation in terms of the ladder operators demonstrates the complete analogy between the various sets of energy eigenfunctions of the two-dimensional HO and the various basis sets of Gaussian paraxial modes. The Hermite–Gaussian modes are analogous to the Cartesian set of eigenfunctions, which are just products of eigenfunctions of the one-dimensional HO in the x- and y-direction. They have vanishing orbital AM. The Laguerre–Gaussian modes are analogous to the HO eigenfunctions of energy and orbital AM. These functions factorize in polar coordinates, with azimuthal dependence $\exp(im\phi)$. The transformation between the two basis sets is identical for modes and for HO eigenfunctions. It is based on the relation in equation (102) between the circular ladder operators and the Cartesian ones. In particular, substitution of the expressions (102) for the circular raising operators in equation (103) for the Laguerre–Gaussian modes directly gives an expansion of these modes in the Hermite–Gaussian modes. In this expansion, only terms occur where the total mode index $n_x + n_y = n_+ + n_-$. This algebraic technique of mode transformation is a convenient alternative for analytical methods [32].

The ladder operators in this section ($\hat{a}_x$, $\hat{a}_y$, $\hat{a}_+$, $\hat{a}_-$) should not be confused with the photon annihilation operators $\hat{a}_{mps}(\omega)$ that occurred in Section 2.6. The description in the present section is completely classical, and the operators transform one mode function into another one.

2.8.4 Orbital Angular Momentum and the Hermite–Laguerre Sphere

The relation (equation (102)) between circular and Cartesian raising operators is obviously the same as the relation between circular and linear polarization vectors. In fact, it is well known that the polarization of light is characterized by the Stokes parameters [33], which determine a point on a sphere with radius 1. This is termed the *Poincaré sphere*. It is analogous to the Bloch sphere of a spin 1/2, where each point on the sphere defines the direction of the expectation value $2\langle \mathbf{S} \rangle$ of the spin vector, which uniquely fixes the spin state. The analogy is defined by the statement that the circular polarization vectors $\mathbf{e}_\pm$ are analogous to the states with spin up and down. The point on the Poincaré sphere defined by the spherical angles θ and ϕ then represents the polarization vector

$$\mathbf{e}(\theta, \phi) = \mathbf{e}_+ e^{-i\phi/2} \cos\frac{\theta}{2} + \mathbf{e}_- e^{i\phi/2} \sin\frac{\theta}{2}. \tag{105}$$

It has been noted that the two first-order Hermite–Gaussian modes $|u_{10}\rangle$ and $|u_{01}\rangle$ span a basis in which every linear superposition can likewise be represented by a point on a sphere [34]. Hermite–Gaussian modes with respect to rotated axes are represented by points on the equator, while the poles represent the two Laguerre–Gaussian modes $|u_{mp}\rangle$ with $p = 0$ and $m = \pm 1$.

In view of the transformation in equation (102), it is natural to generalize the Poincaré sphere representation to the raising operators rather than to the mode functions themselves. Hence, for each point on the sphere defined by θ and ϕ, we introduce the HO raising operator

$$\hat{a}^\dagger(\theta, \phi) = \hat{a}_+^\dagger e^{-i\phi/2} \cos\frac{\theta}{2} + \hat{a}_-^\dagger e^{i\phi/2} \sin\frac{\theta}{2}. \tag{106}$$

The complementary raising operator $\hat{b}^\dagger$ is then represented by the opposite point on the sphere

$$\hat{b}^\dagger(\theta, \phi) = -\hat{a}_+^\dagger e^{-i\phi/2} \sin\frac{\theta}{2} + \hat{a}_-^\dagger e^{i\phi/2} \cos\frac{\theta}{2}. \tag{107}$$

This pair of operators obeys the bosonic commutation rules $[\hat{a}, \hat{a}^\dagger] = 1$, $[\hat{a}, \hat{b}^\dagger] = 0$, and so on. For such a pair of raising operators, we find a complete orthonormal basis

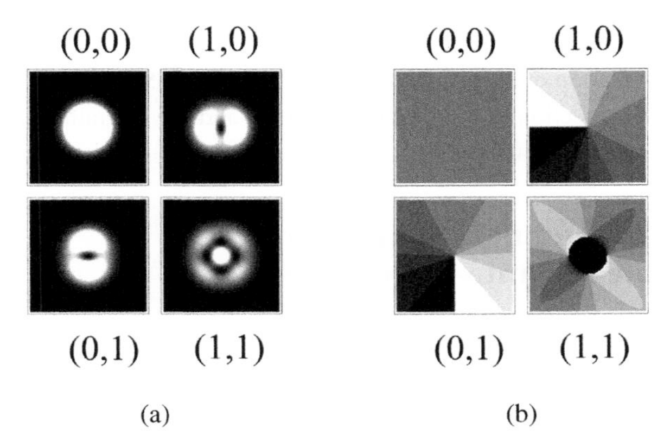

Figure 2.6 Intermediate modes with $\theta = \pi/3$, $\phi = 0$. (a) Intensity distribution. (b) Phase distribution. The phase is indicated by discrete gray tones, with increasing phase shown by brighter tones.

of HO energy eigenstates, by the definition

$$|\Psi_{n_a n_b}\rangle = \frac{1}{\sqrt{n_a! n_a!}} \left(\hat{a}^\dagger\right)^{n_a} \left(\hat{b}^\dagger\right)^{n_b} |\Psi_{00}\rangle. \tag{108}$$

In accordance with equation (97), a mode $|u_{n_a n_b}(z)\rangle$ corresponds to each one of these HO eigenstates. On the poles ($\theta = 0$ or π), the Laguerre–Gaussian modes are reproduced. On the equator ($\theta = \pi/2$), the raising operators $\hat{a}^\dagger$ and $\hat{b}^\dagger$ are linear combinations of $\hat{a}_\pm^\dagger$ with equal strength, and rotated versions of the Hermite–Gaussian modes are obtained. In between the equator and the poles, these basis sets form a continuous transition between the Hermite–Gaussian modes and the Laguerre–Gaussian modes. The fundamental mode $|u_{00}\rangle$ is the same in all basis sets. Again, the expression (108) immediately defines the transformation between this basis set and the Laguerre–Gaussian (or Hermite–Gaussian) modes. In this way, each point on the sphere represents a basis set of modes. It is natural to call this sphere the *Hermite–Laguerre sphere*.

The set of modes corresponding to a point on the equator ($\theta = \pi/2$) contains no phase singularities, but a pattern of orthogonal straight nodal lines with zero intensity that separate regions with opposite phase. For points away from the equator, phase singularities arise. For instance, for $\phi = 0$, the mode with $n_a = 1, n_b = 0$ near the axis is proportional to $(x + iy)\cos(\theta/2) + (x - iy)\sin(\theta/2)$. On the Northern hemisphere ($\theta < \pi/2$), this is a vortex with charge 1. However, the contours of constant intensity near the axis are not circles, but ellipses. This type of elliptical vortices are called noncanonical [35]. The intensity and phase pattern for the lowest-order modes is shown in Figure 2.6.

It is easy to obtain the orbital AM per photon in the modes that we have now defined. The inversion of equations (106) and (107) gives

$$\hat{a}_+ = \left(\hat{a}\cos\frac{\theta}{2} - \hat{b}\sin\frac{\theta}{2}\right)e^{i\phi/2}, \qquad \hat{a}_- = \left(\hat{a}\sin\frac{\theta}{2} + \hat{b}\cos\frac{\theta}{2}\right)e^{-i\phi/2}. \tag{109}$$

When we substitute these expressions in equation (104) and use equation (99), we find that for the orbital AM per photon in the modes (108),

$$\langle u_{n_a n_b}|\frac{\hbar}{i}\frac{\partial}{\partial\phi}|u_{n_a n_b}\rangle = \hbar(n_a - n_b)\cos\theta. \tag{110}$$

The factor $\cos\theta$ resulting from the polar angle on the Hermite–Laguerre sphere also occurs in the expression for the helicity $\hbar\sigma = \hbar\cos\theta$ for the spin per photon for a beam with a polarization represented by (θ, ϕ) on the Poincaré sphere.

2.9 DYNAMICS OF OPTICAL VORTICES

In this section, we derive a number of general properties of the dynamics of vortices in paraxial beams. The relation in equation (97) between a beam $u(R, z)$ and a wave function $\Psi(\xi, \eta, \chi)$ of the HO shows that we may just as well derive the behavior of vortices in the harmonic-oscillator wave function. In fact, since the harmonic oscillator is invariant for a time translation, a single solution of the harmonic oscillator generates a whole class of paraxial beams, since we can make the focal plane of the beam correspond to any instant during the cycle of oscillation. In this section, we demonstrate the advantage of the harmonic-oscillator analogy of paraxial modes when analyzing the propagation properties of mode patterns. Two simple cases are considered. First we will give a simple treatment of stable vortex-carrying configurations. Then we will treat a case of two vortices with opposite charge that can be created or annihilated.

2.9.1 Invariant Mode Patterns

A stationary state of the HO corresponds to a mode with a transverse field pattern that is invariant during propagation. In each transverse plane, the mode pattern has the same shape and only the scale varies, as determined by the width $\gamma(z)$. The most general stationary state is a superposition of states with the same excitation number $N = n_a + n_b$, which is conserved when one changes the basis of states over the Hermite–Laguerre sphere. One may also ask for the set of modes (or HO states) that retain their shape, apart from a rotation. Since rotations are generated by the operator (104), with the Laguerre–Gaussian mode index $m = n_+ - n_-$ as eigenvalue, these

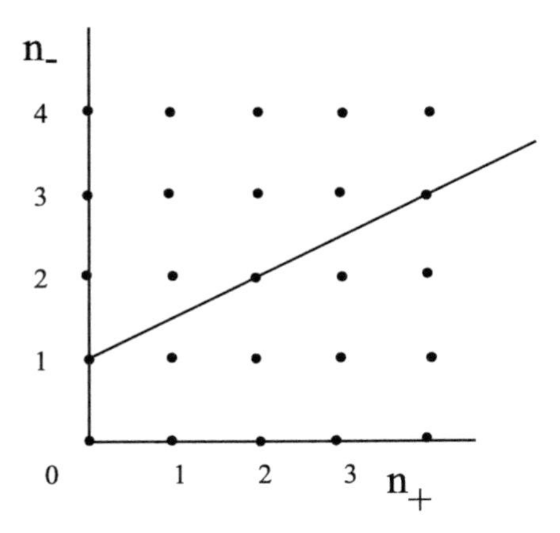

Figure 2.7 Diagram of the modes $|u_{n_+n_-}\rangle$. A superposition of the modes on the line will rotate with increasing value of the Gouy phase χ over an angle 3χ.

states are best described on the basis of Laguerre–Gaussian modes $|u_{n_+n_-}\rangle$ (or the corresponding circular HO eigenstates $|\psi_{n_+n_-}\rangle$). These modes can be arranged in a square lattice, in a diagram with the indices n_+ and n_- along the axis.

Consider a set of pairs of indices on a line in this lattice. Then there is a linear relation between n_+ and n_-, or, equivalently, a linear relation between N and m. This relation can be represented as

$$AN = Bm + C \quad \text{or} \quad (A - B)n_+ + (A + B)n_- = C. \tag{111}$$

The evolution of a state of the HO as a function of χ arises from the phase factor $\exp(-i\chi(N+1))$ for each stationary state with eigenvalue $N+1$. For a superposition of states with indices n_+ and n_- obeying the relation in equation (111), this phase factor is equivalent to an overall phase factor multiplied by the m-dependent term $\exp(-i\chi mB/A)$. This phase factor is equivalent to a rotation of the pattern over an angle $\chi B/A$.

This leads to a remarkable simple conclusion. An arbitrary superposition of HO states $|\psi_{n_+n_-}\rangle$, where the indices are related by equation (111), produces a wave function that rotates at an angular velocity that is B/A times the oscillator frequency. For the corresponding modes, the mode pattern performs a rotation between the focal plane and the far field over an angle $\pi B/2A$. As an example Figure 2.7 shows a line in the diagram of the states $|\psi_{n_+n_-}\rangle$ (or, equivalently, of the Laguerre–Gaussian modes $|u_{n_+n_-}\rangle$), such that all states on the line obey the relation in equation (111) with $A = 1$, $B = 3$, and $C = 4$. An arbitrary superposition of the states on this line will rotate over an angle $3\Delta\chi$, when $\Delta\chi$ is the increase of the phase χ.

This behavior was recently discovered by an analysis of the functional behavior of the Laguerre–Gaussian modes, in terms of the mode indices p and m [36]. The present argument shows the power of the harmonic-oscillator analogy, as well as the advantage of using mode indices n_+ and n_- as labels of the Laguerre–Gaussian modes.

2.9.2 Rotating Patterns of Vortices with Same Orientation

A simple way to describe vortices on a Gaussian background is with multiplication by a vortex factor. For instance, when the HO ground-state wave function is multiplied by the factor $(\xi - \xi_0) + i(\eta - \eta_0)$, a vortex with charge 1 is created at the location $\xi = \xi_0$, $\eta = \eta_0$. Multiplication with the complex conjugate factor creates a vortex with charge -1. Repeating such a multiplication several times at the same location creates a vortex of higher charge. Now suppose that the ground-state wave function is multiplied by a number of factors of this type, all with positive charge. In order to expand the resulting wave function on the basis of the states $|\psi_{n_+n_-}\rangle$, it is convenient to express the factor in terms of the ladder operators in equation (102) as

$$(\xi - \xi_0) + i(\eta - \eta_0) = \hat{a}_- + \hat{a}_+^\dagger - \xi_0 - i\eta_0. \tag{112}$$

Repeated multiplication of the ground-state wave function with factors of this type at different locations is equivalent to the action of a polynomial in the lowering operators $\hat{a}_-$ and $\hat{a}_+^\dagger$. When acting on the ground state, the lowering operator gives zero. In the language of modes, the conclusion is that the resulting mode (after imposing vortices with positive charge) produces a linear combination of Laguerre–Gaussian modes $|u_{n_+0}\rangle$, with $n_- = 0$. In the mode diagram, these modes lie on the line $N = m$, which is equivalent with equation (111) with $A = B$, $C = 0$.

The conclusion of the previous subsection is then that the vortex pattern rotates over an angle $\pi/2$ between focus and the far field. This conclusion has been obtained previously by analytical techniques [37].

2.9.3 Vortex Creation and Annihilation

Two vortices of opposite charge can be spontaneously created or annihilated in the HO wave function, and therefore also in a paraxial light beam. The problem of two vortices with opposite charge in the focus of a paraxial beam has been studied in [37] by analytical techniques. Here we demonstrate the advantage of an algebraic treatment in the HO picture, which is immediately translated into the case of a paraxial beam.

A pair of vortices with charge ± 1 at positions $(\xi, \eta) = (\pm\xi_0, 0)$ imprinted on the background of the Gaussian ground state is described the action of the operator equation (112) and its Hermitian conjugate on the ground state. The (non-normalized) result is found as an expansion in the states $|\psi_{n_+n_-}\rangle$ in the form

$$|\Psi(0)\rangle = (\hat{a}_+ + \hat{a}_-^\dagger + \xi_0)(\hat{a}_- + \hat{a}_+^\dagger - \xi_0)|\psi_{00}\rangle$$
$$= (1 - \xi_0^2)|\psi_{00}\rangle + \xi_0(|\psi_{10}\rangle - |\psi_{01}\rangle) + |\psi_{11}\rangle. \tag{113}$$

This state evolves as a function of χ, which gives a phase factor $\exp(-\chi(N+1))$ to each term. Each wave function has the form of the ground-state wave function multiplied by a simple polynomial. After evaluating the wave functions, one finds

$$\Psi(\chi) = \left[(1-\xi_0^2)e^{-i\chi} + 2i\xi_0\eta e^{-2i\chi} + (\xi^2+\eta^2-1)e^{-3i\chi}\right]\psi_{00}. \tag{114}$$

The location of the vortices are found as the zeros of this function. After multiplication with $\exp(2i\chi)$, this gives the condition

$$(1-\xi_0^2)e^{i\chi} + 2i\xi_0\eta + (\xi^2+\eta^2-1)e^{-i\chi} = 0. \tag{115}$$

The real part of this equation gives the requirement

$$\xi^2 + \eta^2 = \xi_0^2, \tag{116}$$

which implies that the two vortices lie on a circle with radius ξ_0. The imaginary part gives the η-component of the vortex location as

$$\eta = \frac{\xi_0^2 - 1}{\xi_0} \sin\chi. \tag{117}$$

This implies that the two vortices have the same η-value and opposite ξ-values. As a function of χ, they move both upward toward positive η-values when $\xi_0 > 1$, and downward when $\xi_0 < 1$, while remaining on the circle with radius ξ_0.

For large values $\xi_0 \gg 1$, we find $\eta \approx \xi_0 \sin\chi$, each one of the two vortices rotates over an angle χ in opposite directions along the circle, just as if the other vortex were absent. When the factor $(\xi_0^2 - 1)/\xi_0$ is smaller than $-\xi_0$ (or $\xi_0 < 1/\sqrt{2}$), the vortices meet at the point $\xi = 0$, $\eta = -\xi_0$ and disappear. This is an example of annihilation of two vortices of opposite charge. We denote the χ-value of annihilation as χ_1, so that

$$\chi_1 = \arcsin\frac{\xi_0^2}{1-\xi_0^2}. \tag{118}$$

This means that the vortex pair exists during a part of the cycle of the HO. The pair is created at $\chi = -\chi_1$ at the position $\xi = 0$, $\eta = \xi_0$ and then moves downward on opposite sides of the circle to its annihilation point, which is reached when $\chi = \chi_1$. This motion downward from the creation position of the vortex pair toward its annihilation point is shown in Figure 2.8.

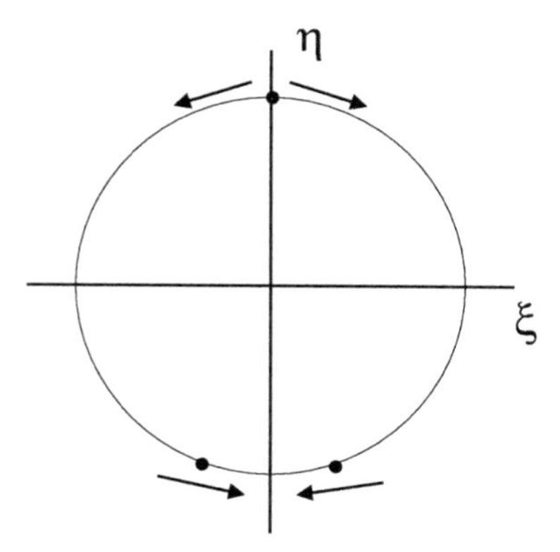

Figure 2.8 Path of the vortex pair along a circle in the $\xi\eta$-plane. The pair is created at the top of the circle and then moves downward on opposite sides to its annihilation position at the lowest point.

We see from equation (113) that the corresponding (non-normalized) paraxial mode expressed as a sum of Laguerre–Gaussian modes $u_{n_+n_-}$ is

$$|u\rangle = \left(1 - \xi_0^2\right)|u_{00}\rangle + \xi_0\left(|u_{10}\rangle - |u_{01}\rangle\right) + |u_{11}\rangle. \tag{119}$$

This mode has two vortices in the focal plane at the locations $x = \pm\xi_0\gamma_0$, $y = 0$. When the distance between the vortices is less than $\gamma_0\sqrt{2}$, they have disappeared in the far field at $z = \pm\infty$. However, we are free to choose the focal plane of the mode to correspond to an arbitrary location during the cycle of the HO. Then we can obtain a mode without vortices in the focal plane. The mode corresponding to the HO state in equation (113) but with the focal plane taken at the phase χ_0 of the oscillator is

$$|u\rangle = \left(1 - \xi_0^2\right)e^{i\chi_0}|u_{00}\rangle + \xi_0 e^{2i\chi_0}\left(|u_{10}\rangle - |u_{01}\rangle\right) + e^{3i\chi_0}|u_{11}\rangle. \tag{120}$$

When the shift in the Gouy phase is taken as $\chi_0 = \pi/2$, the mode has a vortex pair in the far field at $z = \pm\infty$ along the ξ-axis. The pair moves into the lower half-plane ($y < 0$) and is annihilated before focus. After focus, it reappears in the upper half-plane at the top of the circle and then moves down on opposite sides of the circle toward the ξ-axis in the far field at $z = \infty$.

The examples in this section demonstrate the usefulness of the analogy between the two-dimensional harmonic oscillator and paraxial beams.

2.10 CONCLUSION

Angular momentum (AM) is an important and delicate quantity of light. It arises both from circulating phase gradients and from rotating vector properties of the field.

This suggests the possible separation of optical AM into different types, with the flavor of orbital and spin AM. We discuss the possibility and limitations of such a separation, both for a classical and a quantum description of light. For an arbitrary Maxwell field, such a separation is possible. The separate parts have a well-defined significance and generate useful transformations. However, they do not have all the properties of AM. There is a number of special modes of light where the presence of AM is rather obvious. These are also the fields that contain vortices in the form of phase or polarization singularities. Examples of exact solutions of Maxwell's equations are multipole fields with a spherical or cylindrical structure. For these fields, the polarization and phase contributions to AM are inseparable. For both types of multipole fields, the z-component of the AM has a well-defined value per photon, and for the spherical multipole fields the same statement holds for the total AM. For paraxial beams, AM in the propagation direction is separated in a natural way in an orbital and a polarization part. This is not only true for the AM integrated over space, but also for the densities of spin and orbital AM. This remains true for nonmonochromatic beams. We consider the appearance of AM both for stationary beams and for beams in which the light pattern is rotating.

We point out that there is a one-to-one correspondence between monochromatic paraxial beams and solutions of the two-dimensional isotropic harmonic oscillator. This analogy is applied to the description of the dynamics of vortices in paraxial beams in a number of simple examples. We wish to stress, however, that orbital AM in light beams does not require the presence of vortices. A counterexample is a light beam with general astigmatism, where the elliptical wave front has a different orientation than the elliptical intensity pattern in each transverse plane [38]. In this case, the rotation of the ellipses during propagation gives rise to AM [39,40].

REFERENCES

[1] J.D. Jackson, Classical Electrodynamics, Wiley, New York, 1975.

[2] J.W. Simmons, M.J. Guttmann, States, Waves and Particles, Addison–Wesley, Reading, MA, 1970.

[3] J.M. Jauch, F. Rohrlich, The Theory of Photons and Electrons, Springer, Berlin, 1976.

[4] C. Cohen-Tannoudji, J. Dupont-Roc, G. Grynberg, Photons et Atomes, InterEditions, Paris, 1987.

[5] L. Allen, M.W. Beijersbergen, R.J.C. Spreeuw, J.P. Woerdman, Orbital angular momentum of light and the transformation of Laguerre–Gaussian laser modes, *Phys. Rev. A 45* (1992) 8185.

[6] L. Allen, M.J. Padgett, M. Babiker, The orbital angular momentum of light, in: E. Wolf (Ed.), *Prog. Opt. 39* (1999) 291.

[7] J.F. Nye, M.V. Berry, Dislocations in wave trains, *Proc. R. Soc. London, Ser. A 336* (1974) 165.

[8] M. Soskin, M.V. Vasnetsov, Singular optics, in: E. Wolf (Ed.), *Prog. Opt. 42* (2001) 219.

[9] I. Bialynicki-Birula, Z. Bialynicka-Birula, C. Śliwa, Motion of vortex lines in quantum mechanics, *Phys. Rev. A 61* (2000) 032110.

[10] M.R. Dennis, Polarization singularities in paraxial vector fields: morphology and statistics, *Opt. Commun. 213* (2002) 201.

[11] A.Y. Bekshaev, M.S. Soskin, M.V. Vasnetsov, Angular momentum of a rotating light beam, *Opt. Commun. 249* (2005) 367.

[12] S.J. van Enk, G. Nienhuis, Commutation rules and eigenvalues of spin and orbital angular momentum of radiation fields, *J. Mod. Opt. 41* (1994) 963.

[13] J.J. Sakurai, Modern Quantum Mechanics, Addison–Wesley, Reading, MA, 1994.

[14] D. Lenstra, L. Mandel, Radially and azimuthally polarized beams generated by space-variant dielectric subwavelength gratings, *Phys. Rev. A 26* (1982) 3428.

[15] S.J. van Enk, G. Nienhuis, Spin and orbital angular momentum of photons, *Europhys. Lett. 25* (1994) 497.

[16] J. Durnin, Exact solutions for nondiffracting beams. I. The scalar theory, *J. Opt. Soc. Am. A 4* (1987) 651.

[17] J. Durnin, J.J. Miceli, J.H. Eberly, Diffraction-free beams, *Phys. Rev. Lett. 58* (1987) 1499.

[18] M. Lax, W.H. Louisell, W.B. McKnight, From Maxwell to paraxial wave optics, *Phys. Rev. A 11* (1975) 1365.

[19] H.A. Haus, Waves and Fields in Optoelectronics, Prentice Hall, Englewood Cliffs, NJ, 1984.

[20] S.J. van Enk, G. Nienhuis, Eigenfunction description of laser beams and orbital angular momentum of light, *Opt. Commun. 94* (1992) 147.

[21] A.E. Siegman, Lasers, University Science Books, Mill Valley, CA, 1986.

[22] M. Stalder, M. Schadt, Linearly polarized light with axial symmetry generated by liquid crystal polarization converters, *Opt. Lett. 21* (1996) 1948.

[23] Z. Bomzon, G. Biener, V. Kleiner, E. Hasman, Radially and azimuthally polarized beams generated by space-variant dielectric subwavelength gratings, *Opt. Lett. 27* (2002) 285.

[24] A. Niv, G. Biener, V. Kleiner, E. Hasman, Formation of linearly polarized light with axial symmetry by use of space-variant subwavelength gratings, *Opt. Lett. 28* (2003) 510.

[25] I.H. Deutsch, J. Garrison, Paraxial quantum propagation, *Phys. Rev. A 43* (1991) 2498.

[26] G. Nienhuis, Polychromatic and rotating beams of light, *J. Phys. B 39* (2006) 529.

[27] J. Courtial, D.A. Robertson, K. Dholakia, L. Allen, M.J. Padgett, Rotational frequency shift of a light beam, *Phys. Rev. Lett. 81* (1998) 4828.

[28] D. Stoler, Operator methods in physical optics, *J. Opt. Soc. Am. 71* (1981) 334.

[29] H. Kogelnik, T. Li, Laser beams and resonators, *Appl. Opt. 5* (1966) 1550.

[30] G. Nienhuis, L. Allen, Paraxial wave optics and harmonic oscillators, *Phys. Rev. A 48* (1993) 656.

[31] G. Nienhuis, J. Visser, Angular momentum and vortices in paraxial beams, *J. Opt. A: Pure Appl. Opt. 6* (2004) S248.

[32] E. Abramochkin, V. Volostnikov, Beam transformations and nontransformed beams, *Opt. Commun. 83* (1991) 123.

[33] M. Born, E. Wolf, Principles of Optics, Pergamon, New York, 1980.

[34] M.J. Padgett, J. Courtial, Poincaré sphere equivalent for light beams containing orbital angular momentum, *Opt. Lett. 24* (1999) 430.

[35] F.S. Roux, Coupling of noncanonical optical vortices, *J. Opt. Soc. Am. B 21* (2004) 664.

[36] F.S. Roux, The symmetry properties of stable polynomial Gaussian beam, *Opt. Commun. 268* (2006) 196.

[37] G. Indebetouw, Optical vortices and their propagation, *J. Mod. Opt. 40* (1993) 73.

[38] J.A. Arnaud, H. Kogelnik, Gaussian light beams with general astigmatism, *Appl. Opt. 8* (1969) 1687.
[39] J. Visser, G. Nienhuis, Orbital angular momentum of general astigmatic modes, *Phys. Rev. A 70* (2004) 013809.
[40] S.J.M. Habraken, G. Nienhuis, Modes of a twisted optical cavity, *Phys. Rev. A 75* (2007) 033819.

Chapter 3

Singular Optics and Phase Properties

Enrique J. Galvez

Colgate University, USA

Hurricanes, tornadoes, and water spouts evoke a special sense of wonder, not only by their majestic size but by the rotational motion that they entail. "The eye of the storm," the vortex, is particularly mysterious. It represents a place where there is a void; a place that is at the center of something unique; a channel that takes you to an unknown place.

Smaller vortices pose no lack of awe. The swirls in fluids or the turbulent wakes off airplane wing-tips are examples of a ubiquitous phenomenon. Physicists and scientists alike are no less impressed by such rotational displays; magnetic vortices, DNA, and other natural twirls are topics of constant study. Thus, it is no surprise that the discovery of optical vortices a couple of decades ago generated so much interest and still does. Optical vortices are no less spectacular than their atmospheric counterparts. The angular momentum that they carry around them in the optical field is also impressive; that something as ephemeral as light can put objects into rotation is startling to say the least. Vortices constitute an important topic in the study of optical singularities, the topic of this chapter. Like many physical topics involving complex media, the study of optical vortices has revealed interesting results that are not apparent otherwise. The results of these studies are the focus of important applications, such as manipulation of matter and the storage of information in the optical field.

Optical singularities encompass a large range of phenomena. We do not cover the important area of interest on singularities of amplitude, where ray theory predicts infinites: caustics [1]. That they never materialize due to diffraction is no less of an interesting feature of wave fields; they are still responsible for some of nature's most

STRUCTURED LIGHT AND ITS APPLICATIONS

precious optical displays: rainbows. Coexisting with caustics of sunlight at the bottom of a pool, for example, in a seemingly perpetual dance are the voids of darkness: phase singularities. These are places where the phase of the light field is multiply defined, and so the amplitude vanishes. This form of singularities is of two types: shear, where the phase jumps by π due to the sign change of the amplitude of the field and vortices, the point where the phase is multiply defined and about which the phase advances by a multiple of 2π when following a closed path around it. The notion that singularities are an exception is quite a misconception. Quite contrary, they are ubiquitous. They are just masked by our simple forms of perception and by the vastly complex forms that they can take, which we have just started to unravel. Furthermore, the dancing patterns of sunlight at the bottom of a pool are only the two-dimensional slices of a three-dimensional continuum where vortices' paths, the "threads of darkness," are intermingled and stubbornly persistent.

In this chapter, we will discuss the basic properties of phase singularities and optical vortices. We will avoid a review of singular optics as there is a comprehensive one already in the literature [2]. We will limit ourselves to simple cases, threaded by our laboratory experience at Colgate University. This chapter covers vortices in fundamental beams as a form of introduction. We will underscore a growing view that these topics deserve a rightful place in textbooks on optics instruction. The discussion follows with multiple vortices in the optical field, followed by nonintegral vortices as an important case of the latter. This is a simple and controlled evolution in complexity of the cases of vortices in fundamental solutions to the wave equation. A fourth section makes a shy entry into a topic that has only started to be studied: the dynamics of vortices in a propagating beam of light. We will not cover an important and modern topic in singular optics of polarization singularities [3].

3.1 FUNDAMENTAL PHASE SINGULARITIES

The natural place to start describing phase singularities is with the beams where they appear naturally and in their simplest form: Laguerre–Gaussian beams. These are solutions of the paraxial wave equation in cylindrical coordinates. The light in these beams has amplitude that depends on three terms:

$$E(r) \propto r^l \exp\left(-\frac{r^2}{w^2}\right) L_p^l\left(\frac{2r^2}{w^2}\right), \tag{1}$$

where the exponential accounts for the Gaussian fall-off of the amplitude of the light in a narrow beam of half-width w. The last term is an associate Laguerre polynomial of the transverse coordinate r, which has p radial zero crossings. Thus, a Laguerre–Gaussian beam projected on a screen consists of a series of concentric rings. It is

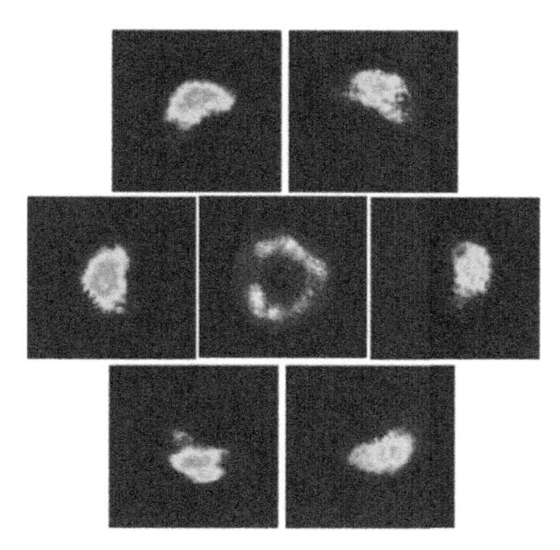

Figure 3.1 False-color image showing the interference of two collinear beams: a Laguerre–Gaussian beam with $l = 1$ (at the center) with a zero-order beam. The surrounding frames correspond to the interference patterns for phase differences in steps of $\pi/3$. See color insert.

noteworthy to observe that $L_0^l(2r^2/w^2) = 1$, so when $p = 0$ the shape of the light beam is a single ring of radius proportional to $\sqrt{l}$ giving the beam its hallmark shape: a "doughnut." For a beam traveling along the z-direction, the phase of the field in cylindrical coordinates is

$$\phi(r, \theta, z) = kz + \frac{kr^2}{2R(z)} + l\theta - \psi(z), \tag{2}$$

where k is the wave-number, $R = \sqrt{(z + z_R^2/z)}$ is the radius of curvature of the wave-front, and $\psi(z)$ is the Gouy phase, given by

$$\psi(z) = \big(2p + |l| + 1\big) \tan^{-1}\left(\frac{z}{z_R}\right), \tag{3}$$

with z_R being the Raleigh range. The third term of equation (2) contains the optical singularity: a phase that depends upon the angular coordinate θ. At $z = 0$, the phase of the beam is simply described by $l\theta$. The field of a beam in the mode with radial index p and angular index l is denoted by LG_p^l.

The interference pattern between a beam in LG_0^l mode and a beam in fundamental mode LG_0^0 consists of an off-axis "blob" that rotates about the beam axis when the phase between the two interfering beams is changed. This type of interference pattern is shown in Figure 3.1. One can easily take images like those in Figure 3.1 with a Mach–Zehnder interferometer [4], where incident light in the fundamental mode going through one arm of the interferometer passes through an optical element that puts

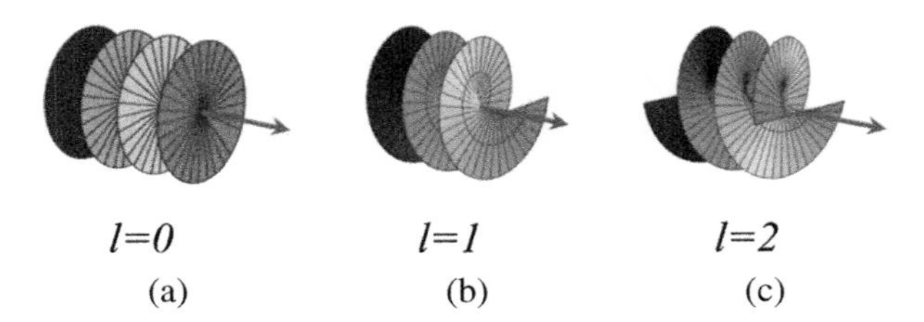

Figure 3.2 Graphic representations of the wave-fronts for Laguerre–Gaussian modes in the first three orders with $p = 0$.

it into a high-order mode (LG_0^1 in the case of the figure). The pattern is then formed at the output of the interferometer when the high-order mode interferes with the light in the LG_0^0 mode that traveled through the other arm. In the case of Figure 3.1, the phase that was inserted was not the usual dynamic phase, due to the path-length difference. Instead, it was due to a geometric phase inserted by transforming the modes of the light in a cyclic way, which we will discuss later [5].

For low values of z, the wave-front (or surface that is the locus of points with the same phase) has the form of l intertwined helical surfaces, as shown in Figures 3.2(b) and 3.2(c). This is in striking contrast to the fundamental Gaussian beam with $l = 0$ shown in Figure 3.2(a), where the wave-front at $z = 0$ consists of parallel planes. When $l \neq 0$, the wave-fronts are tilted in the azimuth direction and curves that follow the gradient of the wave-front are helices. In these modes, the Poynting vector has a transverse component pointing in the azimuth direction. This local distribution of the Poynting vector gives the wave its orbital angular momentum, which is proportional to l [6,7].

We can observe further evidence of the angular dependence of the phase by looking at interference patterns between an $l \neq 0$ Laguerre–Gaussian mode and a fundamental mode when both have different radii of curvature. In this case, the points of constant phase difference are spirals. An example of a typical spiral interference pattern is shown in Figure 3.3. The two forms of collinear interference can be used to identify the phase properties of the mode of the light. Collinear interference ceases to be useful for closely spaced vortices because the collinear interference pattern is unable to resolve them. For this case, noncollinear interference is more useful. In noncollinear interference, an expanded zero-order beam interferes, with the beam we wish to diagnose. Optical vortices appear as forks in the fringe pattern, with the fork denoting the phase dislocation. The number of tines of the imaged fork corresponds to the value of l, the topological charge of the vortex, minus 1 [8].

Computer-generated holograms of fork patterns are a popular way of generating beams with optical vortices in the diffracted orders [9–11], as shown in Figure 3.4.

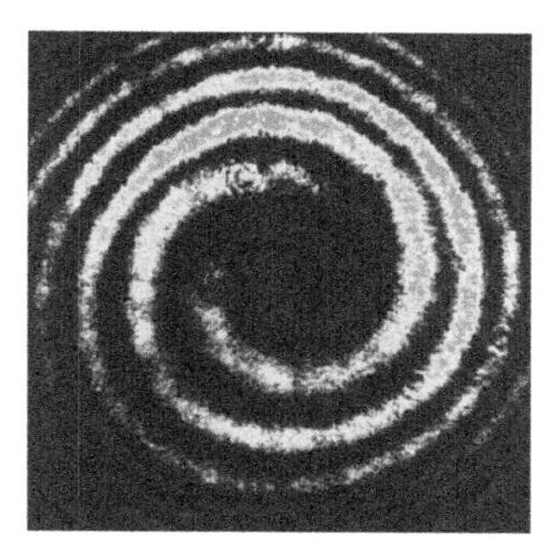

Figure 3.3 False color image of the double spiral interference pattern produced by the interference of a beam with $l = 2$ and an expanded beam in a zero-order (i.e., $l = 0$) mode. See color insert.

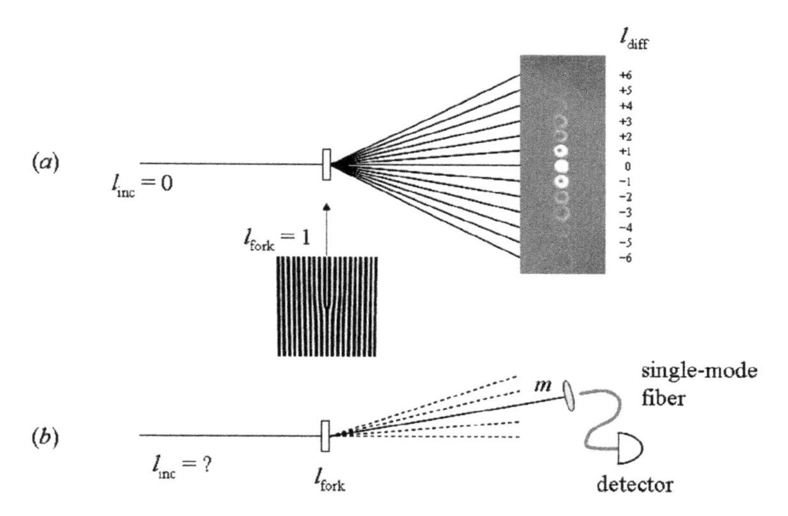

Figure 3.4 (a) Method to generate beams with optical vortices using a computer-generated binary grating with $l = 1$. (b) Method to diagnose the l-component of the state of single photons.

In general, the value of l_{diff} of the mth diffracted order is related to the charge of the incident beam l_{inc} and the charge of the forked grating l_{fork} by [8,12]

$$l_{\text{diff}} = l_{\text{inc}} + m l_{\text{fork}}. \tag{4}$$

Another popular method of producing a beam with vortices is by using a spiral phase plate [13]. The beam that is transmitted when a beam with l_{inc} is incident on a plate with a spiral phase dislocation $2\pi l_{\text{plate}}$ will have charge l_{trans} given by

$$l_{\text{trans}} = l_{\text{inc}} + l_{\text{plate}}. \tag{5}$$

The previous two relations characterize the charge of the output beam.

The forked grating shown in Figure 3.4(a) is an amplitude grating, which disperses the incident energy into a large number of diffracted orders. Because of this, the grating is not efficient at producing any one particular mode. If the objective is to produce a beam in one mode, a more efficient method is necessary. One method uses a phase-blazed grating. This can be done with either a bleached film hologram [14] or a spatial light modulator. The latter produces a phase grating by suitably biasing a pixilated liquid crystal [15,16].

In general, the beams generated by gratings or phase plates may not be pure Laguerre–Gaussian eigenstates. Instead they are a superposition of modes of same l but different p [17]. A mode of higher purity is obtained by either modulating the intensity of the input beam [18], or by modulating the efficiency of the blazed grating [19].

The previous diffractive methods can be extended to produce other families of beams with optical vortices, such as Mathieu beams, a form of Bessel beam that carries orbital angular momentum [20].

The use of equation (4) is the chosen method of detecting the amplitude of the l-component of individual photons in a superposition of orbital-angular-momentum eigenstates [21]. In this method, a single-mode optical fiber is used for detecting only $l = 0$ states. Thus, by choosing a suitable value of l_{fork} and locating the fiber so to detect light coming off the mth diffracted order, one can measure the amplitude of the component of the light that gives $l_{\text{diff}} = 0$. That is, the input mode has $l_{\text{inc}} = -ml_{\text{fork}}$. A schematic of this method is shown in Figure 3.4(b).

As discussed previously, a field of a Laguerre–Gaussian beam carries an orbital angular momentum proportional to l [7]. The higher orbital angular momentum per photon is only one of the features brought by high-order modes, which exist in a space of dimension $N = 2p + |l|$. These higher-dimensional spaces have larger topologies. Following closed paths in the space of modes via mode transformations [22] results in a new form of geometric phase [23]. In these mode transformations, the wave-front of an input mode is disassembled and reassembled into a different mode. The cyclic sculpturing of a mode results in a topological phase. Laboratory measurements of the presence [5] and absence [24] of these geometric phases are also consistent with a correlation between the phases and orbital-angular-momentum exchanges between the light and the optical system.

Optical vortices are not restricted to linear beams. They have been studied extensively in nonlinear media in connection with optical solitons [25]. More recent studies of the phase singularities include the properties of "vortex cores," the sub-wavelength surroundings of the phase singularity [26].

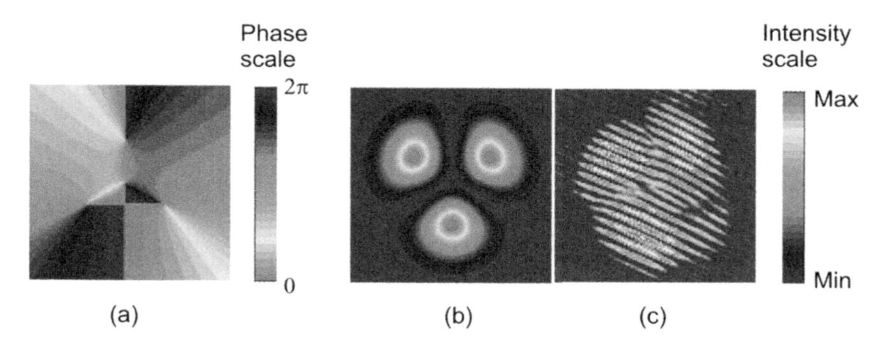

Figure 3.5 Composite vortex beam created by superposition of Laguerre–Gaussian beams with $l = +2$ with $l = -1$ at a ratio of amplitudes $A_{l=+2}/A_{l=-1} = 0.84$. Frames (a) and (b) show the computed phase and intensity of the composite beam. Frame (c) shows a measured interference pattern of the composite beam with a reference plane-wave. See color insert.

We have only mentioned the singularities in monochromatic wave fields. Since the singularities depend on the wavelength of the light, they are dispersive. This has been investigated in white light [16,27,28] as well as in coherent ultra-short pulses [29–31].

3.2 BEAMS WITH COMPOSITE VORTICES

Superposition of eigenmodes of orbital angular momentum provides an entry into a rich area of singular optics. When the component beams of a superposition contain vortices, then new vortices are produced. The number and location of these vortices is determined by the details of the superposition. When the component beams are collinear, we get patterns like the one in Figure 3.5. It was obtained by mixing two $p = 0$ Laguerre–Gaussian modes: one with $l_1 = +2$ and the other with $l_2 = -1$. Frames (a) and (b) in Figure 3.5 show respective calculations of the phase and intensity of the pattern that results when the two beams are combined with nearly equal amplitudes.

Following an analysis of the general situation [32], the angular position ϕ of vortices when the component beams are l_1 and l_2 (and assuming $|l_2| > |l_1|$) is given by

$$\phi = \frac{\delta + n\pi}{l_2 - l_1}, \tag{6}$$

where δ is the phase between the component beams and n is an odd integer. In the example shown in Figure 3.5 one can see that equation (6) predicts three angular positions separated by $2\pi/3$. If δ is changed, the pattern will rotate about its

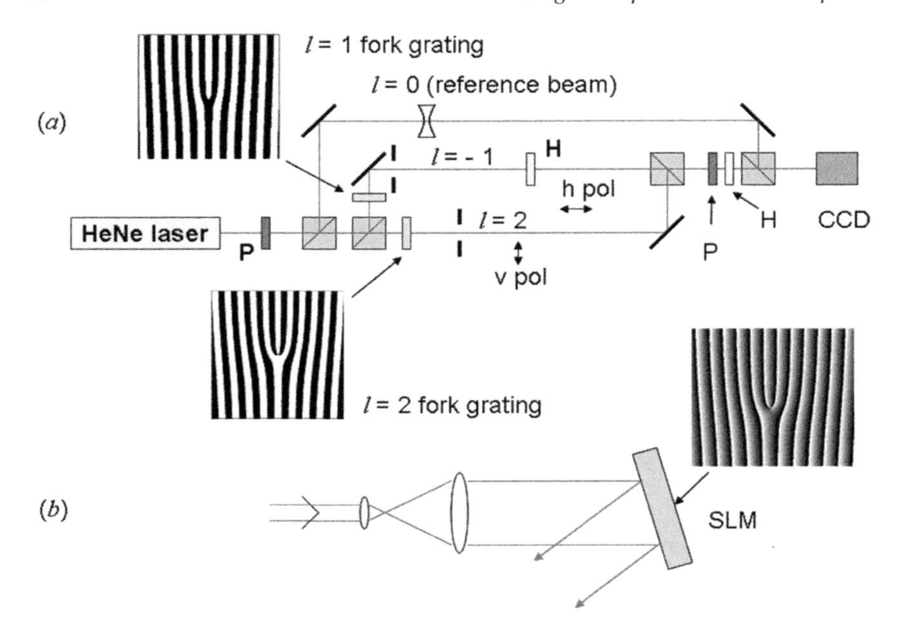

Figure 3.6 Experimental setups used to prepare beams with multiple vortices. In (a), the modes produced by binary gratings are physically combined by interferometers. Half-wave plates (H) and polarizers (P) are used to control the relative weight of the beams. In (b), a phase hologram of the desired beam is programmed in a spatial light modulator.

center. The radial position of the vortices is $r = 0$ for the central one of charge l_1 and

$$r = \frac{w}{\sqrt{2}}\left(\sqrt{\frac{|l_2|!}{|l_1|!}\frac{A_1}{A_2}}\right)^{1/(|l_2|-|l_1|)}, \tag{7}$$

for the $|l_2 - l_1|$ surrounding vortices of charge $l_2/|l_2|$, where w is the beam width or beam spot, and A_1/A_2 is the ratio of the amplitudes of the two modes. Thus, for our example, where $A_1/A_2 = 0.84$, the vortices are located at a radial distance $0.84w$ from the center. The laboratory measurements confirm these predictions [32]: frame (c) of Figure 3.5 shows an interference pattern between the composite beam and a reference plane wave, producing a noncollinear interference pattern in which vortices appear as forks in the interference pattern. Figure 3.5(c) shows three forks in the regions between the lobes with their tines pointing to the upper left. They correspond to the three $l = +1$ vortices that are predicted. The center of the beam has a fork with tines pointing in the opposite direction, revealing an oppositely charged central vortex (i.e., with $l = -1$). These data were taken with nested interferometers like the ones

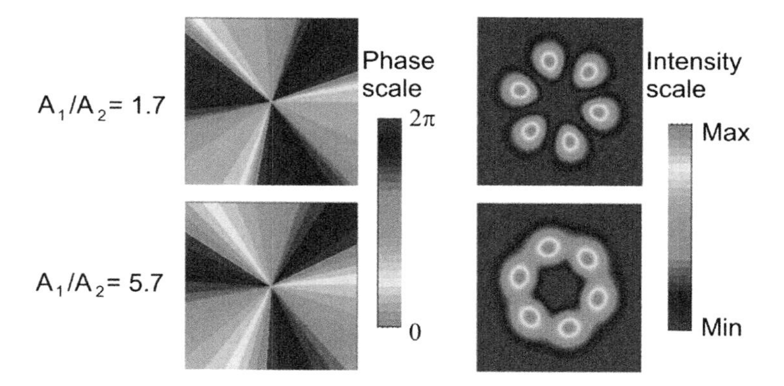

Figure 3.7 Composite beam when $l_1 = +3$ and $l_2 = -3$ for different relative amplitudes A_1 and A_2. See color insert.

shown in Figure 3.6(a). An inner Mach–Zehnder interferometer is used to prepare the composite beam. The amplitudes are adjusted using the polarization of light [32]. Before imaging by a CCD camera, the composite beam is interfered noncollinearly with an expanded beam in the fundamental mode.

An exception to the above description is the superposition of two collinear beams of the same order. When $l_1 = -l_2$, there are no vortices in the periphery of the composite beam. Instead, one gets a central vortex that has the charge of the beam with the greatest amplitude, l_1 or l_2. In the periphery, the field redistributes into $2|l_1|$ symmetric lobes. The phase around the central singularity does not increase linearly with the azimuth angle, as in Laguerre–Gaussian modes. Instead, it increases faster at the angles in between the lobes. The rate of change of the phase in these locations increases as the ratio of the two amplitudes approaches 1, at which point the phases shift by π infinitely fast (i.e., becoming shear singularities). Figure 3.7 shows two examples of this situation for cases with different amplitude ratio. When $A_1/A_2 = 1.7$ (top row), one can see the phase varying rapidly at the radial intensity nodes, with adjacent lobes being roughly out of phase by π. When $A_1/A_2 = 5.7$ (bottom row), one of the modes dominates over the other, and the phase variation approaches a linear relation with the azimuth angle. This case of composite vortices is interesting in that it displays the two types of scalar singularities: vortex and shear.

As is well known, Laguerre–Gaussian beams carry orbital angular momentum in their tilted wave-front. An absorbing object will experience forces all along the beam profile of such a beam. In composite beams, these forces can be tailored. For example, in the case of the beam in Figure 3.6, the pitch of the wave-front increases faster in the region in between the lobes. Objects trapped along the beam profile may experience forces that act nonuniformly along the beam profile, giving the object "kicks" instead of a constant push.

The composite beams get richer when the beams are collinear but displaced. In this case, the displacement between the two component vortices adds a new degree of freedom [33]. The composite beam is no longer a symmetric one. One can then tailor the amplitude and phase of the composite beam by adjusting the relative amplitude, phase, and displacement of the component beams.

The previous approaches involve superposing well-defined beams and studying the resulting composite beam. A different approach is to "program" the vortices into the composite beam then to investigate the mode composition of the programmed pattern. The idea here is to *encode* information in the beam by means of the vortices [34,35]. In this case, one can tailor a beam profile using the versatility of a spatial light modulator [35]. A typical setup that uses a spatial light modulator is shown in Figure 3.6(b). One idea is to use this for classical communication purposes due to the resilience of optical vortices to remain in the beam as it propagates. This, however, must be tempered by the sensitivity of these vortices to perturbations, where a vortex of charge l splits into $|l|$ vortices of charge $l/|l|$ when perturbed by another field. For example, in the case of the collinear composite vortices discussed above, the combination of a beam with charge $l_1 \neq 0$ and a second beam with charge $l_2 = 0$ results in $|l_1|$ vortices distributed around the center of the beam at positions given by equations (6) and (7).

A slightly different purpose of this approach is to create a composite beam for encoding *quantum* information on single photons [34], and thus preparing the quantum state of the light in a superposition of orbital angular momentum eigenstates. Such a system can be used for quantum information purposes [36].

3.3 NONINTEGER VORTEX BEAMS

A unique case of composite vortices is the one that appears when an optical beam is encoded to have a nonintegral phase dislocation. The optical beam is easy to program. When generating beams with passive or active holograms, one has the charge of the beam as an input parameter. If this parameter is set to a noninteger value, the output beam is said to be a noninteger-vortex beam. This poses intriguing questions, such as what is the distribution of vortices in this beam, and what is its orbital angular momentum? Both questions have been addressed in recent studies, and the answers are not as intuitive as one may think. Noninteger vortices do not exist. A beam programmed to have a noninteger phase dislocation has an array of vortices that has been well characterized and measured [37–40]. If the dislocation is set to be a noninteger q that is not a half-integer, the number of vortices in a beam is q rounded to the nearest

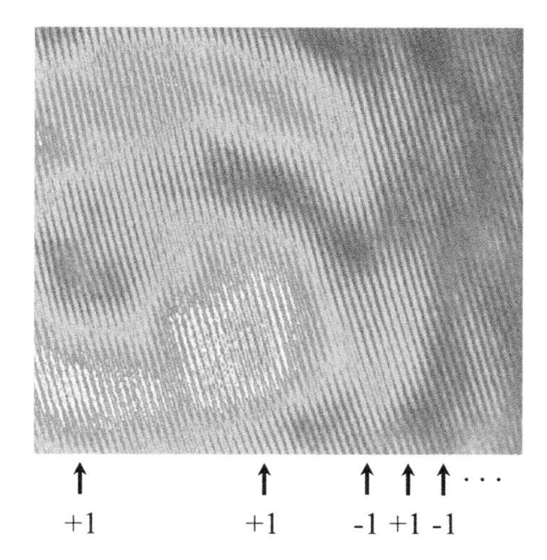

Figure 3.8 False-color image of a noncollinear interference pattern between an expanded fundamental mode and the field of a half-integer vortex beam with $q = 1.5$. The arrows indicate the horizontal positions of easily identified singly charged vortices, denoted by forks in the interference pattern. The direction of the tines denotes the sign of the charge. See color insert.

integer. When q is exactly a half integer, it has p vortices in its profile, where p is the value of q rounded *down* to the nearest integer, *plus* an infinite array of singly charged vortices of alternating sign [38]. This pattern has been measured by two different methods [8,39] and is shown in Figure 3.8. The intensity pattern is characteristic of a noninteger beam, which is that of a broken ring. The array of vortices of alternating sign begins at the broken part and continues in a nearly linear formation radially outward.

What is remarkable about nonintegral vortex beams is that the nonintegral character is encoded in the entire field; should we combine two nonintegral beams with complementary charge, the resulting composites will obey the same rules as they did earlier [32,40]. In the diffraction of integral vortices by forked gratings with nonintegral charge, diffracted beams can come out as integers if the combination of grating charge and diffracted order in equation (4) yields an integer value [8].

Interestingly, the problem of noninteger modes has been studied in more depth at the quantum level. Photon pairs produced by parametric down-conversion can be entangled via their common total orbital angular momentum. By sending each photon of a pair through a rotating noninteger spiral phase plate and by measuring the coincident detections of each pair, one can exploit correlations between the two photons that depend on how noninteger vortices are related [41].

Finally, the orbital angular momentum of a nonintegral beam does not vary linearly with the noninteger value of the dislocation. This is because at the nonintegral value of the dislocation, there is a nonaxial component of the angular momentum [39].

3.4 PROPAGATION DYNAMICS

So far we have discussed vortices in a somewhat static situation, as they appear on a transverse plane. When we consider the distribution of vortices in three dimensions, more interesting effects are revealed. Optical vortices in a composite beam can change their position as the beam propagates [42]. For example, the angular positions of vortices in the composite beams discussed in the previous section depend on the phase δ (as shown in equation (6)) between the component beams. This phase can change due to the Gouy phase, given by equation (3) [43]. A composite beam made up of a superposition of Laguerre–Gaussian beams in modes with vortices of charge l_1 and l_2 would get a Gouy phase shift

$$\Delta\psi = \left(|l_1| - |l_2|\right)\left[\tan^{-1}\left(\frac{z_b}{z_R}\right) - \tan^{-1}\left(\frac{z_a}{z_R}\right)\right] \tag{8}$$

when the beam propagates from positions z_a to z_b. Such a phase shift would manifest as a rotation of the composite vortices. For the case of a composite beam with $l_1 = -1$ and $l_2 = +2$ focused by a lens, its composite image would rotate by $\pi/6$ when going from $z_a = -z_R$ to $z_b = +z_R$.

Other more complex situations can be created by suitable superpositions of high-order modes. These may be torus loops that wrap around the beam axis [44,45]. The loops are vortices that wrap in the transverse direction around the beam axis. The beam axis can contain axial vortices. For suitable values of a perturbing field, the loops can connect with the axial vortices forming "vortex knots." These situations have been created in the laboratory by programming a spatial light modulator with the appropriate combination of modes [19,46].

3.5 CONCLUSIONS

As seen in this chapter, optical beams can have interesting structures due to distributions of the phase in their wave fields. When this phase circulates about points on a transverse plane, rich vortex patterns of wave fields develop. These patterns are robust, as vortices are part of stable solutions of the paraxial wave equation. Superpositions of these fields create interesting situations that can go beyond academic curiosity and can be applied to manipulation of particles or for encoding information.

These singular situations have received wide attention in optics due to the versatility of current optical technology. However, since they involve the properties of wave fields, they are not limited to optics. Interesting applications and phenomena wait to be discovered in other wave contexts. Finally, that the paraxial approximation to the wave equation has the same form as the Schrodinger equation in two dimensions implies that the vortex dynamics of optical wave fields may correspond to interesting dynamics yet to be investigated of quantum waves in two dimensions [45].

ACKNOWLEDGMENTS

We would like to acknowledge contributors to the data and images presented in this chapter: S. Baumann, P. Crawford, N. Fernandes, P.J. Haglin, V. Matos, L. MacMillan, M.J. Pysher, N. Smiley, and H.I. Sztul.

REFERENCES

[1] J.F. Nye, Natural Focusing and Fine Structure of Light, Institute of Physics Publishing, Bristol, 1999.

[2] M.S. Soskin, M.V. Vasnetsov, Singular optics, in: E. Wolf (Ed.), *Progress in Optics*, vol. 42, Elsevier, 2001, pp. 219–276.

[3] I. Freund, Polarization flowers, *Opt. Commun. 199* (2001) 47–63.

[4] E.J. Galvez, Gaussian beams in the optics course, *Am. J. Phys. 74* (2006) 355–361.

[5] E.J. Galvez, P.R. Crawford, H.I. Sztul, M.J. Pysher, P.J. Haglin, R.E. Williams, Geometric phase associated with mode transformations of optical beams bearing orbital angular momentum, *Phys. Rev. Lett. 90* (2003) 2039011–2039014.

[6] L. Allen, M.W. Beijersbergen, R.J.C. Spreeuw, J.P. Woerdman, Orbital angular momentum of light and the transformation of Laguerre–Gaussian laser modes, *Phys. Rev. A 45* (1992) 8185–8189.

[7] L. Allen, M.J. Padgett, The Poynting vector in Laguerre–Gaussian beams and the interpretation of their angular momentum density, *Opt. Commun. 184* (2000) 67–71.

[8] S. Baumann, E.J. Galvez, Non-integral vortex structures in diffracted light beams, *Proc. SPIE 6483* (2007) 64830T.

[9] V.Yu. Bazhenov, M.V. Vasnetsov, M.S. Soskin, Laser beams with screw dislocations in their wavefronts, *JETP Lett. 52* (1990) 429–431.

[10] N.R. Heckenberg, R. McDuff, C.P. Smith, A.G. White, Generation of optical phase singularities by computer-generated holograms, *Opt. Lett. 17* (1992) 221–223.

[11] N.R. Heckenberg, R. McDuff, C.P. Smith, H. Rubinsztein-Dunlop, M.J. Wegener, Laser beams with phase singularities, *Opt. Quantum Electron. 24* (1992) S951–S962.

[12] G.F. Brand, Phase singularities in beams, *Am. J. Phys. 67* (1999) 55–60.

[13] G.A. Turnbull, D.A. Robertson, G.M. Smith, L. Allen, M.J. Padgett, The generation of free-space Laguerre-Gaussian modes at millimeter-wave frequencies by use of a spiral phaseplate, *Opt. Commun. 127* (1996) 183–188.

[14] H. He, N.R. Heckenberg, H. Rubinztein-Dunlop, Optical particle trapping with higher-order doughnut beams produced using high efficiency computer generated holograms, *J. Mod. Opt. 42* (1995) 217–223.
[15] J.E. Curtis, D.E. Grier, Structure of optical vortices, *Phys. Rev. Lett. 90* (2003) 1339011–1339014.
[16] J. Leach, M.J. Padgett, Observation of chromatic effects near a white-light vortex, *New J. Phys. 5* (2003) 154.1–154.7.
[17] M.W. Beijersbergen, R.P.C. Coerwinkel, M. Kristensen, J.P. Woerdman, Helical-wavefront laser beams produced with a spiral phaseplate, *Opt. Commun. 112* (1994) 321–327.
[18] G. Machavariani, N. Davidson, E. Hasman, S. Blit, A.A. Ishaaya, A.A. Friesem, Efficient conversion of a Gaussian beam to a high purity helical beam, *Opt. Commun. 209* (2002) 265–271.
[19] J. Leach, M.R. Dennis, J. Courtial, M.J. Padgett, Vortex knots of light, *New J. Phys. 7* (2005) 55–66.
[20] S. Chavez-Cerda, M.J. Padgett, I. Allison, G.H.C. New, J.C. Gutierrez-Vega, A.T. O'Neil, I. MacVicar, J. Courtial, Holographic generation and orbital angular momentum of high-order Mathieu beams, *J. Opt. B 4* (2002) S52–S57.
[21] A. Mair, A. Vaziri, G. Welhs, A. Zeilinger, Entanglement of the orbital angular momentum states of photons, *Nature 412* (2001) 313–316.
[22] M.W. Beijersbergen, L. Allen, H.E.L.O. van der Veen, J.P. Woerdman, Astigmatic laser mode converters and transfer of orbital angular momentum, *Opt. Commun. 96* (1993) 123–132.
[23] S.J. van Enk, Geometric phase, transformations of Gaussian light beams and angular momentum transfer, *Opt. Commun. 102* (1993) 59–64.
[24] E.J. Galvez, M. O'Connell, Existence and absence of geometric phases due to mode transformations of high-order modes, *Proc. SPIE 5736* (2005) 166–172.
[25] A.S. Desyatnikov, Y.S. Kivshar, L. Torner, Optical vortices and vortex solitons, in: E. Wolf (Ed.), *Progress in Optics*, vol. 47, Elsevier, 2005, pp. 291–391.
[26] M.V. Berry, M.R. Dennis, Quantum cores of optical phase singularities, *J. Opt. A 6* (2004) S178–S180.
[27] M.V. Berry, Coloured phase singularities, *New J. Phys. 4* (2002) 66.1–66.14.
[28] M.V. Berry, Exploring the colours of dark light, *New J. Phys. 4* (2002) 74.1–74.14.
[29] I. Mariyenko, J. Strohaber, C. Uiterwaal, Creation of optical vortices in femtosecond pulses, *Opt. Exp. 13* (2005) 7599–7608.
[30] I. Zeylikovich, H.I. Sztul, V. Kartazaev, T. Le, R.R. Alfano, Ultrashort Laguerre–Gaussian pulses with angular and group velocity dispersion compensation, *Opt. Lett. 32* (2007) 2025–2027.
[31] K. Bezuhanov, A. Dreischuh, G.G. Paulus, M.G. Schätzel, H. Walther, D. Neshev, W. Królikowski, Y. Kivshar, Spatial phase dislocations in femtosecond laser pulses, *J. Opt. Soc. Am. B 23* (2006) 26–35.
[32] E.J. Galvez, N. Smiley, N. Fernandes, Composite optical vortices formed by collinear Laguerre–Gauss beams, *Proc. SPIE 6131* (2006) 19–26.
[33] I.D. Maleev, G.A. Swartzlander Jr., Composite optical vortices, *J. Opt. Soc. Am. B 20* (2003) 1169–1176.
[34] G. Molina-Terriza, J.P. Torres, L. Torner, Management of the angular momentum of light: Preparation of photons in multidimensional vector states of angular momentum, *Phys. Rev. Lett. 88* (2002) 0136011–0136014.
[35] G. Gibson, J. Courtial, M.J. Padgett, M. Vasnetsov, V. Pas'ko, S.M. Barnett, S. Franke-Arnold, Free-space information transfer using light beams carrying orbital angular momentum, *Opt. Exp. 12* (2004) 5448–5456.

[36] G. Molina-Terriza, J.P. Torres, L. Torner, Twisted photons, *Nature Phys. 3* (2007) 305–310.
[37] I.V. Basistiy, M.S. Soskin, M.V. Vasnetsov, Optical warefront dislocations and their properties, *Opt. Commun. 119* (1995) 604–612.
[38] M.V. Berry, Optical vortices evolving from helicoidal integer fractional phase steps, *J. Opt. A 6* (2004) 259–268.
[39] J. Leach, E. Yao, M.J. Padgett, Observation of vortex structure of a noninteger vortex beam, *New J. Phys. 6* (2004) 71–78.
[40] E.J. Galvez, S.M. Baumann, Composite vortex patterns formed by component light beams with non-integral topological charge, *Proc. SPIE 6905* (2008) 69050D.
[41] S.S.R. Oemrawsingh, A. Aiello, E.R. Eliel, G. Nienhuis, J.P. Woerdman, How to observe high-dimensional two-photon entanglement with only two detectors, *Phys. Rev. Lett. 92* (2004) 2179011–2179014.
[42] I.V. Basistiy, V. Yu Bazhenov, M.S. Soskin, M.V. Vasnetsov, Optics of light beams with screw dislocations, *Opt. Commun. 103* (1993) 422–428.
[43] J. Courtial, Self-imaging beams and the Gouy effect, *Opt. Commun. 151* (1998) 1–4.
[44] M.V. Berry, M.R. Dennis, Knotted and linked phase singularities in monochromatic waves, *Proc. R. Soc. London A 457* (2001) 2251–2263.
[45] M.V. Berry, M.R. Dennis, Knotting and unknotting of phase singularities: Helmholtz waves, paraxial waves in $2 + 1$ spacetime, *J. Phys. A: Math. Gen. 34* (2001) 8877–8888.
[46] J. Leach, M.R. Dennis, J. Courtial, M.J. Padgett, Knotted threads of darkness, *Nature 432* (2004) 165.

Chapter 4

Nanoscale Optics: Interparticle Forces

Luciana C. Dávila Romero and David L. Andrews

University of East Anglia, UK

4.1 INTRODUCTION

The nature and variety of optical forces that operate on particles of atomic, molecular, nanoscale or microscale dimensions are in principle similar to those that relate to the effect of light on larger particles. When compared to the latter, however, a significant difference in practical terms is the greater ease, in the case of microparticles, in overcoming gravity. This feature facilitates the study of suspensions or surface layers, for example, in systems comprising micron or nanometer sized particles. The distinct advantage of the nanoscale, in this respect, is nonetheless offset against much more influential levels of thermal motion. The latter problem is particularly acute in the case of atomic samples. This problem is commonly overcome by the use of cold atom traps and optical molasses instrumentation, which utilize atomic cooling through momentum exchange with absorbed and emitted photons. In any such context, conventional optical tweezers and Maxwell–Bartoli mechanisms represent the operation of optomechanical forces whose origins are well understood, and which characteristically operate on individual particles of matter. Further distinctions in behavior can then be drawn on the basis of material composition, the salient response functions being cast in terms that reflect atomic, molecular, dielectric, or metallic constitution, for example. In the latter example, the distinctively complex refractive index represents a quality admitting further opportunities to tailor dispersive optical forces, often supplemented by an exploitation of plasmonic effects.

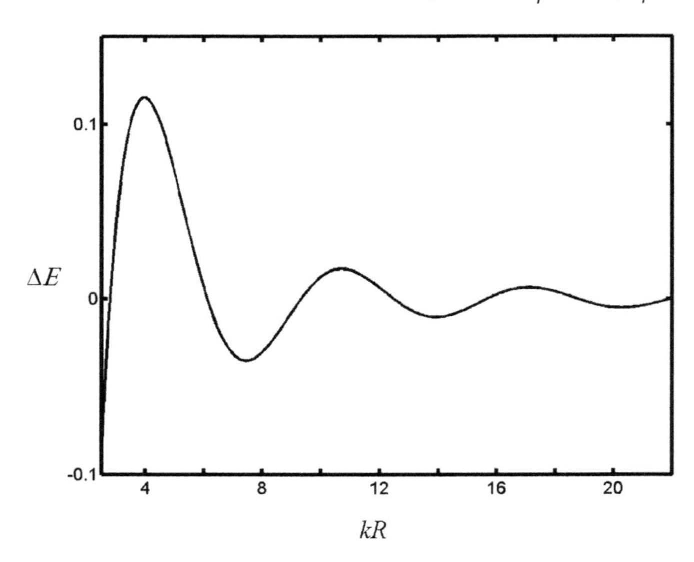

Figure 4.1 Dependence of optically induced potential energy for a pair of particles separated by distance R, plotted against kR, where $k = 2\pi/\lambda$, and λ is the laser wavelength. The interparticle axis is aligned with the electric field of the radiation; the locations of the energetically stable minima depend upon dispersion properties (see later). Graphs of similar form, for example, those shown in references [2,8], provide a compelling motif for the subject.

A relatively recent flurry of activity has been prompted by the discovery and verification of something quite different: an optomechanical force that operates *between* particles at nanoscale separations. The first theoretical proof that intense laser light can produce an optically modified potential energy surface for particle interactions was provided by Thirunamachandran almost thirty years ago [1], but the laser intensities that appeared necessary then represented a significant deterrent. However, before the end of the decade, a landmark paper by Burns and colleagues [2] verified the effect experimentally. This latter work also provided the first graphs for the simplest case of two identical spherical particles, of energy against separation. These graphs exhibited striking landscapes of rolling potential energy maxima and minima, as shown in Figure 4.1. Recognition of the enormous potential for practical applications quickly came about, and the prospects were almost immediately flagged in an influential futurology of chemistry [3]. Subsequent studies have shown that optically induced interparticle forces offer a number of highly distinctive features that can be exploited for the controlled optical manipulation of matter. The terms *optical binding* and *optical matter*, which have gained some currency for such forces, highlight the possibilities for a significant interplay with other interactions, such as chemical bonding and dispersion forces. Exploiting such interactions, new opportunities for creating optically ordered matter have already been demonstrated both theoretically and experimentally [4–11].

At this juncture, progress in theory is developing along several fronts, with many studies invoking essentially classical descriptions of the radiation field. Some of the most adventurous studies relate to perhaps the most demanding experimental challenge—the possibility of engaging off-resonant laser light with Bose–Einstein condensates to achieve "superchemistry", i.e., the coherent manipulation and assembly of atoms and molecules [12,13]. In several treatments, paraxial wave equations have been adopted to describe optical binding between micron-sized spherical particles in the presence of counterpropagating beams [14,15]. The results were analyzed in terms of the relative refractive indices of the spheres and the surrounding medium. Such studies are valid in the Mie size regime, i.e., where the sphere diameter exceeds the wavelength, and input fields are well approximated as paraxial. Considering particles of like dimensions, Chaumet and Nieto-Vesperinas [8] have derived results both for isolated spheres and for spheres near a surface. In the isolated case, they have found that interparticle forces depend significantly upon the polarization and wavelength of the incident light, and upon the particle size. Furthermore, Ng and Chan [16] have determined the equilibrium positions in an array of evenly spaced particles aligned in parallel with the wave-vector of the optical input. Extending the range of applications, studies of optical trapping and binding of cylindrical particles have been carried out by Grzegorczyk and colleagues [17,18].

Further opportunities for application and other readily achievable areas of relevance are now being identified with the benefit of a comprehensive theory based upon quantum electrodynamics (QED) [19,20]. Based upon this theory, calculations on carbon nanotubes, for example, already have indicated dependences on particle orientation. This suggests possibilities for optically modifying the morphology of deposited nanotube films [21], while applications to other dielectric nanoparticles in optical vortex fields have identified opportunities for new forms of optical patterning and clustering [22]. Some of the most recent work has established other more exotic effects, such as an optically induced shift in the equilibrium bond length of van der Waals dimers (molecular pairs held together by weak hydrogen bonds) and, in molecular solids, bulk optomechanical deformation [23]. In the following section, we will first rehearse and explain the state of the art QED theory, placing the various representations within a single consistent framework. With reference to the key equations and against this background, the next section will provide a concise overview of the applications. We will conclude the chapter with a look to the future.

4.2 QED DESCRIPTION OF OPTICALLY INDUCED PAIR FORCES

In the perturbative derivation of optically induced pair forces, as with more common interparticle coupling forces, calculations are generally performed on a system in which each particle resides in its lowest-energy, stable state. For the development of a QED theory, the system state has to be more precisely specified as one in which both particles and the radiation field are in the ground state. This system state couples with other short-lived states in which the electromagnetic field has a nonzero occupation number for one or more radiation modes. The *dispersion interaction*, traditionally interpreted as a coupling between mutually induced moments, emerges from a fourth-order perturbative calculation based on the exchange of two virtual photons, each created at one particle and annihilated at the other. The two virtual quanta may (but need not) overlap in time as they propagate between the two units. Cast in such terms, the theory delivers a result (the Casimir–Polder formula) that is valid for all distances and correctly accounts for the retardation features that lead to long-range R^{-7} asymptote dependence on the pair separation R [24–30]. The virtual photon interpretation also lends a fresh perspective to the physics involved in the more familiar R^{-6} range dependence known as the *van der Waals interaction*—the attractive part of the Lennard–Jones potential, which operates at shorter distances and is largely responsible for the cohesion of condensed phase matter [31].

The photonic basis for this dispersion interaction strongly suggests that other effects may be manifest when intense light is present, i.e., when calculations are performed on a basis state for which the occupation number of at least one photon mode is nonzero. Indeed, it is the same fourth order of perturbation theory that gives the leading result: the annihilation and creation of one photon from the occupied radiation mode in principle substitute for the paired creation and annihilation events of one of the two virtual photons involved in the Casimir–Polder calculation. It is clear that the result of any such calculation on optically conferred pair energies will exhibit linear dependence on the photon number of the occupied mode. Cast in terms of experimental quantities, this will be manifest as an energy shift ΔE_{ind} with a corresponding proportionality to the irradiance of throughput radiation. The corresponding laser-induced coupling forces can be determined from the potential energy result, as the spatial derivative.

4.2.1 Quantum Foundations

To begin, consider the coupling between two particles, with no assumed symmetry, whose laser-induced interactions involve the absorption of a real input photon at one particle and the stimulated emission of a real photon at the other one, with a virtual photon acting as a messenger between the two. The throughput radiation suffers no overall change in its state. Following the Power–Zienau–Woolley approach [32–35], writing the interactions of the vacuum electromagnetic fields with particle ξ in the electric–dipole approximation, we have the interaction Hamiltonian

$$H_{\text{int}}^{\xi} = -\varepsilon_o^{-1} \sum_{\xi} \mu(\xi) \cdot \mathbf{d}^{\perp}(\mathbf{R}_{\xi}), \tag{1}$$

where $\mu(\xi)$ and $\mathbf{R}_{\xi}$, respectively, denote the electric–dipole moment operator and the position vector of dielectric nanoparticles labeled ξ. The operator $\mathbf{d}^{\perp}(\mathbf{R}_{\xi})$ represents the transverse electric displacement field, expressible in the following general mode-expansion

$$\mathbf{d}^{\perp}(\mathbf{R}_{\xi}) = i \sum_{\mathbf{k},\lambda} \left(\frac{\hbar c k \varepsilon_0}{2V} \right)^{1/2} \big[\mathbf{e}^{(\lambda)}(\mathbf{k}) a^{(\lambda)}(\mathbf{k}) \exp(i\mathbf{k} \cdot \mathbf{R}_{\xi}) - \bar{\mathbf{e}}^{(\lambda)}(\mathbf{k}) a^{\dagger(\lambda)}(\mathbf{k}) \exp(-i\mathbf{k} \cdot \mathbf{R}_{\xi}) \big]. \tag{2}$$

In equation (2), V is the quantization volume, and summation is taken over modes indexed by wave-vector $\mathbf{k}$ and polarization λ; a and $a^{\dagger}$ are annihilation and creation operators, respectively, and $\mathbf{e}$ represents the electric field unit vector, with $\bar{\mathbf{e}}$ being its complex conjugate. For present purposes, the distinction between $\mathbf{e}$ and $\bar{\mathbf{e}}$ can be dropped on the assumption that only plane polarizations are to be entertained, which is consistent with experimental practice. Since the laser-induced coupling involves four matter-photon interactions, it requires the application of fourth-order perturbation theory (within the electric–dipole approximation), and the energy is explicitly given by

$$\Delta E_{\text{ind}} = \text{Re}\left[\sum_{t,s,r} \frac{\langle i|H_{\text{int}}|t\rangle \langle t|H_{\text{int}}|s\rangle \langle s|H_{\text{int}}|r\rangle \langle r|H_{\text{int}}|i\rangle}{(E_i - E_t)(E_i - E_s)(E_i - E_r)} \right]. \tag{3}$$

In general, an arbitrary ket $|\varepsilon\rangle$ here refers to a member of the set of basis states of the unperturbed Hamiltonian, such that we have

$$|s\rangle = |\text{mol}\rangle_s \otimes |\text{rad}\rangle_s = |\text{mol}_s; \text{rad}_s\rangle, \tag{4}$$

where $|\text{mol}\rangle_s$ and $|\text{rad}\rangle_s$ respectively, define the status of all particles and radiation states involved. Specifically, $|i\rangle$ is the unperturbed system state and the kets $|r\rangle$, $|t\rangle$, $|s\rangle$ are virtual states.

From equations (1) and (2), it follows that each Dirac bracket in the numerator of (3) is associated with the creation or annihilation of a photon. Details emerge on application to a specific system; here we consider two chemically identical particles A and B, the latter displaced from A by vector $\mathbf{R}$. Assuming neither particle possesses a

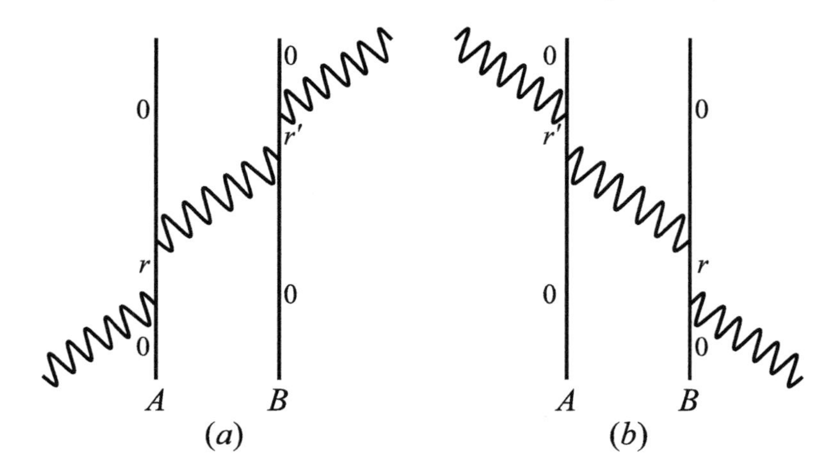

Figure 4.2 Two typical Feynman diagrams (each with 23 further permutations) are used to calculate dynamic contributions to laser-induced interaction energy. Vertical lines denote world-lines of the two particles; wavy lines outside vertical lines denote real (laser) photons, and those inside denote virtual photons; time progresses upward. Adapted from [19].

permanent electric dipole moment, it is readily shown that each must suffer two dipole transitions and that 48 different cases arise with each case generating a dynamic contribution to the energy shift. As a calculational aid, these contributions are typically represented in the form of nonrelativistic Feynman diagrams, as shown in Figure 4.2. In the complete set, 24 entail absorption of the laser photon at A, and in the other 24 the same process occurs at B. The latter may be deduced on the basis of mirroring the former, in the sense that A exchanges with B and that $\mathbf{R}$ changes sign. Accordingly, we denote as $\Delta E_{\text{ind}}^{A\to B}$ the energy shift resulting from orderings in which the absorption of laser light occurs at A, and stimulated emission at B, with $\Delta E_{\text{ind}}^{B\to A}$ denoting the converse. Note that the direction indicated by the superscript does *not* determine the direction of virtual photon propagation; for example, among the contributions to $\Delta E_{\text{ind}}^{A\to B}$, half involve virtual photon propagation toward B, but the other half involve propagation toward A.

Hence, we have the following expression for the total induced energy shift, ΔE_{ind}

$$\begin{aligned}\Delta E_{\text{ind}} &= \Delta E_{\text{ind}}^{A\to B} + \Delta E_{\text{ind}}^{B\to A} = \Delta E_{\text{ind}}^{A\to B}(\mathbf{R}) + \Delta E_{\text{ind}}^{A\to B}(-\mathbf{R}) \\ &= \Delta E_{\text{ind}}^{A\to B}(\mathbf{R})\big|_{\text{even}} + \Delta E_{\text{ind}}^{A\to B}(\mathbf{R})\big|_{\text{odd}} + \Delta E_{\text{ind}}^{A\to B}(\mathbf{R})\big|_{\text{even}} \\ &\quad - \Delta E_{\text{ind}}^{A\to B}(\mathbf{R})\big|_{\text{odd}} \\ &= 2\Delta E_{\text{ind}}^{A\to B}(\mathbf{R})\big|_{\text{even}},\end{aligned} \tag{5}$$

where *even* and *odd* denote the corresponding parts of the function $\Delta E_{\text{ind}}^{A\to B}(\mathbf{R})$ with respect to $\mathbf{R}$. After using expression (3) and then performing a sequence of calculational steps (detailed elsewhere [19]), the induced energy shift $\Delta E_{\text{ind}}^{A\to B}$ emerges as follows, using the convention of implied summation over repeated subscript (Cartesian) indices

$$\Delta E_{\text{ind}}^{A\to B}(k,\mathbf{R}) = \left(\frac{n\hbar ck}{\varepsilon_0 V}\right)\text{Re}\big[e_i^{(\lambda)}\alpha_{ij}^{A}(k)V_{jk}^{\pm}(k,\mathbf{R})\alpha_{kl}^{B}(k)e_l^{(\lambda)}\exp(-i\mathbf{k}\cdot\mathbf{R})\big]. \quad (6)$$

Here n is the number of laser photons within a quantization volume V, and we have introduced the well-known dynamic polarizability tensor α_{ij}^{ξ} and the fully retarded resonance dipole–dipole interaction tensor of the general form

$$V_{jk}^{\pm}(k,\mathbf{R}) = \frac{\exp[\mp ikR]}{4\pi\varepsilon_0 R^3}\big\{(1\pm ikR)(\delta_{jk}-3\hat{R}_j\hat{R}_k)-(kR)^2(\delta_{jk}-\hat{R}_j\hat{R}_k)\big\}. \quad (7)$$

Given that the analytically arbitrary choice of sign has no physical consequence [36], we shall stick with the negative sign (as generally assumed without comment in older work) and drop it from our notation henceforth, i.e., $V_{jk}^{-}(k,\mathbf{R}) \equiv V_{jk}(k,\mathbf{R})$.

4.2.2 Defining the Geometry

As a convenient starting point for an exploration of various geometries and degrees of rotational freedom, we will begin by considering the coupling of two fixed particles, with no assumed symmetry. The geometry for the pair is specified as follows: Particle A is at the origin ($\mathbf{R}_A = 0$), and particle B is on the z-axis ($\mathbf{R}_B = R\hat{\mathbf{z}}$) such that the separation between the two particles is given by $\mathbf{R} \equiv \mathbf{R}_B - \mathbf{R}_A = R\hat{\mathbf{z}}$; the angles ϕ and θ denote the orientations of the optical polarization vector respect to $\mathbf{R}$, as shown in Figure 4.3. The figure depicts a case in which particles have the same orientation; however, in a more general case, this will not necessarily apply.

In order to fully describe the system with due regard to its internal degrees of freedom, it is necessary to consider frames of reference:

(a) A *fixed frame* (or *laboratory frame*), denoted by $(\hat{x},\hat{y},\hat{z})$ as shown in Figure 4.3
(b) For each particle, $\xi = A$ and B, a *particle frame*, chosen with regard to the particle symmetry such that the corresponding polarizability tensor is diagonalized in three nonzero components $(\hat{x}'^{\xi},\hat{y}'^{\xi},\hat{z}'^{\xi})$.

The latter frames enter the calculations at a later stage; for the present stage, we can refer all vectors and tensors to the fixed frame. Thus, for example, from equation (7),

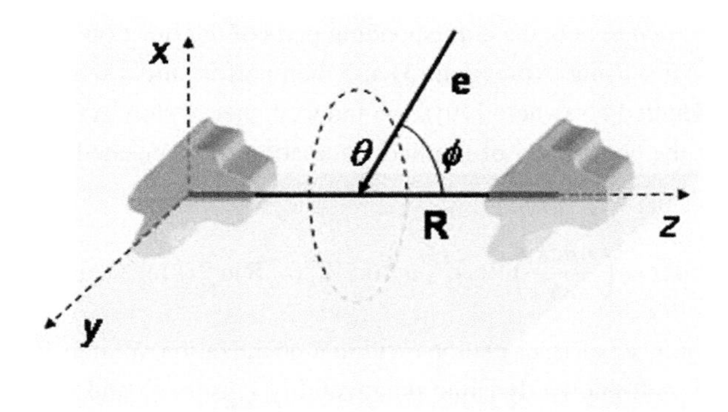

Figure 4.3 Geometry of the particle pair and the polarization vector **e** of an electromagnetic field. For simplicity, both particles are shown with the same orientation. In the electric–dipole approximation, the direction of the optical propagation vector is irrelevant, serving only as a constraint on possible directions of **e**.

the components of the tensor $V_{jk}(k, \mathbf{R})$ are explicitly

$$V_{jk}(k, \mathbf{R}) = \begin{cases} \frac{\exp[ikR]}{4\pi\varepsilon_0 R^3}\{(1 - ikR) - (kR)^2\} & \text{for } (jk) = \{xx, yy\}, \\ -\frac{\exp[ikR]}{2\pi\varepsilon_0 R^3}(1 - ikR) & \text{for } (jk) = \{zz\}, \\ 0 & \text{in any other case.} \end{cases} \tag{8}$$

From equations (6) and (8), and using the relationship $I = n\hbar c^2 k/V$ for the laser irradiance I, we have

$$\Delta E_{\text{ind}}^{A\to B}(k, \mathbf{R}) = \left(\frac{I}{4\pi\varepsilon_0^2 c}\right) \text{Re}\Bigg[\left[e_i^{(\lambda)}\mathbf{Z}_{il}^{(1)}e_l^{(\lambda)}(1 - ikR) - e_i^{(\lambda)}\mathbf{Z}_{il}^{(2)}e_l^{(\lambda)}(kR)^2\right] \times \frac{\exp(ikR)\exp(-i\mathbf{k}\cdot\mathbf{R})}{R^3}\Bigg], \tag{9}$$

where we have introduced two pair response tensors, $\mathbf{Z}_{il}^{(1)}$ and $\mathbf{Z}_{il}^{(2)}$. The latter are defined in the pair-fixed frame as

$$\begin{aligned} \mathbf{Z}_{il}^{(1)} &= \mathbf{Z}_{il}^{(2)} - 2\alpha_{iz}^A\alpha_{zl}^B, \\ \mathbf{Z}_{il}^{(2)} &= \alpha_{ix}^A\alpha_{xl}^B + \alpha_{iy}^A\alpha_{yl}^B. \end{aligned} \tag{10}$$

When the real part of the square bracket is taken in expression (9), the induced energy shift is expressible as

$$\Delta E_{\text{ind}}^{A\to B}(k,\mathbf{R})$$
$$= \left(\frac{I}{4\pi\varepsilon_0^2 cR^3}\right)\{e_i^{(\lambda)}\mathbf{Z}_{il}^{(2)}e_l^{(\lambda)}$$
$$\times\left[\cos(kR-\mathbf{k}\cdot\mathbf{R})+kR\sin(kR-\mathbf{k}\cdot\mathbf{R})-(kR)^2\cos(kR-\mathbf{k}\cdot\mathbf{R})\right]$$
$$-2e_i^{(\lambda)}\alpha_{iz}^{A}\alpha_{zl}^{B}e_l^{(\lambda)}\left[\cos(kR-\mathbf{k}\cdot\mathbf{R})+kR\sin(kR-\mathbf{k}\cdot\mathbf{R})\right]\}. \tag{11}$$

Securing the complete result, using expression (5), it is apparent that the induced energy shift is given by

$$\Delta E_{\text{ind}}(k,\mathbf{R})$$
$$= \left(\frac{I}{2\pi\varepsilon_0^2 cR^3}\right)\{\left[e_i^{(\lambda)}\mathbf{Z}_{il}^{(2)}e_l^{(\lambda)}-2e_i^{(\lambda)}\alpha_{iz}^{A}\alpha_{zl}^{B}e_l^{(\lambda)}\right]\left[\cos(kR)+kR\sin(kR)\right]$$
$$-e_i^{(\lambda)}\mathbf{Z}_{il}^{(2)}e_l^{(\lambda)}k^2R^2\cos(kR)\}\cos(\mathbf{k}\cdot\mathbf{R}). \tag{12}$$

The interparticle force can be found by simply taking the derivative of the energy shift with respect to the separation of the particles

$$\mathbf{F}_{\text{ind}} = -\frac{\partial\Delta E_{\text{ind}}}{\partial\mathbf{R}}$$
$$= \left(\frac{I}{2\pi\varepsilon_0^2 cR^4}\right)\{\left[e_i^{(\lambda)}\mathbf{Z}_{il}^{(2)}e_l^{(\lambda)}-2e_i^{(\lambda)}\alpha_{iz}^{A}\alpha_{zl}^{B}e_l^{(\lambda)}\right]$$
$$\times\{\left[3\cos(kR)+3kR\sin(kR)-k^2R^2\cos(kR)\right]\cos(\mathbf{k}\cdot\mathbf{R})$$
$$+\left[k_zR\cos(kR)+k_zkR^2\sin(kR)\right]\sin(\mathbf{k}\cdot\mathbf{R})\}$$
$$-e_i^{(\lambda)}\mathbf{Z}_{il}^{(2)}e_l^{(\lambda)}\{\left[k^2R^2\cos(kR)+k^3R^3\sin(kR)\right]\cos(\mathbf{k}\cdot\mathbf{R})$$
$$+k_zk^2R^3\cos(kR)\sin(\mathbf{k}\cdot\mathbf{R})\}\}. \tag{13}$$

We now analyze particular cases, deriving explicit results for other systems of physical interest. We will assume particles of cylindrical symmetry to accommodate the more usual case of spherical symmetry and to deliver results that have validity for nanotubes and most other significantly anisotropic nanoparticles.

4.2.3 Tumbling Cylindrical Pair

First we shall address a system in which two particles freely rotate in the incident light as a binary system, with each particle maintaining a fixed distance and orientation with respect to its counterpart. In this case, we not only have the angles that define the direction of the polarization vector respect to the pair, as shown in Figure 4.3; it is

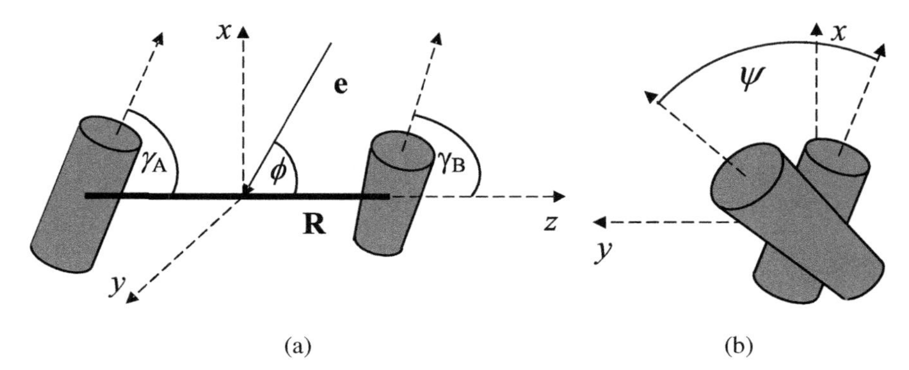

Figure 4.4 Geometry of the tumbling pair system for a pair of cylindrical particles.

also necessary to introduce three angles that will determine the relative orientations of the two components. As shown in Figure 4.4, these internal angles are defined as $(\gamma_A, \gamma_B, \psi)$; γ_A is the angle between particle A and the z-axis (assuming that the $\hat{z}'^A$ lies on the xz-plane); γ_B is the angle between the z-axis and the molecular principal axis z'^B (which may or may not lie on the xz-plane); while angle ψ is the angle between $\hat{z}'^A$ and z'^B axes projected onto the xy-plane.

As noted earlier, the polarization of the incoming and outgoing radiation is considered to be linear. Representing the polarization in the laboratory frame, generally we have

$$\mathbf{e}_i^{(\lambda)} = \sin\phi\cos\theta\hat{\mathbf{x}} + \sin\phi\sin\theta\hat{\mathbf{y}} + \cos\phi\hat{\mathbf{z}}. \tag{14}$$

In the case of the tumbling cylindrical pair, the polarizability tensor of each particle is diagonal when expressed with respect to the corresponding particle's reference frame

$$\boldsymbol{\alpha}^{\xi} = \begin{pmatrix} \alpha_{\perp}^{\xi} & 0 & 0 \\ 0 & \alpha_{\perp}^{\xi} & 0 \\ 0 & 0 & \alpha_{\parallel}^{\xi} \end{pmatrix}. \tag{15}$$

However, this frame rotates with the tumbling pair. It makes more sense to refer all vector and tensor components in the general energy expression (6) to a laboratory-fixed frame in which the polarization components are static, and to this end the polarizability for each particle must be recast in the laboratory-fixed frame. By appropriate unitary transformations, we find the following results for particles A and B:

$$\alpha^A\Big|_{\substack{\text{Fixed}\\\text{frame}}} = \alpha_{\parallel}^A \begin{pmatrix} 1-\eta^A\cos^2\gamma_A & 0 & \eta^A\cos\gamma_A\sin\gamma_A \\ 0 & 1-\eta^A & 0 \\ \eta^A\cos\gamma_A\sin\gamma_A & 0 & 1-\eta^A\sin^2\gamma_A \end{pmatrix} \tag{16a}$$

and

$$\alpha_{ij}^{B}\Big|_{\substack{\text{Fixed}\\ \text{frame}}} = \alpha_{\parallel}^{B}\begin{pmatrix} (1-\eta^{B}\cos^{2}\gamma_{B})\cos^{2}\psi+(1-\eta^{B})\sin^{2}\psi & \eta^{B}\sin^{2}\gamma_{B}\cos\psi\sin\psi & \eta^{B}\cos\gamma_{B}\sin\gamma_{B}\cos\psi \\ \eta^{B}\sin^{2}\gamma_{B}\sin\psi\cos\psi & (1-\eta^{B}\cos^{2}\gamma_{B})\sin^{2}\psi+(1-\eta^{B})\cos^{2}\psi & \eta^{B}\cos\gamma_{B}\sin\gamma_{B}\cos\psi \\ \eta^{B}\cos\gamma_{B}\sin\gamma_{B}\cos\psi & \eta^{B}\cos\gamma_{B}\sin\gamma_{B}\sin\psi & (1-\eta^{B}\sin^{2}\gamma_{B}) \end{pmatrix}. \tag{16b}$$

Here we have introduced the anisotropy factors $\eta^{\xi} \equiv (\alpha_{\parallel}^{\xi} - \alpha_{\perp}^{\xi})/\alpha_{\parallel}^{\xi}$ to simplify the expressions. Given the complexity that ensues, we restrict consideration to that of an isotropic average with respect to the incoming light, the calculation of which requires the use of a phase-average method [37]. In this case, the induced energy shift is given by the following, in which the j_m are spherical Bessel functions:

$$\begin{aligned}\langle\Delta E_{\text{ind}}\rangle = \frac{I}{\varepsilon_0 c}\operatorname{Re}\bigg[\bigg\{\frac{1}{3}j_0(kR) - \frac{1}{6}j_2(kR)\bigg\}\big\{V_{xx}\big(\alpha_{xx}^{A}\alpha_{xx}^{B} + \alpha_{yy}^{A}\alpha_{yy}^{B} + \alpha_{zx}^{A}\alpha_{xz}^{B}\big) \\ + V_{zz}\big(\alpha_{xz}^{A}\alpha_{zx}^{B} + \alpha_{zz}^{A}\alpha_{zz}^{B}\big)\big\} + \frac{1}{2}j_2(kR)\big\{V_{xx}\alpha_{zx}^{A}\alpha_{xz}^{B} + V_{zz}\alpha_{zz}^{A}\alpha_{zz}^{B}\big\}\bigg],\end{aligned} \tag{17}$$

where we have used $\alpha_{ij}^{\xi}\big|_{\substack{\text{Fixed}\\ \text{frame}}} \equiv \alpha_{ij}^{\xi}$ to simplify notation, with the explicit components for each particle determined by equation (2.16). The resulting expression can be explicitly calculated by using equation (8), giving

$$\begin{aligned}\langle\Delta E_{\text{ind}}\rangle \\ = \frac{I}{4\pi\varepsilon_0^2 cR^3} \\ \times\bigg[\bigg[\bigg(-kR + \frac{3}{kR} - \frac{1}{k^3R^3}\bigg)\frac{1}{4}\sin 2kR - \bigg(1 - \frac{1}{k^2R^2}\bigg)\frac{1}{2}\cos 2kR\bigg] \\ \times\big(\alpha_{xx}^{A}\alpha_{xx}^{B} + \alpha_{yy}^{A}\alpha_{yy}^{B}\big) \\ + \bigg[-\bigg(\frac{2}{kR} - \frac{1}{k^3R^3}\bigg)\frac{1}{2}\sin 2kR + \bigg(1 - \frac{1}{k^2R^2}\bigg)\cos 2kR + \sin^2 kR\bigg]\alpha_{zx}^{A}\alpha_{xz}^{B} \\ + \bigg[-\bigg(\frac{2}{kR} - \frac{1}{k^3R^3}\bigg)\frac{1}{2}\sin 2kR + \bigg(1 - \frac{1}{k^2R^2}\bigg)\cos 2kR - \cos^2 kR\bigg]\alpha_{xz}^{A}\alpha_{zx}^{B} \\ + \bigg[\bigg(\frac{1}{kR} - \frac{1}{k^3R^3}\bigg)\sin 2kR + \frac{2}{k^2R^2}\cos 2kR\bigg]\alpha_{zz}^{A}\alpha_{zz}^{B}\bigg].\end{aligned} \tag{18}$$

The corresponding laser-induced force, directly deducible from this expression, is explicitly given in the original papers [19,21].

4.2.4 Collinear Pair

In this case, we shall consider two cylindrically symmetric particles aligned collinearly, i.e., their principal axes of symmetry coincide, serving to define the axis z as shown in Figure 4.5. Owing to the symmetry of the system, it is readily seen that the induced energy shift is independent of the angle θ shown in Figure 4.3. Therefore, the polarization vector now takes the simpler form

$$\mathbf{e} = \sin\phi\hat{\mathbf{x}} + \cos\phi\hat{\mathbf{z}}. \tag{19}$$

In this case, the polarizability tensor for each particle is given by

$$\alpha_{ij}^{(\xi)} = \begin{pmatrix} \alpha_{\perp}^{(\xi)} & 0 & 0 \\ 0 & \alpha_{\perp}^{(\xi)} & 0 \\ 0 & 0 & \alpha_{\parallel}^{(\xi)} \end{pmatrix}, \tag{20}$$

where we chose the molecular frame to coincide with that of the particles $(x, y, z) \equiv (x_A, y_A, z_A) \equiv (x_B, y_B, z_B)$. From equations (19) and (20), the induced energy shift can be expressed as

$$\begin{aligned}\Delta E_{\text{ind}}(k, \mathbf{R}) &= \left(\frac{I}{2\pi\varepsilon_0^2 c R^3}\right)\left\{\left[\alpha_{\perp}^{A}\alpha_{\perp}^{B}\sin^2\phi - 2\alpha_{\parallel}^{A}\alpha_{\parallel}^{B}\cos^2\phi\right]\left[\cos(kR) + kR\sin(kR)\right]\right. \\ &\quad \left. - \alpha_{\perp}^{A}\alpha_{\perp}^{B}\sin^2\phi k^2R^2\cos(kR)\right\}\cos(\mathbf{k}\cdot\mathbf{R}),\end{aligned} \tag{21}$$

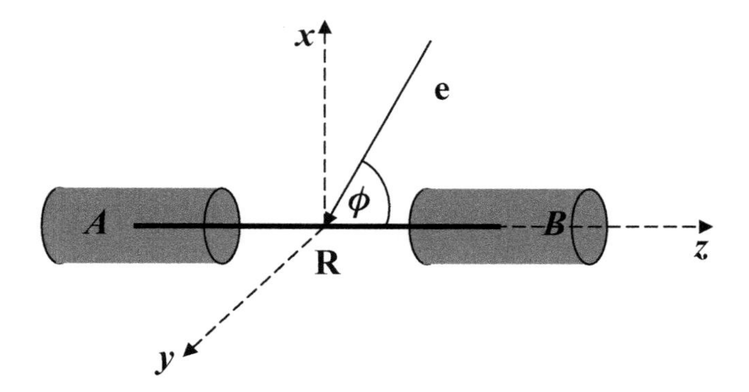

Figure 4.5 Geometry for a pair of collinear and cylindrical symmetric particles.

and the induced force as

$$
\begin{aligned}
\mathbf{F}_{\text{ind}} &= -\frac{\partial \Delta E_{\text{ind}}}{\partial \mathbf{R}} = -\frac{\partial \Delta E_{\text{ind}}}{\partial R}\hat{\mathbf{z}} \\
&= \left(\frac{I}{2\pi\varepsilon_0^2 c R^4}\right)\Big\{\big(\alpha_\perp^A \alpha_\perp^B \sin^2\phi - 2\alpha_\parallel^A \alpha_\parallel^B \cos^2\phi\big) \\
&\quad \times \big[\big(3\cos kR + 3kR\sin kR - k^2R^2\sin kR\big)\cos(\mathbf{k}\cdot\mathbf{R}) \\
&\quad + \big(k_z R\cos kR + k_z kR^2 \sin kR\big)\sin(\mathbf{k}\cdot\mathbf{R})\big] \\
&\quad - \alpha_\perp^A \alpha_\perp^B \sin^2\phi\big[\big(k^2R^2\cos kR + k^3R^3\sin kR\big)\cos(\mathbf{k}\cdot\mathbf{R}) \\
&\quad + k_z k^2 R^3 \cos kR \sin(\mathbf{k}\cdot\mathbf{R})\big]\Big\}. \qquad (22)
\end{aligned}
$$

If the particles have spherical symmetry, then $\alpha_\perp^\xi = \alpha_\parallel^\xi = \alpha_0^\xi$, and the induced energy shift and the induced force are more simply expressible as

$$
\begin{aligned}
\Delta E_{\text{ind}}^{\text{symm}}(k,\mathbf{R}) &= \left(\frac{I}{2\pi\varepsilon_0^2 c R^3}\right)\alpha_0^A \alpha_0^B \\
&\quad \times \big\{\big[\sin^2\phi - 2\cos^2\phi\big]\big[\cos(kR) + kR\sin(kR)\big] \\
&\quad - \sin^2\phi k^2R^2\cos(kR)\big\}\cos(\mathbf{k}\cdot\mathbf{R}), \qquad (23a)
\end{aligned}
$$

$$
\begin{aligned}
\mathbf{F}_{\text{ind}}^{\text{symm}} &= \left(\frac{I}{2\pi\varepsilon_0^2 c R^4}\right)\alpha_0^A \alpha_0^B\big\{\big(\sin^2\phi - 2\cos^2\phi\big) \\
&\quad \times \big[\big(3\cos kR + 3kR\sin kR - k^2R^2\sin kR\big)\cos(\mathbf{k}\cdot\mathbf{R}) \\
&\quad + \big(k_z R\cos kR + k_z kR^2\sin kR\big)\sin(\mathbf{k}\cdot\mathbf{R})\big] \\
&\quad - \sin^2\phi\big[\big(k^2R^2\cos kR + k^3R^3\sin kR\big)\cos(\mathbf{k}\cdot\mathbf{R}) \\
&\quad + k_z k^2R^3\cos kR\sin(\mathbf{k}\cdot\mathbf{R})\big]\big\}. \qquad (23b)
\end{aligned}
$$

In this case, it is interesting to demonstrate the considerable simplification that can be effected if we consider $kR = 1$, and therefore $\cos(\mathbf{k}\cdot\mathbf{R}) \approx 1$, $\cos kR \approx 1$, and $\sin kR \approx kR$. Then the above expressions reduce to

$$
\Delta E_{\text{ind}}^0(k,\mathbf{R}) = \left(\frac{I}{2\pi\varepsilon_0^2 c R^3}\right)\big[\alpha_\perp^A \alpha_\perp^B \sin^2\phi - 2\alpha_\parallel^A \alpha_\parallel^B \cos^2\phi\big], \qquad (24a)
$$

$$
\mathbf{F}_{\text{ind}}^0 = \left(\frac{3I}{2\pi\varepsilon_0^2 c R^4}\right)\big(\alpha_\perp^A \alpha_\perp^B \sin^2\phi - 2\alpha_\parallel^A \alpha_\parallel^B \cos^2\phi\big), \qquad (24b)
$$

which in turn become even more compact for the spherically symmetric case

$$\Delta E_{\text{ind}}^{0,\text{symm}} = \frac{I}{2\pi \varepsilon_0^2 c R^3} \alpha_0^A \alpha_0^B \left(3\sin^2\phi - 2\right), \tag{25a}$$

$$\mathbf{F}_{\text{ind}}^{0,\text{symm}} = \frac{3I}{2\pi \varepsilon_0^2 c R^4} \alpha_0^A \alpha_0^B \left(3\sin^2\phi - 2\right). \tag{25b}$$

Finally, consider a pair that can freely tumble while retaining a fixed collinear orientation of its component particles. Then, by averaging over all possible directions for the radiation, we have

$$\begin{aligned} \langle \Delta E_{\text{ind}} \rangle = \frac{I}{3\pi \varepsilon_0^2 c R^3} \big\{ \big[\alpha_\perp^A \alpha_\perp^B - \alpha_\parallel^A \alpha_\parallel^B \big] \big[\cos(kR) + kR\sin(kR) \big] \\ - \alpha_\perp^A \alpha_\perp^B k^2 R^2 \cos(kR) \big\} \cos(\mathbf{k} \cdot \mathbf{R}), \end{aligned} \tag{26}$$

and in the short range

$$\langle \Delta E_{\text{ind}}^0 \rangle = \left(\frac{I}{3\pi \varepsilon_0^2 c R^3} \right) \times \big[\alpha_\perp^A \alpha_\perp^B - \alpha_\parallel^A \alpha_\parallel^B \big]. \tag{27}$$

The laser-induced force can be calculated in a similar manner. Clearly, in the spherical case, $\alpha_\perp^\xi = \alpha_\parallel^\xi$, giving vanishing results for both the energy shift and the force. However, this short-range asymptote is not representative of the complex patterning of energy and force observed at longer distances. The behavior beyond the short-range is itself of considerable interest, and it is a subject we explore in due detail later in the chapter.

4.2.5 Cylindrical Parallel Pair

Another interesting case is when the two cylindrically symmetric particles are parallel to each other and perpendicular to their relative displacement vector **R**, as shown in Figure 4.6. In this case, given the geometry of the system, it is necessary to retain both angular degrees of freedom ϕ and θ in equation (14). The polarizability for this system is given by

$$\alpha_{ij}^{(\xi)} = \begin{pmatrix} \alpha_\parallel^{(\xi)} & 0 & 0 \\ 0 & \alpha_\perp^{(\xi)} & 0 \\ 0 & 0 & \alpha_\perp^{(\xi)} \end{pmatrix}. \tag{28}$$

Note the difference from the previous case, equation (20), due to the effective rotation of the particles in the xz-plane (from the definition of molecular angles in the Tumbling Pair section, we can see that in this case the polarizability can be obtained from

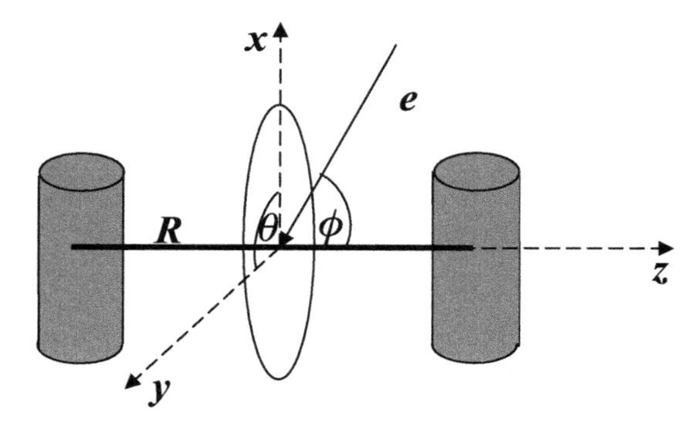

Figure 4.6 Geometry for a pair of parallel and cylindrical symmetric particles.

expressions (16a) and (16b) for $\gamma_A = \gamma_B = \pi/2$, $\psi = 0$). The induced energy shift $\Delta E_{\text{ind}}(k, \mathbf{R})$ is now given by

$$\begin{aligned}
&\Delta E_{\text{ind}}(k, \mathbf{R}) \\
&= \left(\frac{I}{2\pi\varepsilon_0^2 c R^3}\right)\Big\{\big[\alpha_\parallel^A \alpha_\parallel^B \sin^2\phi \cos^2\theta + \alpha_\perp^A \alpha_\perp^B \sin^2\phi \sin^2\theta - 2\alpha_\perp^A \alpha_\perp^B \cos^2\phi\big] \\
&\quad \times \big[\cos(kR) + kR\sin(kR)\big] \\
&\quad - \big[\alpha_\parallel^A \alpha_\parallel^B \sin^2\phi \cos^2\theta + \alpha_\perp^A \alpha_\perp^B \sin^2\phi \sin^2\theta\big]k^2R^2\cos(kR)\Big\}\cos(\mathbf{k}\cdot\mathbf{R}), \qquad (29)
\end{aligned}$$

and the induced force is

$$\begin{aligned}
\mathbf{F}_{\text{ind}} &= -\frac{\partial \Delta E_{\text{ind}}}{\partial \mathbf{R}} \\
&= \left(\frac{I}{2\pi\varepsilon_0^2 c R^4}\right)\Big\{\Big\{\big[\alpha_\parallel^A \alpha_\parallel^B \sin^2\phi \cos^2\theta + \alpha_\perp^A \alpha_\perp^B \sin^2\phi \sin^2\theta - 2\alpha_\perp^A \alpha_\perp^B \cos^2\phi\big] \\
&\quad \times \big[3\cos(kR) + 3kR\sin(kR) - k^2R^2\cos(kR)\big] \\
&\quad - \big[\alpha_\parallel^A \alpha_\parallel^B \sin^2\phi \cos^2\theta + \alpha_\perp^A \alpha_\perp^B \sin^2\phi \sin^2\theta\big] \\
&\quad \times \big[k^2R^2\cos(kR) + k^3R^3\sin(kR)\big]\Big\}\cos(\mathbf{k}\cdot\mathbf{R}) \\
&\quad + \Big\{\big[\alpha_\parallel^A \alpha_\parallel^B \sin^2\phi \cos^2\theta + \alpha_\perp^A \alpha_\perp^B \sin^2\phi \sin^2\theta - 2\alpha_\perp^A \alpha_\perp^B \cos^2\phi\big] \\
&\quad \times \big[k_z R\cos(kR) + k_z kR^2\sin(kR)\big] \\
&\quad - \big[\alpha_\parallel^A \alpha_\parallel^B \sin^2\phi \cos^2\theta + \alpha_\perp^A \alpha_\perp^B \sin^2\phi \sin^2\theta\big] \\
&\quad \times \big[k_z k^2 R^3\cos(kR)\big]\Big\}\sin(\mathbf{k}\cdot\mathbf{R})\Big\}. \qquad (30)
\end{aligned}$$

In the short-range approximation ($kR \ll 1$), the corresponding expressions are

$$\Delta E_{\text{ind}}^{0} = \frac{I}{2\pi\varepsilon_0^2 c R^3}\{[\alpha_{\parallel}^{A}\alpha_{\parallel}^{B}\cos^2\theta + \alpha_{\perp}^{A}\alpha_{\perp}^{B}\sin^2\theta]\sin^2\phi - 2\alpha_{\perp}^{A}\alpha_{\perp}^{B}\cos^2\phi\} \tag{31a}$$

$$F_{z,\text{ind}}^{0} = \left(\frac{3I}{2\pi\varepsilon_0^2 c R^4}\right)[\alpha_{\perp}^{A}\alpha_{\perp}^{B}(\sin^2\phi(2+\sin^2\theta) - 2) + \alpha_{\parallel}^{A}\alpha_{\parallel}^{B}\sin^2\phi\cos^2\theta]. \tag{31b}$$

Again, if the parallel pair freely tumbles with respect to the electromagnetic field, it is necessary to consider the isotropic average case, and we can see from equation (27) that

$$\langle\Delta E_{\text{ind}}^{0}\rangle = \frac{I}{6\pi\varepsilon_0^2 c R^3}[\alpha_{\parallel}^{A}\alpha_{\parallel}^{B} - \alpha_{\perp}^{A}\alpha_{\perp}^{B}], \tag{32}$$

$$\langle F_{z,\text{ind}}^{0}\rangle = \left(\frac{I}{2\pi\varepsilon_0^2 c R^4}\right) \times [\alpha_{\parallel}^{A}\alpha_{\parallel}^{B} - \alpha_{\perp}^{A}\alpha_{\perp}^{B}]. \tag{33}$$

In a case in which the particles have spherical symmetry, it is readily verified that the above results reduce to the same limiting expressions as those given in previous sections.

The above results for cylindrical particles in various configurations have been applied to single-walled carbon nanotubes. These particles are of interest not only for their intrinsic properties and applications, but since they are strongly polarizable species they also afford ideal opportunities to exploit the quadratic dependence on polarizability featured in the force equations. Assuming that the $\alpha_{\perp}$ and $\alpha_{\parallel}$ values are consistent with the corresponding static polarizabilities, then for nanotubes 200 nm in length and 0.4 nm in radius, separated by a distance $R = 2$ nm, and with an incident intensity $I = 1 \times 10^{16}$ W m^{-2}, the results deliver forces ranging between 10^{-12} and 10^{-5} N, according to the geometry [21]. Significantly, this full range of values is amenable to determination by atomic force microscopy. However, there may be a more important consequence: the wide variation in values, and the scale of the highest values suggest that there is a realistic possibility for the nanomanipulation of carbon nanotubes, based on laser control of optomechanical forces.

4.2.6 Spherical Particles

For spherical particles, the energy shift may be obtained by setting $\alpha_0^{\xi} = \alpha_{\parallel}^{\xi} = \alpha_{\perp}^{\xi}$ in equations (21) or (29). This shift may be expressed as a function of the geometric

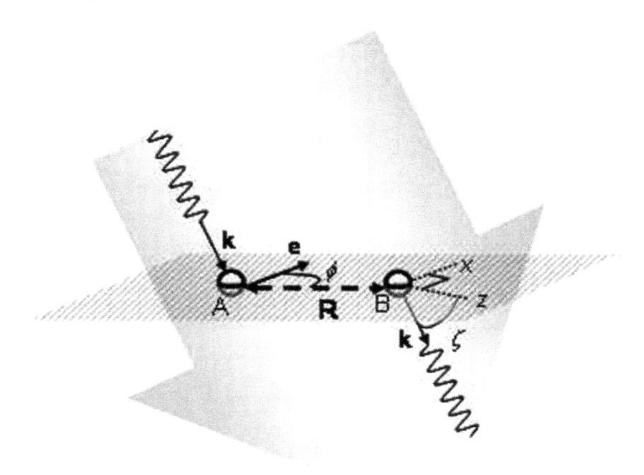

Figure 4.7 Particles A and B, displaced by R, trapped in a polarized laser beam. The polarization vector, **e**, defines the x-axis, forming an angle ϕ with **R**. Together, these vectors define the x, z-plane, the beam propagation vector **k** subtending an angle ζ onto z.

parameters show in Figure 4.7.

$$\Delta E(\mathbf{k},\mathbf{R}) = \left(\frac{2I}{\varepsilon_0 c}\right) \mathrm{Re}\left\{\alpha_0^A V_{xx}(k,\mathbf{R})\alpha_0^B\right\} \cos(kR\sin\phi\cos\zeta) \tag{34}$$

we can obtain contour plots of the energy surface determined in equation (34) to provide detailed information about the location of the system's stability points, as shown in Figure 4.8. A host of interesting features emerge, even from the examples exhibited here [38,39]. In each energy landscape, local minima distinguish optical binding configurations. The contours intersect the abscissa scale orthogonally, reflecting an even dependence upon each angular variable; the variation in the domain $(\pi/2, \pi/2)$ is notionally revealed by unfolding along the distance axis. The physical significance is that a system whose (kR, ζ, ϕ) configuration has $\phi = 0$ or $\zeta = 0$ (but is not situated at a local minimum) is always subject to a force drawing it toward a neighboring minimum without change of orientation. For the same reason, there is no torque when $\phi = \pi/2$ or $\zeta = \pi/2$. However, a system in an arbitrary configuration will generally be subject to forces leading to both forces and torques. For example, inspection of Figure 4.8(a) shows that while a pair in the configuration $(6.0, \pi/4, 0)$ is subject to a torque tending to increase ϕ to $\pi/2$, its trajectory will be accompanied by forces that tend to first increase and then decrease R. The details, which will additionally involve changes in ζ, can, of course, be determined from the total derivative of equation (34). Other features, also exemplified in Figure 4.8(a), are off-axis islands of stability such as the one that can be identified at $(10, \pi/10, 0)$. In general, the optically induced pair

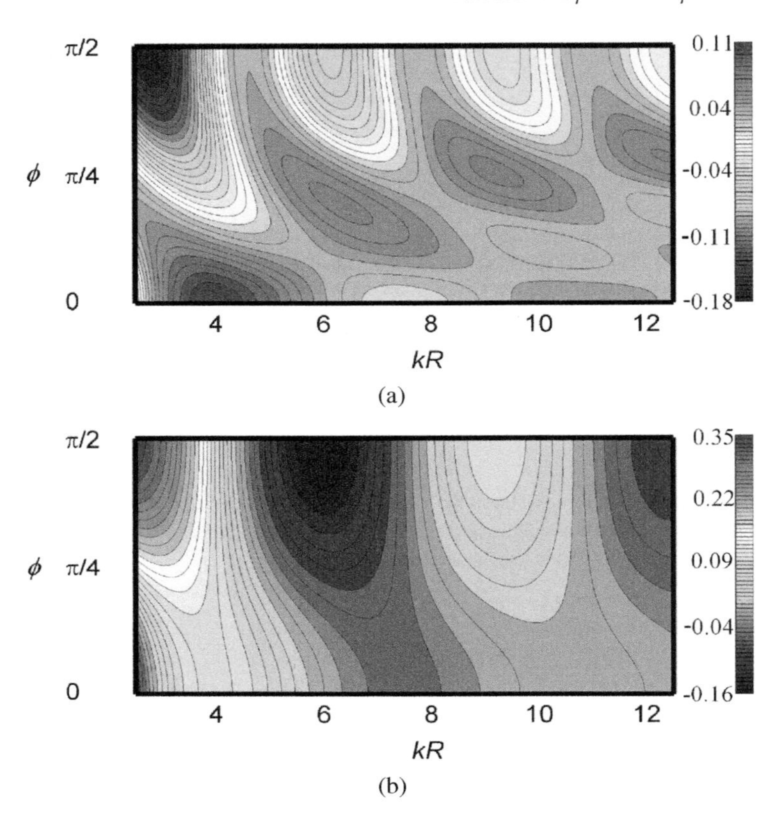

Figure 4.8 Contour maps of optically induced pair energy. Plots of ΔE as a function of ϕ and kR: (a) $\zeta = 0$; (b) $\zeta = \pi/2$. The variation of ΔE with kR along the abscissa, $\phi = 0$, shows its first two maxima at $kR \sim 4.0$, 10.5, and the first (nonproximal) minimum, at $kR \sim 7.5$ (compare to Figure 4.1). The horizontal scale typically spans distances R of several hundred nanometers, depending upon the value of k (see text). The units of the color scale are $\alpha_0^{(A)}\alpha_0^{(B)}2Ik^3/(4\pi\varepsilon_0^2 c)$. Adapted from references [38,39]. See color insert.

potential provides a prototypical template for the optical assembly of larger numbers of particles, facilitating the optical fabrication of structures of molecules, nanoparticles, microparticles, and colloidal particles.

4.2.7 Spherical Particles in a Laguerre–Gaussian Beam

The nature and form of optically induced forces between particles in an optical vortex are of special interest. Here we entertain the possibilities afforded by having

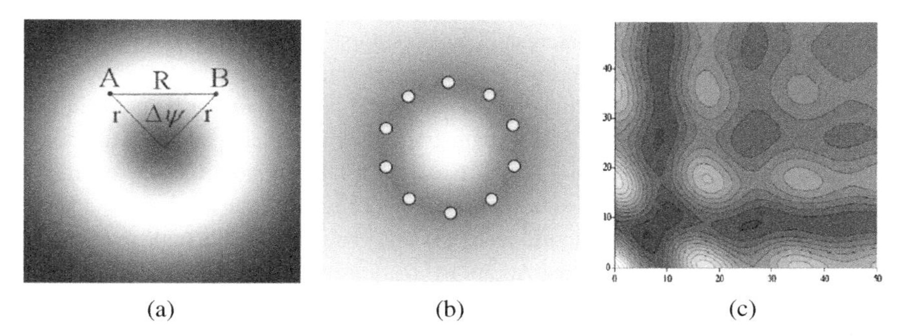

Figure 4.9 (a) Geometry of a particle pair in a Laguerre–Gaussian beam ($p = 0$); (b) Clustering of nanoparticles in a Laguerre–Gaussian beam; (c) Contour graphs of ΔE^{0}_{ABC} against $\Delta\psi_1$ (x-axis) and $\Delta\psi_2$ (y-axis) for three particles in a Laguerre–Gaussian beam with $l = 20$; lighter shading denotes higher values of ΔE^{0}_{ABC}. Adapted from references [22,40,41].

two or more particles (assumed to be spherical for simplicity) trapped in a Laguerre–Gaussian beam, or two such beams, counterpropagating to offset Maxwell–Bartoli forces. First, consider particles A and B trapped in the annular high-intensity region of a Laguerre–Gaussian beam with arbitrary l and $p = 0$, i.e., an optical vortex with one radial node at the beam center. For significant forces to arise, the interparticle distance R will usually be small compared to the radius of the optical trap, and it is helpful to recast the energy and force equations in terms of the angular displacement $\Delta\psi$ between A and B; see Figure 4.9(a). The general result (for arbitrary p) is

$$\Delta E_{\text{ind}} = \left(\frac{I f_{lp}^2 \alpha_0^2}{4\pi\varepsilon_0^2 c A_{lp} R^3}\right)\left\{\cos^2\phi\left(\cos kR + kR\sin kR - k^2R^2\cos kR\right)\right.$$
$$\left. - 2\sin^2\phi(\cos kR + kR\sin kR)\right\}\cos(l\Delta\psi), \tag{35}$$

where f_{lp} and A_{lp} are standard Laguerre–Gaussian beam functions as defined in Chapter 1; usually α_0 is the polarizability of a spherical nanoparticle (the same for A and B); $\hbar ck$ denotes the input photon energy, ϕ is again the angle between the polarization of input radiation and $\mathbf{R}$, and $\Delta\psi = \psi_B - \psi_A$ is the azimuthal displacement angle. In the short-range region ($kR = 1$), the leading term of equation (1) is determined from Taylor series expansions of $\sin(kR)$ and $\cos(kR)$. By the use simple trigonometry, the result ΔE_{ind} can be expressed as [22,40,41];

$$\Delta E^{0}_{\text{ind}} = \left[\frac{I f_{lp}^2 \alpha_0^2 (1 - 3\sin^2\phi)}{8\sqrt{2}\pi\varepsilon_0^2 r^3 c A_{lp}}\right]\frac{\cos(l\Delta\psi)}{(\eta - \cos\Delta\psi)^{3/2}}. \tag{36}$$

Here, η is a damping factor whose introduction, in place of the unity that emerges from simple trigonometry, precludes a singularity at $\Delta\psi = 0$.

The result has a number of interesting features. First, at $l = 0$, that is, for a conventional Hermite–Gaussian laser beam, a single energy minimum occurs at $\Delta\psi = 180°$. This illustrates that the energetically most favorable position of the particles in the beam cross-section is where they are diametrically opposite each other, as might be expected. Second, for odd values of $l > 1$, only a local minimum (not the energetically most favorable) arises for this configuration. Third, for even values of l, a local *maximum* occurs at 180°. The fourth feature is that generally, for $l \neq 0$, there are l angular minima and $(l - 1)$ maxima. Additional features reflect the behavior associated with increasing values of l. Fifth, the number of positions for which the particle pair can be mutually trapped increases, becoming less energetically favorable as the angular disposition increases toward diametric opposition. Sixth, the absolute minima are found at decreasing values of $\Delta\psi$, physically signifying a progression toward particle clustering.

To identify the possibilities for stable formations of more than two particles, as shown in Figure 4.9(b), the two-particle analysis is readily extended to a system of three (or more) particles. In this case, ΔE^0_{ABC} is determined by summing the pairwise laser-induced interactions of the three particles with each other, employing variables $\Delta\psi_1$ and $\Delta\psi_2$ as the azimuthal displacements between particles A–B and B–C, respectively. A typical contour plot of ΔE^0_{ABC} against $\Delta\psi_1$ and $\Delta\psi_2$ is shown in Figure 4.9(c). Such results are indicative of a rich scope for further theoretical and experimental exploration.

4.3 OVERVIEW OF APPLICATIONS

The body of experimental work on optical binding and its related studies is growing apace. In this section, we shall summarize just a fragment of the novel work being done in this area by a number of different research groups. Most experimental research has been stimulated by an interest in applying optical binding to the organization and manipulation of matter at scales comparable to the wavelength of light. In such a context, bear in mind that optical binding forces will have a profound effect that will modify the outcome of all optical manipulation techniques involving more than one particle [42]. Obvious examples are processes involving the assembly of optical structures using holographic optical traps [43]; equally, the two-dimensional assembly of particles in the presence of counterpropagating beams results from the combined effects of optical trapping and binding [44]. Although the systems used and the setup designs vary significantly, these and other such studies have common goals—principally to understand the nature of the binding forces due to the presence of an electromagnetic field, and to develop tools for the noncontact control of matter on the micron and submicron nanoscale.

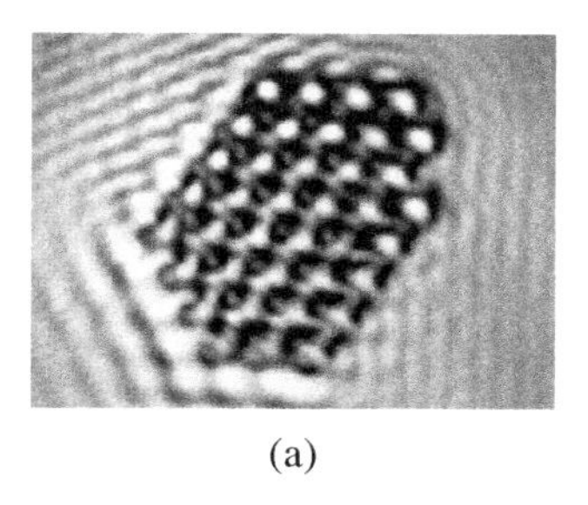

(a)

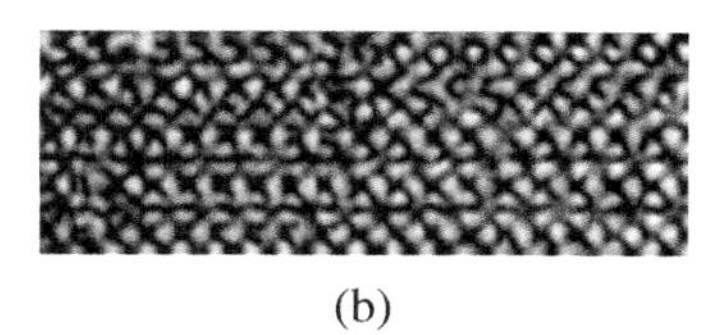

(b)

Figure 4.10 (a) Self-assembled 2D "optical crystal" formed in a 30 μm Gaussian trap generated by a single laser beam. The multiple coherent scattering of the polystyrene spheres (3 μm) generates sinusoidal fringes through its interference with the trap. Adapted from [45]; (b) 1D optical crystals formed in fringes created by interference of two plane waves, 3 μm polystyrene beads self-organize along each trap, optical binding forces promoting a regular, equidistant placement. Adapted from [46].

One of the most common systems used to demonstrate optical binding comprises micron-sized essentially spherical polyethylene beads in a liquid suspension. In two of the first reports by Burns and colleagues [2,4], the authors noted that the relative position of a pair of such spheres were influenced by each other when placed in an optical trap. When the spheres were well separated in the trap, their motions along the trap appeared random; but as they approached each other, they appeared to depart from diffusive behavior, tending to spend more time in relatively close proximity. The relative motion of the pair was recorded by studying the diffraction patterns thereby created in the scattered field. Significantly, these reports showed that there are discrete separations at which the particles are more likely to be found, and that the positions of the inferred neighboring energy minima differ by distances approximately equal to the wavelength of the light.

In later work [45,46], Fournier and colleagues demonstrated that optical binding forces are at least partly responsible for the self-arrangement of optically trapped particles separated by distances ranging up to a few wavelengths. Several cases have been reported, as shown in Figure 4.10, and it has been shown that the binding force can dominate the usual (optical tweezer) gradient trapping force in systems where one expects a large number of particles to arrange according to a trapping template. When a free particle approaches an already formed structure, being subject to a potential energy landscape already patterned by many-body optical interference, the added sphere becomes accommodated within the whole ensemble (which then reorganizes until it reaches a new minimum energy configuration). It is important to emphasize that it is extremely difficult to disentangle optical forces due to gradient and scattering forces in most of these experiments, and that the generation of "optical crystals" as shown in these examples is generally due to contributions from both types of interaction, as is specifically shown in Figure 4.11.

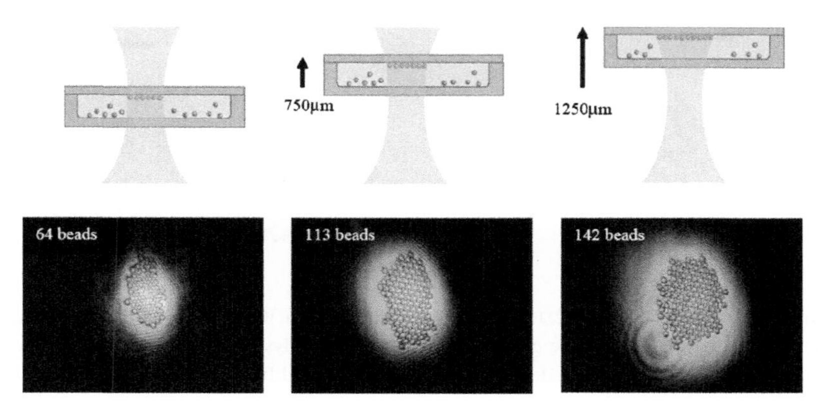

Figure 4.11 2D optical crystals resulting from the combination of binding and trapping in a Gaussian trap produced by 1 Watt laser power at 532 nm. Adapted from [46].

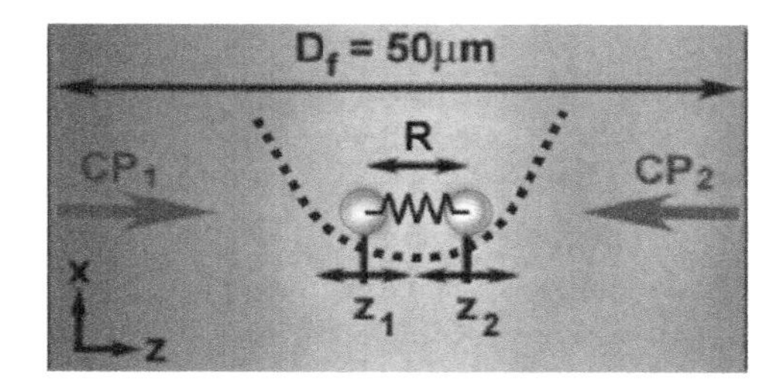

Figure 4.12 Counterpropagating light fields (CP_1 and CP_2: 1070 nm) are delivered by optical fibers with a separation D_f. A pair array forms in the gap between the two fibers, R is the equilibrium separation of the sphere centers, and z_1, z_2 indicate small displacements from equilibrium along the axis. The array center of symmetry coincides with half the fiber separation. The two normal modes of the bound pair are indicated: Dashed line represents the potential related to the center of mass motion of the two-sphere system; zigzag between the two spheres indicates the optically induced pair potential, determining relative motion within the system. Adapted from [51].

It is worth noting that the experimental setup used in the studies whose results are exemplified in this section is commonly referred to as *transverse optical binding* [47], which means that the wave-vector of the electromagnetic field is perpendicular to the plane containing the 2D optical crystal, or to the axis of a 1D optical chain. As shown in Figure 4.11, this setup takes advantage of a container cell to confine the optical crystal. Nonetheless, such a configuration admits numerous possibilities of scattering from the cell, making it difficult to analyze the optical binding contribution. A different approach is to achieve *longitudinal optical binding* [48–51], as shown in Figure 4.12, where two counterpropagating (noncoherent) beams impinge on the system.

Completing the picture, there have also been studies of optical binding between nano-metallic particles trapped in electromagnetic fields [52–55]. Indeed, the detection of light-induced aggregation in 10 nm gold clusters was first reported over ten years ago [52]. At the time, this was attributed to van der Waals-like forces between closely approaching clusters and cluster aggregates. A subsequent theoretical study [53] showed that interparticle interaction energy is a sensitive function of particle size. More recent theoretical work has shown that such optical binding forces are significantly stronger than traditional van der Waals forces, and that there are realistic possibilities to exploit optically induced forces for the noncontact organization of novel metallic structures [54] such as the metallic necklaces reported in [55].

4.4 DISCUSSION

Relating theory to experiment in this field is perhaps more than usually difficult, but it is a challenge that carries a promise of rich rewards in the form of new techniques for the nanomanipulation of matter. Part of the problem is that producing suitable conditions for the sought effects generally necessitates the use of specialized cells or optical traps, each of which can generate additional partly contributory optical effects that usually compete with optical binding. Another difficulty is that many existing descriptions of optical binding mechanism are a little vague, and it is not always clear whether two different descriptions amount to the same or to potentially competing phenomena. In the hope of bringing more clarity and precision to the field, the theoretical methods and results presented in this chapter are based upon a robust and thorough quantum electrodynamical analysis of optically induced interparticle interactions. In this framework, it is understood that laser-induced forces and torques between nanoparticles occur by pairwise processes of stimulated photon scattering. The analysis clarifies the fundamental involvement of quantum interactions with the throughput radiation, and also the form of electromagnetic coupling between particles. It further reveals that additional torque features arise in an optical vortex.

In applying the results to nanoparticles such as polystyrene beads (whose electronic properties are neither those of one large molecule nor those of a chromophore aggregate), the molecular properties that appear in the given equations need to be translated into bulk quantities; for example, the polarizability becomes the linear susceptibility. Moreover, one must account for the optical properties of the medium that supports the particles. In most of the experiments discussed in the previous section, we showed that the relative values of the refractive index between the beads and the surrounding medium significantly influence the optical binding phenomena, thereby modifying the bead positions of stability. In fact, proper registration by the theory of the responsible local field effects is also straightforward; it is already known how the retarded potential of equation (7) is affected [56]. Alongside the incorporation of Lorentz field factors, the dependence on kR changes to a dependence on $n(ck)kR$, where the multiplier is the complex refractive index. For example, in a liquid illuminated by 800 nm radiation, when the refractive index at that wavelength is 1.40, the potential energy minimum registered in Figure 4.1 at $kR \sim 7.5$ signifies a pair separation of 670 nm rather than 960 nm.

Recently, there has been fresh interest in the angular properties of the force fields resulting from optical binding. Multidimensional potential energy surfaces have been derived and shown to exhibit unexpected turning points, producing intricate patterns of local force and torque. Numerous local potential minimum and maximum can be identified, and islands of stability conducive to the formation of rings have been identified [38,39]. The major challenge to be addressed is to account for the effects of particle numbers, namely the additional and distinctive features that must arise when more than two isolated particles are involved. There are three distinct aspects to this. First and simplest, there is a need to identify the effects that will obviously arise as a consequence of the superposition of optically modified pair potentials. Second, there is a need to fully analyze the contributions from *multi*-particle processes of stimulated scattering, involving the entangled near-field interactions of more than two particles. And finally, since stimulated scattering releases throughput radiation essentially unchanged, there is a need to entertain the *multiple processes* of stimulated scattering in order to properly address the kind of 1D arrays and 2D optical crystal structures that experiments have so beautifully revealed. We are confident that these challenges for the future will soon bear the fruit of establishing better and clearer links between theory and experiment.

ACKNOWLEDGMENTS

We are pleased to acknowledge that all of the QED work at UEA has been made possible through funding from the EPSRC. We also gratefully acknowledge many useful discussions of the work, especially with David Bradshaw and Justo Rodríguez at UEA, and Kishan Dholakia at the University of St. Andrews.

REFERENCES

[1] T. Thirunamachandran, Intermolecular interactions in the presence of an intense radiation-field, *Mol. Phys. 40* (1980) 393.

[2] M.M. Burns, J.-M. Fournier, J.A. Golovchenko, Optical binding, *Phys. Rev. Lett. 63* (1989) 1233.

[3] G. Whitesides, What will chemistry do in the next twenty years?, *Angew. Chem. Int. Ed. Engl. 29* (1990) 1209.

[4] M.M. Burns, J.-M. Fournier, J.A. Golovchenko, Optical matter—crystallization and binding in intense optical-fields, *Science 249* (1990) 749.

[5] P.W. Milonni, M.L. Shih, Source theory of the Casimir force, *Phys. Rev. A 45* (1992) 4241.

[6] F. Depasse, J.-M. Vigoureux, Optical binding between two Rayleigh particles, *J. Phys. D: Appl. Phys. 27* (1994) 914.

[7] P.W. Milonni, A. Smith, Van der Waals dispersion forces in electromagnetic fields, *Phys. Rev. A 53* (1996) 3484.

[8] P.C. Chaumet, M. Nieto-Vesperinas, Optical binding of particles with or without the presence of a flat dielectric surface, *Phys. Rev. B 64* (2001) 035422.

[9] M. Nieto-Vesperinas, P.C. Chaumet, A. Rahmani, Near-field photonic forces, *Philos. Trans. R. Soc. London A 362* (2004) 719.

[10] S.K. Mohanty, J.T. Andrews, P.K. Gupta, Optical binding between dielectric particles, *Opt. Exp. 12* (2004) 2746.

[11] D. McGloin, A.E. Carruthers, K. Dholakia, E.M. Wright, Optically bound microscopic particles in one dimension, *Phys. Rev. E 69* (2004) 021403.

[12] D.H.J. O'Dell, S. Giovanazzi, G. Kurizki, V.M. Akulin, Bose–Einstein condensates with $1/r$ interatomic attraction: Electromagnetically induced "gravity", *Phys. Rev. Lett. 84* (2000) 5687.

[13] F. Dimer de Oliveira, M.K. Olsen, Mean field dynamics of Bose–Einstein superchemistry, *Opt. Commun. 234* (2004) 235.

[14] N.K. Metzger, E.M. Wright, K. Dholakia, Theory and simulation of the bistable behaviour of optically bound particles in the Mie size regime, *New J. Phys. 8* (2006) 139.

[15] V. Karásek, K. Dholakia, P. Zemánek, Analysis of optical binding in one dimension, *Appl. Phys. B 84* (2006) 149.

[16] J. Ng, C.T. Chan, Localized vibrational modes in optically bound structures, *Opt. Lett. 31* (2006) 2583.

[17] T.M. Grzegorczyk, B.A. Kemp, J.A. Kong, Trapping and binding of an arbitrary number of cylindrical particles in an in-plane electromagnetic field, *J. Opt. Soc. Am. A 23* (2006) 2324.

[18] T.M. Grzegorczyk, B.A. Kemp, J.A. Kong, Stable optical trapping based on optical binding forces, *Phys. Rev. Lett. 96* (2006) 113903.

[19] D.S. Bradshaw, D.L. Andrews, Optically induced forces and torques: Interactions between nanoparticles in a laser beam, *Phys. Rev. A 72* (2005) 033816, *Phys. Rev. A 73* (2006) 039903, Corrigendum.

[20] A. Salam, On the effect of a radiation field in modifying the intermolecular interaction between two chiral molecules, *J. Chem. Phys. 124* (2006) 014302.

[21] D.L. Andrews, D.S. Bradshaw, Laser-induced forces between carbon nanotubes, *Opt. Lett. 30* (2005) 783.

[22] D.S. Bradshaw, D.L. Andrews, Interactions between spherical nanoparticles optically trapped in Laguerre–Gaussian modes, *Opt. Lett. 30* (2005) 3039.

[23] D.L. Andrews, R.G. Crisp, D.S. Bradshaw, Optically induced inter-particle forces: from the bonding of dimers to optical electrostriction in molecular solids, *J. Phys. B: At. Mol. Opt. Phys. 39* (2006) S637.

[24] H.B.G. Casimir, D. Polder, The influence of retardation on the London–van der Waals forces, *Phys. Rev. 73* (1948) 360.

[25] P.W. Milonni, The Quantum Vacuum: An Introduction to Quantum Electrodynamics, Academic Press, San Diego, CA, 1994, p. 54.
[26] D.L. Andrews, L.C. Dávila Romero, Conceptualization of the Casimir effect, *Eur. J. Phys. 22* (2001) 447.
[27] E.A. Power, Casimir–Polder potential from first principles, *Eur. J. Phys. 22* (2001) 453.
[28] G.J. Maclay, H. Fearn, P.W. Milonni, Of some theoretical significance: Implications of Casimir effects, *Eur. J. Phys. 22* (2001) 463.
[29] B.W. Alligood, A. Salam, On the application of state sequence diagrams to the calculation of the Casimir–Polder potential, *Mol. Phys. 105* (2007) 395.
[30] F. Capasso, J.N. Munday, D. Iannuzzi, H.B. Chan, Casimir forces and torques: physics and applications to nanomechanics, *IEEE J. Select. Topics Quantum Electron. 13* (2007) 400.
[31] A. Altland, B. Simons, Condensed Matter Field Theory, University Press College, Cambridge, 2006, p. 29.
[32] E.A. Power, S. Zienau, Coulomb gauge in nonrelativistic quantum electrodynamics and the shape of spectral lines, *Philos. Trans. R. Soc. A 251* (1959) 427.
[33] R.G. Woolley, Molecular quantum electrodynamics, *Proc. R. Soc. A 321* (1971) 557.
[34] E.A. Power, T. Thirunamachandran, Nature of Hamiltonian for interaction of radiation with atoms and molecules $(e/mc)\mathbf{p}.\mathbf{A}$, $-(\boldsymbol{\mu}\mathbf{E})$ and all that, *Am. J. Phys. 46* (1978) 370.
[35] R.G. Woolley, Charged particles, gauge invariance, and molecular electrodynamics, *Int. J. Quantum Chem. 74* (1999) 531.
[36] R.D. Jenkins, G.J. Daniels, D.L. Andrews, Quantum pathways for resonance energy transfer, *J. Chem. Phys. 120* (2004) 11442.
[37] D.L. Andrews, M.J. Harlow, Phased and Boltzmann-weighted rotational averages, *Phys. Rev. A 29* (1984) 2796.
[38] J. Rodríguez, L.C. Dávila Romero, D.L. Andrews, Optically induced potential energy landscapes, *J. Nanophotonics 1* (2007) 019503.
[39] L.C. Dávila Romero, J. Rodríguez, D.L. Andrews, Electrodynamic mechanism and array stability in optical binding, *Opt. Commun. 281* (2008) 865.
[40] D.S. Bradshaw, D.L. Andrews, Optical forces between dielectric nanoparticles in an optical vortex, in: D.L. Andrews (Ed.), *Nanomanipulation with Light*, in: *SPIE Proceedings*, vol. 5736, 2005, p. 87.
[41] D.S. Bradshaw, D.L. Andrews, Optical ordering of nanoparticles trapped by Laguerre–Gaussian laser modes, in: D.L. Andrews (Ed.), *Nanomanipulation with Light II*, in: *SPIE Proceedings*, vol. 6131, 2006, p. 61310G.
[42] D.G. Grier, A revolution in optical manipulation, *Nature 424* (2003) 810.
[43] D.G. Grier, S.-H. Lee, Y. Roichman, Y. Roichman, Assembling mesoscopic systems with holographic optical traps, in: D.L. Andrews (Ed.), *Complex Light and Optical Forces*, in: *SPIE Proceedings*, vol. 6483, 2007, p. 64830D.
[44] C.D. Mellor, T.A. Fennerty, C.D. Bain, Polarization effects in optically bound particle arrays, *Opt. Exp. 14* (2006) 10079.
[45] J.-M. Fournier, G. Boer, G. Delacrétaz, P. Jacquot, J. Rohmer, R.P. Salathé, Building optical matter with binding and trapping forces, in: K. Dholakia, G.C. Spalding (Eds.), *Optical Trapping and Optical Micromanipulation*, in: *SPIE Proceedings*, vol. 5514, 2004, p. 309.
[46] J.-M. Fournier, J. Rohmer, G. Boer, P. Jacquot, R. Johann, S. Mias, R.P. Salathé, Assembling mesoscopic particles by various optical schemes, in: K. Dholakia, G.C. Spalding (Eds.), *Optical Trapping and Optical Micromanipulation II*, in: *SPIE Proceedings*, vol. 5930, 2005, p. 59300Y.

[47] M. Guillon, Field enhancement in a chain of optically bound dipoles, *Opt. Exp. 14* (2006) 3045.
[48] M. Guillon, Optical trapping in rarefied media: towards laser-trapped space telescopes, in: K. Dholakia, G.C. Spalding (Eds.), *Optical Trapping and Optical Micromanipulation II*, in: *SPIE Proceedings*, vol. 5930, 2005, p. 59301T.
[49] W. Singer, M. Frick, S. Bernet, M. Ritsch-Marte, Self-organized array of regularly spaced microbeads in a fiber-optical trap, *J. Opt. Soc. Am. B 20* (2003) 1568.
[50] N.K. Metzger, E.M. Wright, W. Sibbett, K. Dholakia, Visualization of optical binding of microparticles using a femtosecond fiber optical trap, *Opt. Exp. 14* (2006) 3677.
[51] N.K. Metzger, R.F. Marchington, M. Mazilu, R.L. Smith, K. Dholakia, E.M. Wright, Measurement of the restoring forces acting on two optically bound particles from normal mode correlations, *Phys. Rev. Lett. 98* (2007) 068102.
[52] H. Eckstein, U. Kreibig, Light-induced aggregation of metal-clusters, *Z. Phys. D 26* (1993) 239.
[53] K. Kimura, Photoenhanced van der Waals attractive force of small metallic particles, *J. Phys. Chem. 98* (1994) 11997.
[54] A.J. Hallock, P.L. Redmond, L.E. Brus, Optical forces between metallic particles, *Proc. Natl. Acad. Sci. 102* (2005) 1280.
[55] G. Ramakrishna, Q. Dai, J. Zou, Q. Huo, Th. Goodson III, Interparticle electromagnetic coupling in assembled gold-necklace nanoparticles, *J. Am. Chem. Soc. 129* (2007) 1848.
[56] G. Juzeliūnas, D.L. Andrews, Quantum electrodynamics of resonance energy transfer, *Adv. Chem. Phys. 112* (2000) 357.

Chapter 5

Near-Field Optical Micromanipulation

Kishan Dholakia and Peter J. Reece

University of St. Andrews, UK

5.1 INTRODUCTION

The domain of near-field optics is undoubtedly a buoyant and prolific topic in current photonics. This area, which is typically associated with the notion of evanescent waves and novel hybrid structures, has come to the fore due the prospects of overcoming the "diffraction limit of light"—the bottleneck that is created in all forms of optics applications because we cannot focus a light beam down to a spot size of arbitrarily small dimensions, but rather are restricted to approximately half the wavelength of the light in the host medium. The motivation of studying near-field optics is in the realization of a new generation of optical systems where we are not constrained by the diffraction limit criterion. Optical microscopy and imaging are two areas that stand to benefit from this advance, and it is well documented that the near field has delivered new imaging modalities such as the scanning near-field optical microscope. However the near field's impact is broader than imaging and delivers in a myriad of other optics areas.

In this chapter, we will focus on the emergent theme of near-field optical micromanipulation and in particular emphasize the major experimental advances made in this regime. Micromanipulation via evanescent fields offers several potential advantages as have been elucidated earlier. The nature of evanescent waves means that the manipulated particles are organized near an interface that is appropriate for studies where proximity to a physical substrate is critical. This includes areas such as surface-enhanced Raman spectroscopy, studies of cell adhesion and cell signaling during

STRUCTURED LIGHT AND ITS APPLICATIONS

growth and differentiation, as well as interactions of colloidal aggregates. The near field is not restricted in terms of focusing, thus this opens opportunities for patterning the light field such that we may create optical potential energy landscapes with sub-wavelength periodicity and over large areas. The rapid decay of an evanescent wave assists in creating small trapping regions and high localization of trapped objects close to an interface. The very geometries used for such guiding and trapping permit ease of viewing and also decoupling of observation from the incident-trapping light fields. The aim of the chapter is to give the reader a flavor of this upcoming topic of micro-manipulation and to highlight areas where the field may progress in years to come.

5.1.1 What Is the Near Field?

In the general sense, near-field optics is often associated with interactions and phenomena involving evanescent light fields. The theory governing near-field optics is nontrivial, and ideally a full consideration would be given to the Maxwell equations to elucidate the exact behavior in this regime. Evanescent waves, which are electromagnetic modes with an exponentially decaying component, are prevalent in many areas of photonics from optical wave-guiding to near-field optical microscopy. They become significant when one is dealing with the scattering and diffraction of electromagnetic radiation by microscopic and nanoscopic objects and when the length scales are comparable to the wavelength of the light, λ. Two different examples may help to illustrate the properties of evanescent waves. We may first appreciate what is meant by an evanescent wave by considering the diffraction of an incident ideal plane electromagnetic wave propagating in direction z with a wave vector k incident upon an aperture of width a. Beyond the aperture, the field may be represented as a superposition of plane waves, each satisfying the dispersion relation

$$k_z = \sqrt{\left(\frac{2\pi}{\lambda}\right)^2 - (k_x^2 + k_y^2)},$$

where k_x and k_y represent the in-plane component of the wave vector. From the Fourier transform relation between wave vector and spatial coordinates, we find that $\Delta x \Delta k_x \simeq 2\pi$ [1]. If the aperture is less than the wavelength ($a < \lambda$) of the incident light, some of the plane wave components will have in-plane wave-vectors such that k_z is purely imaginary, and the wave is described in this region with a $e^{-|k_z|z}$ [2]. This shows an exponential decay of the wave very rapidly in the z-direction, and thus these waves do not propagate into the far field: in turn, this means that they exist solely in the near field. One may also consider the role of evanescent waves due to the Fresnel equations and the condition of total internal reflection [3]. Essentially, as one tunes the angle for light propagating from a dense (high refractive index) media to a rarefied (low-index) media, the light becomes trapped within the high index material and undergoes what is termed *total internal reflection*. However, for this to occur we need to have an evanescent wave present in the rare media with a k vector that is parallel to the interface between the two materials. As previously noted, this wave is present only close to the surface and decays away very rapidly from the interface.

5.1.2 Optical Geometries for the Near Field and Initial Guiding Studies

Near-field laser trapping refers to the manipulation of a microparticle by optical forces that originate from an evanescent wave. The strength of an evanescent wave decays rapidly with the distance perpendicular to the boundary where the evanescent field is generated. This concept was first utilized by Kawata and Sugiura for optical guiding near a surface [4]. The radiation pressure force for guiding was generated by an evanescent wave produced at the interface between two media under the total internal reflection condition by illumination with a laser at 1064 nm as shown in Figure 5.1. Microparticles were successfully guided along the surface [4]. This was followed by further demonstrations of guiding along buried waveguide structures [5].

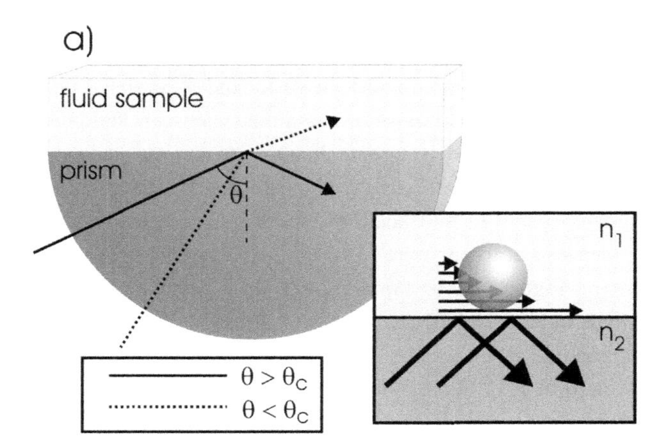

Figure 5.1 Excitation of evanescent waves at the interface between a prism of refractive index (n_2), and an adjacent lower refractive index (n_1) medium using Kretschmann geometry. Light incident on the interface at angles greater than the critical angle is reflected by total internal reflection, producing an exponentially decaying evanescent field in the adjacent medium (*inset*). Particles close to the surface will experience an optical force due to the presence of the evanescent field, directed along the direction of propagation and toward the surface of the prism.

In the first demonstration of near-field optical micromanipulation, Kawata and colleagues showed that micrometer sized colloids in the proximity of a water/glass interface supporting evanescent waves experience a net optical guiding force in the direction of the evanescent field wave vector and also away from the surface [4]. The choice of angles of incidence close to the critical angle enabled the use evanescent fields with large penetration depths, of the order of the size of the particle, which enhanced the degree of guiding. Oetama and Walz studied such guiding in more depth in 2002 [6] and used a ray optics model to try and match theory with experiment. They predicted translation of particles close to the surface, the pulling of particles toward the surface, and even rotation about an axis parallel to the surface. The 5-μm diameter spheres were guided (translated) across the surface at speeds of the order of 1 μm/s using laser light at 514 nm from an argon ion laser and an optical beam diameter of 80 μm at the prism surface. Despite some experimentally observed nonlinear behavior in the power dependence, which was associated with extraneous factors, the ray optics model was seen to deliver good qualitative comparisons to the experimental data.

Following these studies, several proposals appeared to try and advance the near field to a trapping geometry with an emphasis on nanoparticles as opposed to microparticles. A metallic tip illuminated by a laser beam was proposed to produce a localized evanescent field by the surface plasmon effect so that a particle of a few nanometers in size suspended in water [7] or air [8] could be trapped. Numerical calculations showed that a strong field enhancement from light scattering created a sufficiently strong trapping potential that overcomes the Brownian motion of the particle and achieve confinement. Okamoto and Kawata looked at a related technique where a subwavelength sphere was held in water using the evanescent field near a subwavelength aperture [9]. Naturally, these are challenging from an experimental viewpoint due to Brownian motion for such small particles, issues with loading the trap, and deleterious radiation pressure and heating effects. Chaumet and colleagues further developed the idea of a trapping system with an apertureless near-field microscope probe, as shown in Figure 5.2, to capture and hold a particle in air or vacuum [8]. Though Brownian motion is reduced, there is no sample medium for any form of particle damping in this system.

In such geometries, it is difficult to control the distance between the probe and samples, which is in the range of tens of nanometers due to the evanescent nature of illumination. And although a metallic tip leads to an enhanced evanescent field, the heating generated by surface plasmon may result in thermal forces that reduce the stability of the optical trap. We must also consider the light throughput of near-field tips or apertures, which is quite low. As we shall see later in the chapter, the main techniques developed to perform optical trapping and micromanipulation at the interface have been the use of the Kretchsmann geometry and the use of microscope objectives

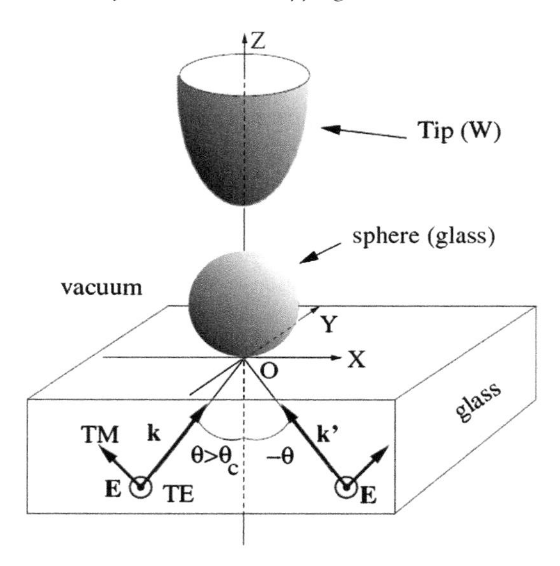

Figure 5.2 Apertureless scanning near-field optical microscope is an alternative approach for near-field optical trapping, which may provide additional compositional information on the trapped object. The schematic diagram shows the trapping arrangement. The substrate is illuminated from below under condition of total internal reflection generating evanescent waves in the adjacent sample medium. A metallic probe brought close to the surface will generate a scattering field that results in the creation of an optical trap near the apex of the tip. Reprinted with permission from [8]. Copyright 2002 by the American Physical Society.

specifically designed for total internal reflection. Additionally, several successful experiments have employed a waveguide geometry for evanescent wave trapping. These will be described later, but first we will give an overview of how we theoretically consider optical trapping forces in the near field.

5.2 THEORETICAL CONSIDERATIONS FOR NEAR-FIELD TRAPPING

The ability to theoretically model optical forces on a trapped particle continues to be a contemporary topic of immense interest. While a number of models have been developed to explain the principles of optical forces acting on a trapped particle in the Rayleigh and geometric optics regimes [10,11], the majority of optical trapping experiments are realized in the more theoretically challenging Lorenz–Mie–Debye region, where the size of the trapped particle is of the order of the wavelength of the trapping beam. Solutions to these problems typically require the full rigors of the

Maxwell equations and considerable computational effort, as an analytical solution is seldom available. Issues that complicate the problem include the description of the tightly focused light if one is using a high numerical aperture objective, as in gradient force optical tweezers, and the optical properties of the material that should include any absorption or resonances that may be present.

The determination of optical forces is generally split into two parts: (1) calculation of the optical field in the presence of the trapped particle and (2) calculation of the force due to this field distribution. Calculation of the electromagnetic fields in a specified geometry is a generic problem of photonics and, as such, numerous techniques are available, each with its own merits and drawbacks. Some of the more popular methods have been developed from calculations of the scattering field of small particles in arbitrary fields using Mie theory, generalized Lorenz–Mie theory, or vector diffraction theory. They include the multiple–multipole method [7], coupled–dipole method [12], integral method [13], and finite element methods [9]. Calculation of the optical forces typically follow either the Maxwell stress tensor [14] or distributed Lorentz force approach [15], which have been shown to be theoretically equivalent [16].

Theoretical modeling of the light field in the near field creates additional difficulties due to the complex nature of the evanescent fields and the close proximity of the trapped particle to the supporting dielectric surface. Following initial experiments by Kawata and Sugiura [4], Almaas and Brevik used Arbitrary Beam Theory (ABT) to attempt to explain the main experimental results, i.e., the repulsive force normal to the surface and the polarization dependent guiding velocities [17]. They found that there were a number of contradictions between theory and experiment, namely the normal force was predicted to be toward the surface, and p-polarization was estimated to give larger guiding velocities. These results were supported by Walz, who used geometric optics arguments to predict the emergent directions of rays passing through the sphere from the surface and the exchange of momentum and resulting forces and torques upon the particle [18]. These models agreed qualitatively within the range of parameters studied, but Walz also showed that for larger particles there was a crossover where the s-polarization produced a stronger guiding force; they also predicted the presence of torques for larger particles. For both theories, it should be noted that multiple reflections between the particle and interface are not considered.

Later work by Almaas and colleagues extended their theory to look at radiation force on an absorbing micrometer-sized dielectric sphere placed in an evanescent field [19]. They generalized their previous studies to include absorption, i.e., of a complex refractive index, and found general expressions for the vertical force for this case. Whereas the horizontal force (parallel to the interface and key for guiding and trapping) can be well accounted for within this constraint, there is no possibility of describing the repulsiveness of the vertical force that was seen in experiments by

Kawata. They concluded that the presence of surfactants, which make the surfaces of the spheres partially conducting, was responsible for this experimentally observed effect. Interestingly, Lester and Nieto-Vesperinas showed that the inclusion of the underlying substrate can have a significant influence on the resulting forces [20], and a recent analysis of Mie particles guided along a waveguide by Jaising and Helleso showed that morphology-dependent resonances may play a significant role in determining the magnitude and direction of the forces [21]. Both effects were shown to potentially cause repulsion from the surface. Also, Arias-Gonzalez and Nieto-Vesperinas showed that for metallic nanoparticles, the force normal to the surface could be either attractive or repulsive, depending on the particle size relative to the wavelength and the polarization of the incident beam [22]. They also investigated the role of plasmon resonances in these systems. This was verified experimentally by Gaugiran and colleagues for guiding of gold nanoparticles on a silicon nitride waveguide [23].

A number of different near-field geometries that aim to exploit localized fields have been investigated theoretically, and many of these studies focus on trapping nanoparticles. Novotny and colleagues used the multiple–multipole method to calculate the field and forces acting on nanometer sized particles close to an illuminated metallic tip [7]. Okamata and Kawata used finite-difference-time-domain calculations to estimate the forces on a particle next to a nanoaperture [9]. Chaumet and colleagues computed optical forces for their proposed trapping system in Figure 5.2 using the coupled dipole method [8]. In this geometry, a metallic tip is brought into close proximity to a surface supporting two counterpropagating evanescent waves. The interaction of the tip with the evanescent field generates optical forces normal to the surface and close to the tip apex, which are attractive or repulsive depending upon the polarization of the evanescent waves; the forces parallel to the surface are then balanced by the counterpropagating waves. The coupled dipole method (CDM) has the advantage of incorporating retardation and multiple scattering between objects and substrate present. Ganec and colleagues used vector diffraction theory to calculate optical forces for a near-field trapping geometry using a total internal reflection microscope objective [24]. Quidant and colleagues calculated the optical forces acting on a particle above a surface patterned with resonant gold nanostructures using the Green dyadic method [25].

5.3 EXPERIMENTAL GUIDING AND TRAPPING OF PARTICLES IN THE NEAR FIELD

In this section, we will review the major experiments that have been performed in the near field and discuss some of the results obtained. As we will see, not all of the observations made on particle behavior are fully understood or correlate with available theory.

5.3.1 Near-Field Surface Guiding and Trapping

In a conventional gradient force optical trap or optical tweezers, a particle with higher refractive index than its surroundings is attracted to and held at the center of a single highly focused light beam [26]. Multiple optical traps may be generated through the use of technology such as acousto-optic deflectors or spatial light modulators [27]. Typically, a minimum of 1 mW of power is needed for a single trap site, which places excessive power requirements on the generation of very large trap arrays using these techniques. Evanescent wave trapping offers a viable alternative for large-scale organization of microparticles, particularly as this shows the ability to simultaneously generate and manipulate large two-dimensional arrays of particles in close proximity to the interface. An evanescent wave trap is formed using a counterpropagating geometry similar to the dual beam fiber trap [28]. Radiation pressure, acting along the direction of the k-vector of the EW wave, will drive particles in the vicinity of the surface to the center of the illuminated region where radiation pressure from the opposite beam is balanced. Gradient forces attract the particle to the surface and also in the transverse direction due to the focus of the beam. The arrangement for such studies is the Kretschmann geometry; shown graphically in Figure 5.1. Notably, as near-field optics is not subject to the free-space diffraction limit, there are possibilities for enhanced degrees of localization, as compared with conventional approaches to optical micromanipulation. Moreover, in contrast to standard "far-field" optical trapping or tweezing, evanescent-field optical micromanipulation need not use a high-numerical aperture microscope objective lens: more weakly focused light fields may be exploited, and indeed the term *lensless optical trapping* (LOT) has been coined [29].

Garces-Chavez and colleagues explored stable particle trapping over a surface in the Kretschmann geometry [29], extending previous work in such a configuration that had been restricted to solely guiding or propulsion of objects. In this manner, they managed to obtain trapping in arrays on the surface. The experimental system is based upon the Kretschmann geometry and employs a right-angle BK-7 prism. The trapping laser light was a continuous wave ytterbium fiber laser operating at 1070 nm incident at the boundary between glass and water, at an angle of incidence beyond the critical angle to achieve total internal reflection (TIR). Guiding was observed, and when using two equal counterpropagating beams the optical radiation pressure forces along the surface were largely balanced, creating line traps. The area of the light field was large, thus the gradient force was typically weak and not particularly localized. Therefore, the authors elected to image a pattern of linear fringes onto the prism using a Ronchi ruling (500 line pairs per inch) onto this interface. The overall effect was to initiate counterpropagating surface waves to create what is often termed a *potential energy landscape* on the surface.

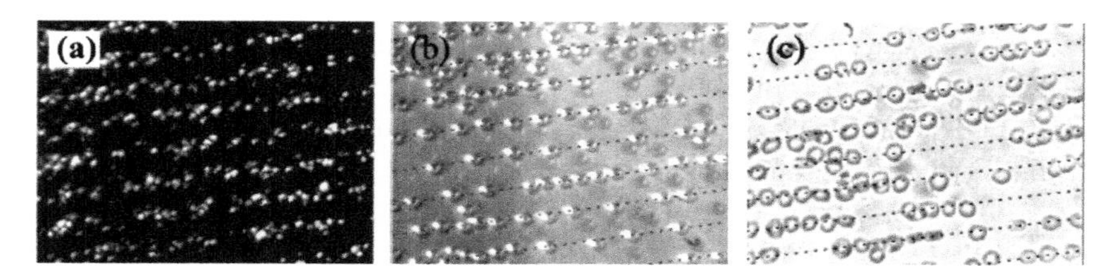

Figure 5.3 Lensess optical trapping (LOT): simultaneous micromanipulation of large numbers of particles using patterned evanescent waves. (a) Light scattered by a dense suspension of 1 μm particles; (b) bright field images of 5 μm polystyrene spheres; and (c) red blood cells. With only one beam, microobjects assemble into and are transported along optically defined channels. The lines show the location of each one of the fringes of light. The width of the image is 150 μm. Reprinted with permission from [29]. Copyright 2005 by the American Institute of Physics.

The periodic potential energy landscape formed on the surface by introduction of the ruling generated linear fringes in each beam path. These were overlapped at the sample plane such that linear potential wells, with a 12 mm period, result. They showed guiding and transport of particles over areas of the order of 1 mm^2 with typical images capturing just a fraction of the area under illumination, as shown in Figure 5.3. In the case of guiding, as expected, the velocity of the colloidal objects increased linearly with power, though it is to be noted that use of high powers (of the order of several hundreds of milliwatts) were required to attain velocities of 1 mm/s. As might be expected, the microparticles aligned and followed trajectories commensurate to that of the ruling. The rulings created lateral traps for the objects. From the perspective of biology, the authors further showed the guiding and trapping of red blood cells, typically as biconcave disks (7 mm of size), suspended in phosphate buffered saline (PBS) along fringe structures created in an evanescent field. It is very interesting to note that subsequent work in a similar geometry showed a marked deviation from this behavior observed for colloidal particles due to the presence of optical binding interactions. This very intriguing observation by Mellor and Bain [30] showed that the particle–particle interaction could result in the formation of 2D lattices of hexagonal or other order; this will be described in more detail later.

5.3.1.1 Enhancement in surface optical trapping

While the experiments described above indeed show the potential for large area micromanipulation and guiding in a surface evanescent trap, the evanescent field as stated earlier is naturally weak: typical velocities observed are quite slow ($\sim$1 μm/s) for relatively high powers. Thus attention in the last few years has turned to enhancement techniques for the evanescent field while retaining the advantages of this Kretschmann geometry. Two key methods of enhancement are described here: the use of surface plasmon enhancement, and the use of a dielectric resonator. These methods have been employed in other areas of optics for enhancement, e.g., sensors or atomic physics, but their use for near-field micromanipulation of mesoscopic objects is relatively new. Here, we will briefly explore both of these schemes.

5.3.1.2 Surface plasmons

A metal–dielectric interface may support resonant electromagnetic modes corresponding to the collective oscillation of free electrons within the metal. These oscillations, termed *surface plasmon polariton* (SPP), have been the topic of intensive research as they possess interesting optical properties that are inherently linked to advances in near-field studies. These include aperture-less optical nanolithography with resolution comparable to electron beam lithography [31], compact integrated optical/electrical nanostructures with subwavelength dimensions (plasmonics) [32], single molecule detection using Surface Enhanced Raman Scattering (SERS) [33], super-lensing effects for subdiffraction limited imaging [34], and metamaterials [35].

Due to the presence of an associated electric field enhancement, SPP modes have the potential to benefit near-field optical micromanipulation. Any enhancement to the evanescent field trapping we described earlier should yield a system where optical forces can trap or localize particles more strongly and over a significantly larger surface area. In the simplest geometry, propagating SPP modes are excited by coupling light through a thin gold layer ($\sim$40 nm) coated on the top surface of a prism (Kretschmann) via attenuated total reflection—an approach compatible with existing techniques for near-field guiding and trapping. The degree of enhancement depends strongly upon the type of structure that is supporting the modes. In localized modes, excited in metallic nanostructured substrates, the field enhancements may be orders of magnitude larger than the incident beam [36]. In addition, the use of patterned substrates may lead to greater flexibility in trapping arrangements for multiple-trapping applications. This was demonstrated theoretically by Quidant and colleagues, who showed that periodically patterned substrates of resonant gold nanostructures should yield an optical potential landscape with individual trapping sites between the gold structures [25]. Finally, the functionality afforded by the use of SPP modes may lead to future synergies that incorporate near-field optical trapping with other plasmon-based applications. It should be noted that the use of SPPs is established in atom optics, where a dipole interaction for deflection of an atomic beam has been seen with SPP-based atomic mirrors [37].

A study of the radiation forces generated by propagating SPP modes using a photonic force microscope was published by Volpe and colleagues [38]. In this experiment, the perturbation of an optically trapped dielectric sphere (optical tweezers) in close proximity to a gold–fluid interface supporting SPP excitations was used to probe local optical forces. They showed that as the trapped object was brought close to the interface, it experienced force directed toward the surface as well as a guiding force acting along the direction of propagation of the SPP mode. Importantly, the forces were shown to be enhanced by the presence of the SPP through a correspondence between the maximum optical force and optimal coupling efficiency of the incident light to plasmon mode.

In addition to the enhanced optical field, SPP modes dissipate within the metallic layer and in turn the energy imparted to the surrounding media generates local heating. Optically induced thermal effects can be used to locally control forces within the medium and may be an effective tool for manipulating particles on the microscopic scale [40,41]. As an example of such surface plasmon-induced thermal gradients, liquid droplets on a surface may be moved using the Marangoni effect that generates differences in surface tension across the droplet [42]. In general thermal gradients, regardless of origin, at the microscopic scale are known to induce convection and thermophoresis [43]. In the presence of a thermal gradient, colloids migrate to the colder region by thermal diffusion (thermophoresis). This effect is not well understood and is sometimes termed the *Ludwig–Soret effect.* Convection is the more well-known and dominant mechanism when considering thermal gradients, and it creates a circulation of fluid around a volume in order to transfer heat to the surrounding medium and typically drives colloids toward the hotter regions within a chamber. The geometry of the system may dictate the relative influence of these phenomena. In this manner, it has been shown in related studies that convective and thermophoretic effects can accumulate and trap colloids and macromolecules using the interplay between these two effects [43].

Garces-Chavez and colleagues conducted a study of the collective dynamics of large numbers of micrometer-sized particles in the presence of SPP [39]. The experimental setup was similar to that in previous work, as shown in Figure 5.1. However, in this instance, the surface was coated with a thin (40 nm) layer of gold. The incident trapping light was tuned around the critical angle, and the reduction in reflectance at a certain angle showed the successful excitation of the surface plasmon. The observations showed initially that in a "thick" sample chamber, the dissipation of the SPP would typically result in heating that would lead to convective effects. In itself, this could be usefully utilized: a mere 50 mW of laser power was used in this system to accumulate several thousand particles in an elliptical area where the two beams were incident upon the prism surface, as shown in Figure 5.4. Analysis of the particle positions could show liquid-solid-like behavior. By reducing the depth of the sample chamber, the authors were able to suppress convective effects and show that optical forces from the evanescent field dominated; in this case, an enhancement of a factor of three was observed. By increasing the power gently, still within a thin chamber,

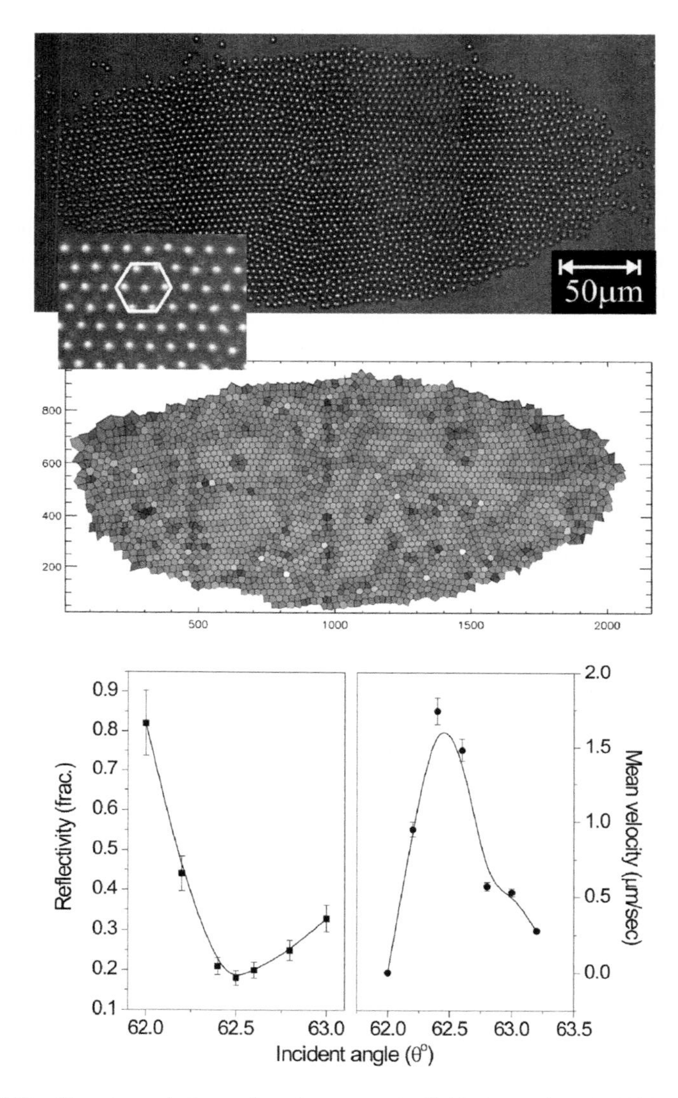

Figure 5.4 *Top*: Experimental observation of an array of colloids (approximately 2800 particles) accumulated in the presence of a surface plasmon polariton excited by attenuated total reflection. *Middle*: Voronoi plot showing the Wigner–Seitz cell for each colloid in the array indicates that at the center there is a predominantly close-packed crystalline arrangement (hexagonal). At the periphery, particles are more fluidlike with no preferred nearest neighbor arrangement. *Bottom*: Reflectivity of the incident light and mean velocity (curve is a guide to the eye) of the particles as a function of incident angle. Copyright 2006 by the American Physical Society [39]. See color insert.

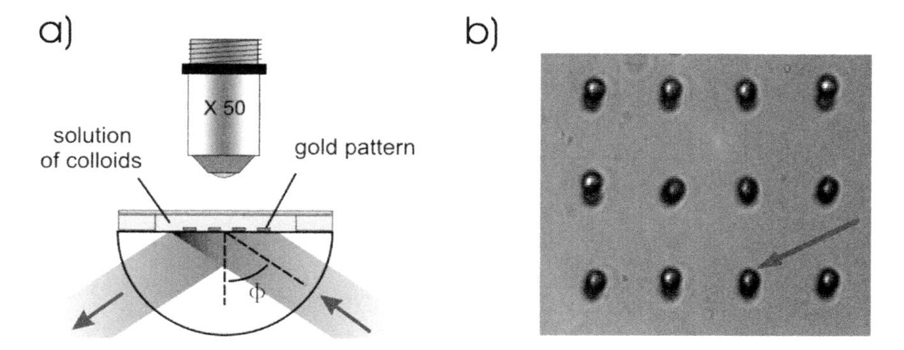

Figure 5.5 Localized surface plasmon modes generated on periodic patterned gold films can be used to obtain greater field enhancement and more localized trapping volumes. (a) Excitation of the modes may be achieved using a similar geometry to the propagating surface plasmons. (b) Demonstration of parallel trapping of an array of 4.88 μm polystyrene spheres on a patterned substrate. Reprinted by permission from Macmillan Publishers Ltd: *Nature Physics* [44], Copyright 2007.

thermophoretic behavior ensued with the particle migrating away from the hot central region to the beam periphery. Interestingly, under conditions where optical forces dominate, particles were observed to form ordered linear arrays; this ordering was associated with optical binding interactions, discussed later in the chapter.

Work by Righini and colleagues combined the effects of both surface-plasmon–induced optical forces and associated thermal forces to create a new near-field optical sorting geometry (see Figure 5.5a) [44]. By patterning the metal film, they were able to create local SPP excitations in predefined positions, resulting in an optical potential landscape able to influence the dynamic of the colloid flow. The strong size dependence of the optical trapping at the gold structures was used to selectively filter out a polydispersed mixture of colloids; species that strongly interact with the plasmon traps were held in the traps, and the optically induced convective forces were used to transport the weakly interacting particles out of the area of interest, as in Figure 5.5b.

5.3.1.3 Cavity enhancement for surface trapping

The demonstrated evanescent field enhancement achieved using both propagating and localized surface plasmon polaritons present a promising approach; however, the unresolved issue of thermal dissipation of the plasmon modes and the resulting thermal effects may limit their potential application. An alternative method for achieving a field enhancement in the Kretschmann configuration, which avoids the problem of heating, is through the use of a dielectric resonator. This approach was first applied to atom optics by Labeyrie and colleagues [45] as an evanescent wave atomic mirror.

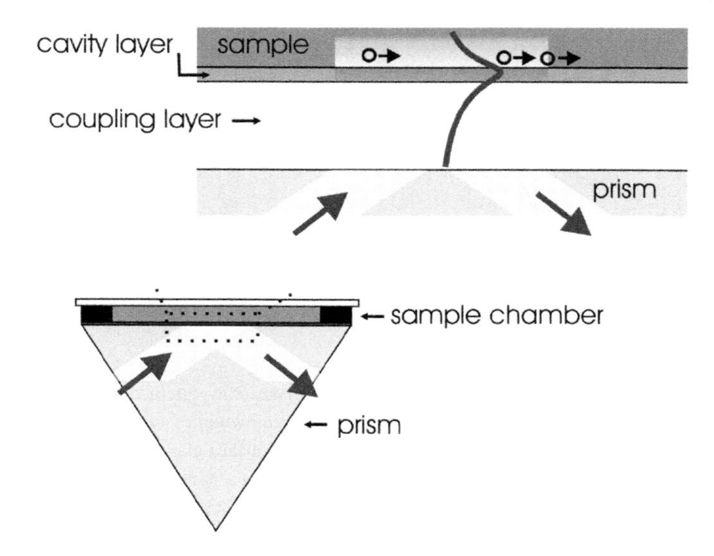

Figure 5.6 Principle of evanescent wave field enhancement in a resonant dielectric waveguide. Under resonance conditions, achieved by tuning the angle of incidence, light incident at the prism-coupling layer interface will couple to the waveguide mode via frustrated total internal reflection. The field enhancement of the waveguide mode is determined by the thickness of the coupling layer and the waveguide losses. The waveguide mode extends into the sample volume and interacts with colloidal particles near the surface.

Proposals existed for application to trapping [46] and were experimentally realized by Reece and colleagues [47] for extended-area micromanipulation of colloidal aggregates.

As shown in Figure 5.6, the dielectric resonator consists of a high refractive index prism coated with a low refractive index coupling layer and a high refractive index cavity layer. A microfluidic chamber containing the sample solution is placed in direct contact with the top cavity layer. The coupling layer and colloidal dispersion act as cladding layers for a waveguide mode that is supported by the cavity layer and excited through the prism by frustrated total internal reflection. This arrangement is similar in structure to a prism coupled planar waveguide and has been exploited in studying the nonlinear optical properties of colloidal aggregates, as discussed later in the chapter. The key design features of the dielectric resonator that make it amenable to optical micromanipulation are (1) the electric field enhancement is determined by the optical properties of the structure rather than the material properties, and (2) the guided mode is weakly confined by the cavity layer and has a large evanescent wave component extending into the sample medium, which may be utilized for optical micromanipulation.

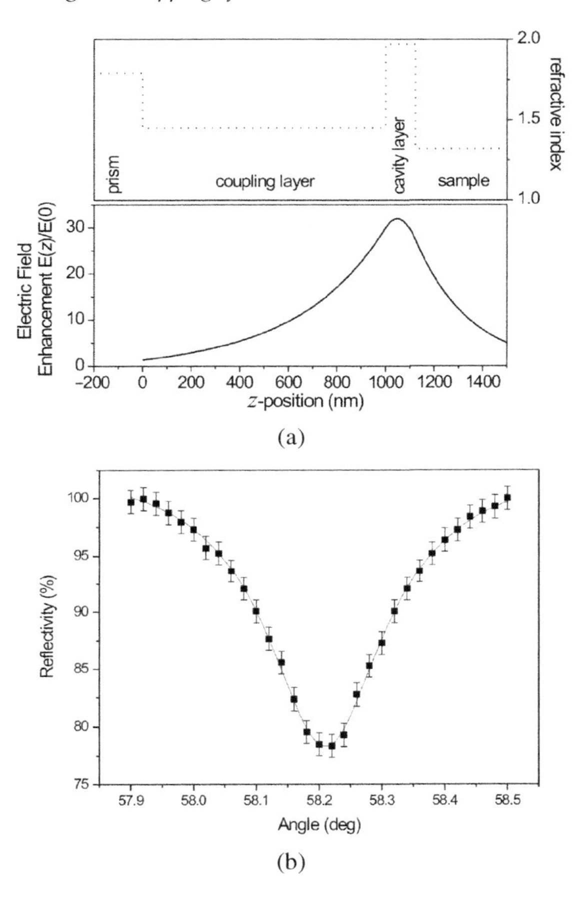

Figure 5.7 (a) Example of the refractive index profile (*top*) and the calculated resonant waveguide mode (*bottom*) for the dielectric resonator structure used by Reece and colleagues [47]. The electric field enhancement is calculated as the ratio of the electric field at the sample cavity layer interface to the incident beam. (b) Experimentally determined angular dependent reflectivity around the coupling angle showing the finite acceptance angle of the resonant structure.

The origin of the field enhancement at the sample surface is achieved by virtue of the resonant coupling with the incident laser beam, a technique that is used widely in photonics to excite resonant modes. It may be easily understood by considering the structure as analogous to a Fabry–Perot optical resonator (e.g., a laser cavity), where the coupling layer acts as the input/output coupler and the waveguide mode is equivalent to a cavity mode. The enhancement of the evanescent field is then simply related to the finesse of the resonator and can be several orders of magnitude greater than the incident field (see Figure 5.7a). Importantly, the very low optical losses of the dielectric waveguide means that the light that is not scattered by the colloidal particles will be reradiated from the waveguide mode back to the prism rather than absorbed, as is the case for surface plasmons. In principle, the limit on the field enhancement will be dictated by a combination of the optical losses of the sample and waveguide layers (e.g., interface scattering, absorption), however as the acceptance angle of the resonant coupling reduces with increasing finesse, this will place a practical limit on degree of enhancement for a specified beam waist.

Using the dielectric resonator approach, Reece and colleagues [47] demonstrated a tenfold increase in the guiding velocity of 5 μm particles in the presence of resonantly coupled waveguide modes, compared with an evanescent wave generated under total internal reflection. They also showed that extended area trapping of large numbers of particles could be achieved. The ability to organize large arrays of particles over extended areas, coupled with the reduced power requirement afforded by the cavity enhancement, provides much potential for the realization of massively parallel processing (such as sorting) of micrometer-sized particles, such as cells. Ultimately, the resonantly coupling approach described here may be readily applied to other resonant optical structures to achieve greater levels of evanescent field enhancement in one or more dimension. These may include the use of photonic crystal point defects [48], toroidal resonators [49], or microsphere resonators cavities [50].

5.3.1.4 Sorting of particles in a surface trap

In addition to the more ubiquitous and well-known phenomena of guiding and trapping particles, the concept of separation and sorting of objects has become a prominent feature in optical micromanipulation. This has been spurred on by the need to develop compact, microfluidic versions of the more common bulk fluorescence activated cell sorters. A variety of sorting devices and systems using optical forces have been implemented in the last few years, and broadly they may be classed as either active or passive devices: An active sorter requires an input signal (e.g., fluorescence from an appropriately tagged cell) that then triggers another part of the system, which for the purposes of our discussion would be an optical switch that either guides or drags the particle of interest to a suitable reservoir. Sorting may also be achieved passively, which means that no tags are required and that the sorting is achieved solely on the differing affinity of the object or cell to the periodic light pattern upon which it is placed. This means that we do not need markers, and the sorting may be readily extended over a large area and is thus an attractive option [51]. However, such passive sorting is difficult and, for cells, the discrimination between different cell types is not sufficient to ensure separation.

In the evanescent wave or near-field trapping area, some sorting has been achieved, as we will see in the example here. In relation to this section, we will look at the

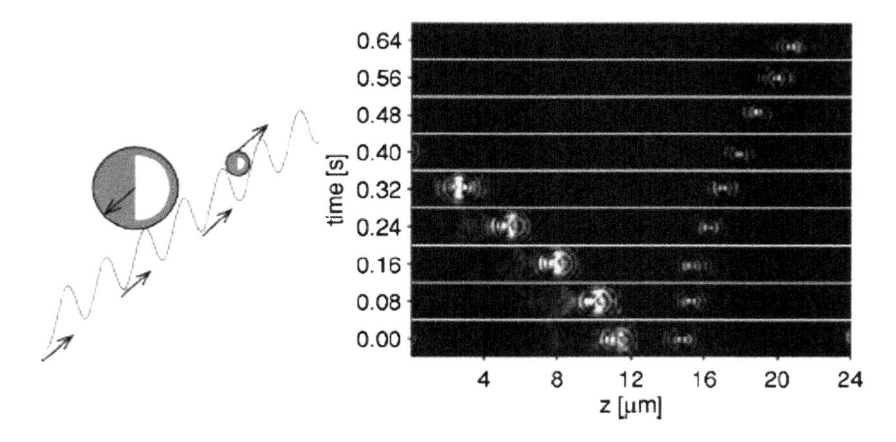

Figure 5.8 Optical sorting based upon interfering evanescent waves. The interference of two coherent counterpropagating beams at the top surface of a prism will create evanescent waves with a periodic fringe pattern; the position of the fringes can be controlled by changing the relative phase of the beams. By mismatching the beam intensities, a tilted optical potential can be produced (*left*). For particles that interact weakly with the fringes, the mismatch in radiation pressures will cause the particles to move down the potential. Strongly interacting particles will be localized by the fringes and can be made to move up the potential by sweeping the fringes. This is realized experimentally with polystyrene colloids sized 750 nm and 350 nm (*right*). Copyright 2006 by the American Physical Society [52].

particle motion upon an interferometric pattern on the surface of a prism in the Kretschmann geometry. Cizmar and colleagues [52] used interference optical traps to study the dynamics of particles in this geometry. In their study interference patterns were formed by coherent counter-propagating evanescent fields and the relative intensity of two fields was used to create a tilted washboard potential [53]. They observed selective trapping based upon particle size, where where particles that weakly interact with the interference fringes freely move down the tilted potential, whereas strongly interacting particles remained confined to the fringes (and strongly interacting particles) could be made to move up the tilted potential, as illustrated in Figure 5.8. The work also included a novel method to determine particle position within the extended interference pattern.

5.3.2 Trapping Using TIR Objectives

Recently, a new near-field trapping and tweezing geometry that utilizes focused evanescent wave illumination has been proposed and demonstrated by Gu and colleagues [54]. The technique makes use of a high numerical aperture total internal reflection fluorescence (TRIF) microscope objective with a circular obstruction placed

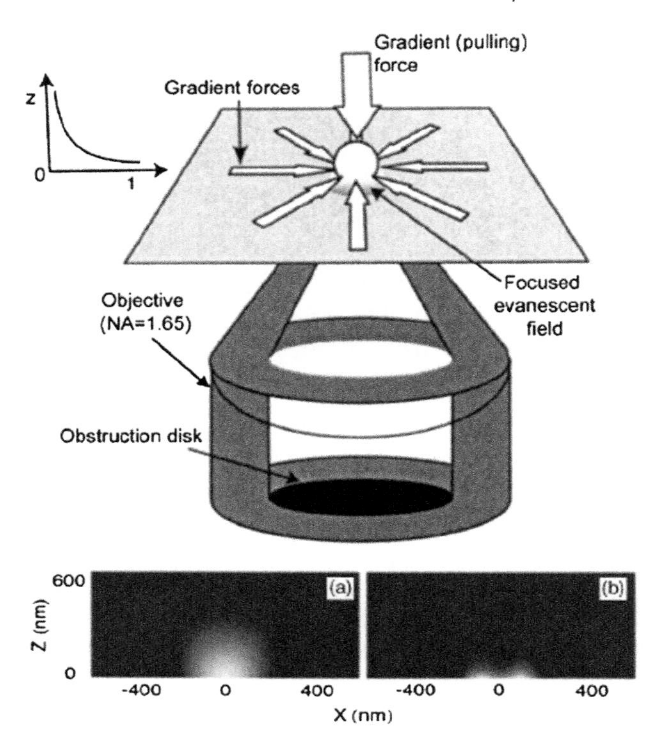

Figure 5.9 Near-field optical trapping using a total internal reflection fluorescence microscope (TIRF) objective. An expanded beam incident on the back aperture of the TIRF objective is partially blocked by an obstruction to allow only rays that will reach the sample plane under total internal reflection conditions to pass. The result is to produce a single beam gradient force optical trap composed entirely of focused evanescent waves; the associated forces are indicated in the diagram. A comparison of the electric field at the focal plane in the absence (a) and presence (b) of the obstruction highlights the changes the axial trapping volume. Reprinted with permission from [54]. Copyright 2004, American Institute of Physics.

at its back aperture, which allowed only light reaching the sample plane under TIR conditions to be passed through the objective. This has advantages over previously discussed geometries in that it provides a method for focusing the evanescent waves to achieve a localized near-field trap while avoiding the need for metallic components that generate undesirable heating.

The schematic diagram in Figure 5.9 illustrates the principle of the trap; the circular obstruction produces an annular beam that enters the back of the high NA objective ($NA = 1.65$) and is focused down to the sample plane. The beam forms a tightly focused, circularly symmetric spot in the trapping plane and has an exponentially decaying component in the axial direction. The optical field under these trapping conditions has been simulated using vector diffraction theory [24]. A comparison of the field with and without the obstruction is shown in Figure 5.9(a) and 5.9(b), which represent the conditions for trapping under evanescent waves and propagating waves, respectively. As is evident, the axial component of the field is substantially reduced, and, consequently, the axial size of the trapping volume (which the authors defined by the position where the intensity drops to 50% of that at the interface) is reduced to approximately 60 nm [6]—approximately one order of magnitude shorter than that in the far field case. Interestingly, the transverse profile of the evanescent wave trap has two peaks.

Gu and colleagues verified the presence of evanescent wave trapping by studying the trapping efficiencies (Q-values) for different circular obstruction sizes. They showed that for the case of the axial component, the trapping efficiency reduced as the size of the obstruction was increased to the point where no axial translation could be achieved; they associated this with the reduction in the propagating part of the beam. For the in-plane trapping component, trapping was achieved for all obstruction sizes, including size where the propagating rays are completely occluded from the trap. In the case of the purely evanescent wave trap, the trapping efficiency also reduced substantially, compared with no obstruction.

Gu and colleagues have recently advanced these experiments and explored the use of a near-field trap to stretch, fold, and rotate red blood cells (erythrocytes) [55]. The study used near-infrared light at 1064 nm with a laser of power up to 2 W. To support their observations, they also looked at numerical simulations of the stresses upon a red blood cell using both finite domain time difference and Maxwell stress tensor approaches. Experimentally, rotation of a red blood cell was seen with rotation rates of 1.5 rpm for a power of just less than 20 mW. The rotation was induced by the alignment of the cell with the polarization of the incident beam. The trapping geometry allowed them to create a near-field optical stretcher, deforming the cell by up to 23% from its original size. Folding of a trapped cell was also seen with higher laser powers (>18 mW). The technique merits further study for elucidating the cell's mechanical properties. Near-field manipulation may provide a new method by which membrane elasticity, viscoelastic properties, and cell deformation may be quantified.

5.3.3 Micromanipulation Using Optical Waveguides

A key feature of integrated optics and telecommunication is the use of optical waveguides for transmission of signals. Such waveguides have refractive index differences and specific modes that propagate within them. The light mode coupling is

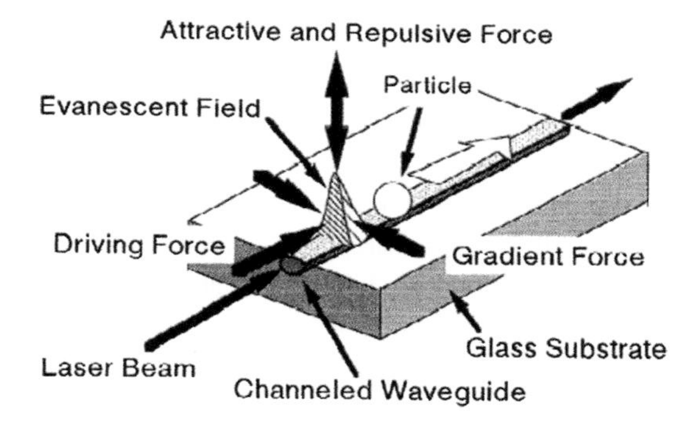

Figure 5.10 Channel waveguides with a fluid cladding (*top*) layer provide addition gradient forces in the transverse plane due to the confinement of the waveguide mode. Optical forces resulting from waveguide mode are indicated, particles in the vicinity of the waveguide will be confined to the top of the waveguide and guided along the direction of propagation due to radiation pressure. Reprinted with permission from [5]. Copyright 1996, Optical Society of America.

an issue, but naturally the light is typically confined within the higher refractive index region and, as might be expected, the light may leak into the rarified medium quite readily. Importantly, the use of such waveguides may provide interesting new geometries for transport, confinement, and sorting of microparticles with the ability to use modern micro- and nanofabrication procedures to tune the interaction as well as develop potential observations into real devices. In this section, we will review the geometries for optical waveguides used with particular emphasis on experiments performed.

Propulsion and trapping of particles along waveguides originated in 1996 when Kawata and Tani propelled particles in the Mie size regime along a channeled waveguide, as indicated in Figure 5.10. The experiment shows the transport of polystyrene spheres of size 1 μm to 5 μm in diameter along the waveguide surface due to the presence of an evanescent wave. The gradient force localizes the particles to the waveguide region laterally. The laser used operated at 1047 nm with a power in excess of 2 W and was able to propel particles along the waveguide at up to 14 μm/s. Metal microparticles (~500 nm in diameter) were also propelled along the guide, and their lateral confinement was inferred as due to a surface current and near-field scattering events.

Gold nanoparticles are highly polarizable and have numerous applications, as stated earlier, and many are due to the exploitation of the plasmon resonance. Hole and colleagues explored the behavior of gold nanoparticles (250 nm) on a waveguide [56]. Gold has a complex refractive index, and thus absorption needs to be considered. Caesium ion-exchanged waveguides were used as systems in these experiment. Photolithography was used to fabricate in a soda-lime glass substrate by defined straight channels (diameter 2.5 μm to 11 μm) in an aluminium film evaporated onto the surface of the substrate and a subsequent high-temperature process. The sample chamber was formed in a molded polydimethylsiloxane (PDMS) elastomer situated on the surface of the waveguides. The particles were propelled with a light field at 1066 nm but they coupled to the end of a waveguide by means of a single-mode fiber delivery system.

Axial propulsion forces, due to particle absorption as well as the more common scattering, resulted in a rapid motion of these particles along the waveguide. Several key points were obtained from their data. First, the particles showed a velocity distribution that was dependent on their three-dimensional position on the waveguide, and with careful data analysis they tracked the modal beat pattern on the surface of a dual-mode waveguide. The peak intensity in the evanescent field of the waveguide was estimated to be 9 GW m^{-2} per Watt of propagating modal power, using beam propagation analysis. The trapping and propulsion of large numbers of gold nanoparticles in the evanescent fields of the optical waveguides was observed, however the particles did not move at constant velocity, but their speed profile had a broad distribution corresponding to the evanescent field intensity profile. For the optimal waveguides in their systems, high velocities were recorded: a velocity of 500 μm/s was achieved for a modal power of 140 mW.

The separation (sorting) of particles has already been described in relation to surface traps where bidirectional motion of submicron objects of differing size was observed in relation to studies in the standard Kretschmann geometry. In the waveguide systems, sorting can also be initiated and seen and constitute another form of near-field optical sorting system. Grujic and colleagues used a Y-branched optical waveguide for microparticle sorting [57]. They used polystyrene microparticles optically guided in the waveguide's evanescent field, which could be directed down the desired, more strongly illuminated, output branch. The evanescent light used for guiding and sorting was a fiber laser operating at a wavelength of 1066 nm. This was butt-coupled directly to the waveguide. It is important to note that this form of sorting is rather more akin to "active" sorting rather than the passive sorting schemes we described in earlier in this chapter. A diagram of the experimental setup is shown in Figure 5.11.

The power distribution between the two output branches is selected by the relative position of the fiber to the waveguide input facet. Microspheres can be efficiently sorted down the two waveguide branches by changing the field distribution in the multimode input trunk. This provides a simple method for reliable particle sorting with very high probability of success under appropriate conditions. The extension of this sorting technology to the biological domain would need bio-matter with a sufficiently high refractive index mismatch relative to the buffer medium to induce a sufficient gradient force for localization. When considering biological macromolecules, these could

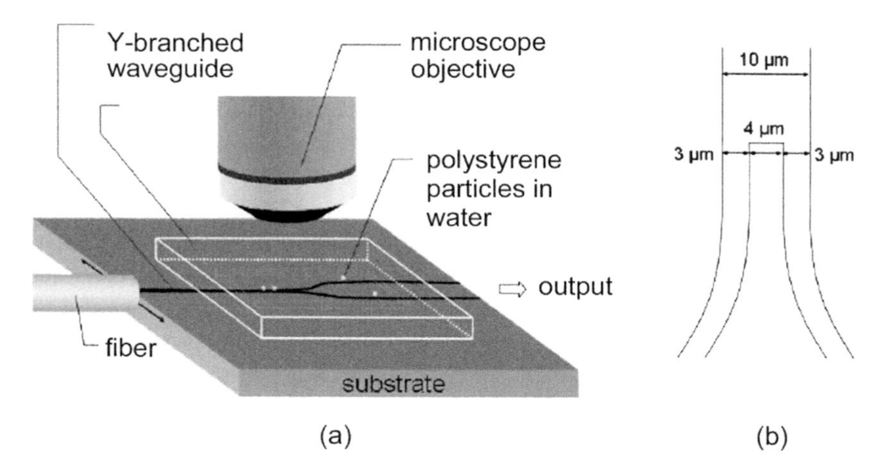

Figure 5.11 Additional functionality such as sorting and routing may be achieved with the use of optical circuits. This is illustrated in the demonstration of optical sorting of polystyrene spheres on a Y-branched waveguide. By controlling the coupling laser into the waveguide, particles may be switched between the upper and lower branch of the Y-junction. Reprinted with permission from [57]. Copyright 2005, Optical Society of America.

be made to adhere to suitably functionalized latex spheres for subsequent selection in the Y sorter. The method has future potential to be easily other particle manipulation techniques or imaging modalities.

Grujic and Helleso also looked at the formation and propulsion of chains of dielectric objects upon a caesium ion exchange waveguide [58]. Chains were assembled in a counterpropagating beam geometry and were observed to be guided along a higher velocity by approximately 15% (for 7 micron diameter spheres) compared to the velocity of single particles placed on the waveguide. Both hydrodynamic and coupling between the microspheres may play an important role in this behavior. From the hydrodynamic viewpoint, the fluid displaced by one of the spheres entrains the other. Optical binding type interactions may also be present. The study showed how to create long one-dimensional chains of particles in this waveguide geometry.

Recently, Gaugiran and colleagues studied polarization and particle-size dependence of radiation forces acting on gold nanoparticles guided on the surface of silver ion and silicon nitride waveguides [23]. Specifically, they were interested in identifying the conditions under which the force normal to the surface becomes repulsive—an effect that had previously been theoretically predicted by Ariaz-Gonzalez and colleagues [22]. Experimentally, Gaugiran and colleagues found that there was indeed a very large discrepancy between the guiding velocity of 600 nm gold particles for TM and TE polarizations. They found that the guiding velocity for the TM polarization case was significantly larger than for TE. As the guiding velocities were predicted to be similar, they associated this discrepancy to the presence of a repulsive force and an attractive force for the TE and TM cases, respectively. Their work was supported by finite element calculations, which showed that under the experimental conditions they would expect to have forces of opposite signs for the different polarizations.

5.4 EMERGENT THEMES IN THE NEAR FIELD

From the previous discussions, one can see that near-field optical micromanipulation is certainly establishing itself as more than a niche topic in the broader context of optical trapping. In several respects, the field is in its infancy and promises much in terms of new science as well as applications in microfluidics and new insights into mechanisms of colloidal self-assembly. In this section, we will explore some emergent themes on the topic.

5.4.1 Optical-Force-Induced Self-Organization of Particles in the Near Field

Externally modulated optical potentials, generated using adaptive optics, are commonly employed for creating multiple optical trapping sites. While the method of optical landscaping has the flexibility of producing arbitrary trapping arrangements, the precision of predefined trap locations is limited by the resolution of the optics, and large arrays suffer from heavy power requirements and trap-loading issues. Two types of optical-force-induced self-assembly have been observed in near-field optical trapping arrangements, which have the potential to overcome some of these limitations. Also, the inherently coupled nature of the interactions driving these self-assembly processes gives rise to nonlinear effects such as bistability and modulational instability, which may be utilized to create dynamically reconfigurable microstructured arrays.

In 2005, Mellor and Bain [30] showed that large arrays of colloidal particles accumulated at the surface of a counterpropagating evanescent waves trap formed into a highly ordered array with submicrometer precision; where the lattice properties were determined by the polarization of the trap (see Figure 5.12), the underlying interference pattern, and the size of the trapped particles. By switching between polarization states of the trap, they observed real-time transitions between checkerboard and hexagonal lattice patterns [59]. They associated the ordering mechanism with optical binding interactions. Optical binding arises when optically trapped particles locally redistribute the trapping field through refocusing and/or scattering, to create a new

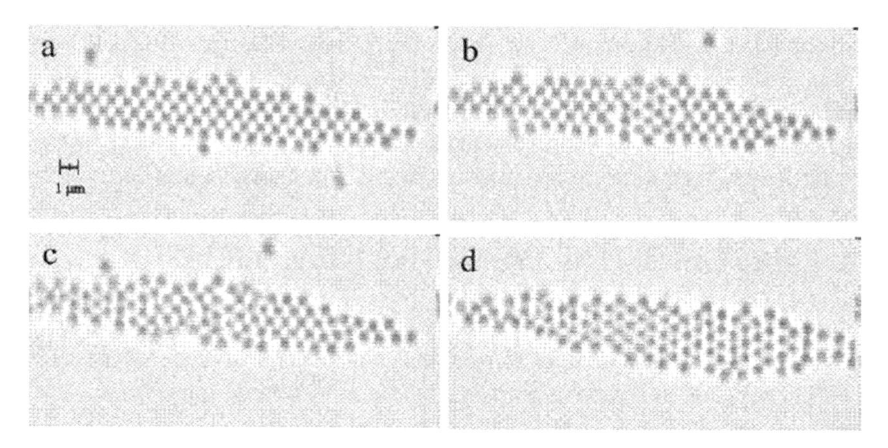

Figure 5.12 Controlling the process of self-organization in optical forces induced colloidal aggregates. (a)–(d) Successive video frames showing the reorientation of a two-dimensional optically bound lattice when the polarization of the optical trap is switched from s- to p-polarization; the lattice geometry is seen to change from a square lattice to a hexagonal lattice. Reprinted with permission from [59]. Copyright 2006, Optical Society of America.

stable trapping position for adjacent particles to occupy. Under favorable conditions, they lead to the formation of ordered arrangement of mutually interacting optically bound states, a self-assembling process that is achieved without the need for external modulation of the incident-trapping field. The concept of optical binding was first demonstrated by Golovchenko and colleagues [60] and led to the proposal of optically bound matter where large crystals of particles were supported by optical forces. To date, the concept of optical binding has been used to interpret a wide range of related observations, including the attraction/repulsion forces between two freestanding semiconductor waveguides, the arrangement of nanoparticles in a Laguerre–Gaussian beam, and the interactions between Mie particles aligned along the propagation axis of an optical trap (longitudinal binding).

Reece and colleagues [61] demonstrated optically induced self-organization of colloidal arrays in the presence of unpatterned counterpropagating waveguide modes of a dielectric resonator. High concentrations ($>0.1\%$ solid) of submicrometer-diameter colloids accumulated at the center of the trap formed regularly spaced linear arrays oriented along the direction of the incident EW wave vector with a lateral separation of several micrometers. The presence of large numbers of particles was necessary to observe the pattern formation. The colloidal arrays formed along the laser propagation-axis were shown to be linked to the breakup of the incident field into optical spatial solitons, the lateral spacing of the arrays being related to modulation instability of the soft condensed matter system. Optical spatial solitons (OSS) are spatially localized, nondiffracting modes that are supported in nonlinear optical media. They are the result of a balance between diffraction and self-phase modulation, which results in self-focusing under intense illumination. For a plane-wave incident field, small wavefront perturbations cause the optical field to break up into periodic arrays of spatial optical solitons or more complex patterns, an effect known as *modulational instability* (MI).

The presence of both OSS and MI is suggestive of the presence of optical nonlinearity in the colloidal dispersion. In fact, colloidal dispersions have long been considered potential candidates for artificial optical nonlinear Kerr materials. Ashkin and colleagues indicated the potential of bulk colloidal suspensions to act as an artificial Kerr medium in a number of nonlinear optical experiments, including self-focusing, optical bistability, and four-wave mixing. The mechanism behind the nonlinear properties of these materials is the electrostrictive effect due to the optical gradient forces experienced by the dielectric particulates, which aggregate at regions of high intensity, thereby locally increasing the refractive index and leading to a self-phase modulation. In the experiments described by Reece and colleagues [61], the dielectric resonator was treated as a prism-coupled nonlinear waveguide, where the colloidal dispersion acted as an artificial Kerr medium with a local nonlinear response (continuum approximation). This was sufficient to predict the main experimental observations: the presence of a threshold power for the onset of the array formation, the magnitude of the lateral array spacing, and its power dependence, as shown in Figure 5.13. For a deeper understanding of the optical nonlinearity of the colloidal dispersions and its resulting effects, the detailed granular nature must be considered, i.e., the nonlocal response. Conti and colleagues [62] have developed a theoretical nonlocal model for the nonlinear optical response of emulsions (including colloid dispersions) that accounts for microscopic particle–particle interactions using a static structure factor. However, this is just one of a number of competing models that aim to describe the detailed dynamics of the nonlinear process.

5.4.2 Near-Field Trapping with Advanced Photonic Architectures

Recent studies have proposed the use of a photonic crystal structure to trap nanoparticles in the near field. Photonic crystals are intricately linked with the controlled propagation and confinement of photons. The very geometry of the photonic crystal can engineer a suitable environment for the guiding or trapping of photons. Confinement of photons is particularly associated with photonic crystal structures embedded within a slab waveguide. The ability to confine light in such structures also

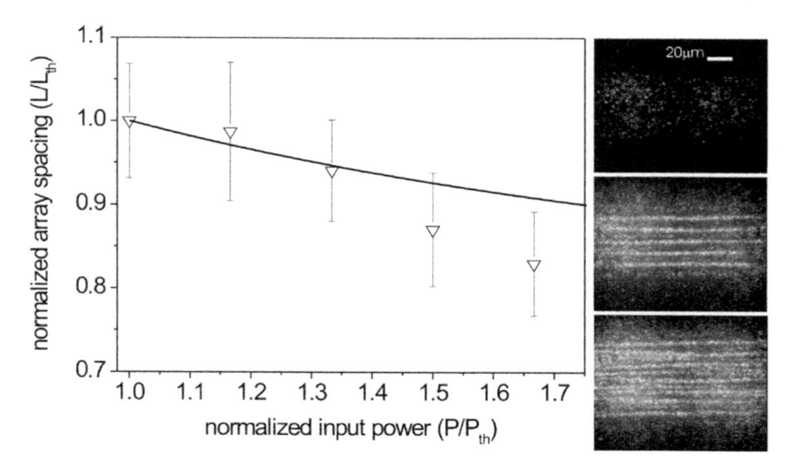

Figure 5.13 Dynamics of linear arrays of colloids formed at the surface of a dielectric resonator. The three panels on the right show the array formations for different illumination intensities. Below the threshold, the OSS are not observed (*top*), even after extended periods of illumination; just above threshold, stable arrays of OSS are formed. The graph on the left shows the period of array spacing as a function of the input power above threshold. The period (L) is normalized with respect to the period at threshold (L_{th}), and the input coupling power (P) is normalized to the power at threshold (P_{th}). The solid line indicates the variation as predicted by the continuum model. Copyright 2007 by the American Physical Society [61].

leads to large-field gradients of field intensity that thus make them ideal for creating either optical landscapes with inherently strong forces. This is not true for just microscopic or nanoparticles, the topic of this chapter, but indeed may be extended to the trapping of atomic species. Rahmani and Chaumet considered the near field of a photonic crystal slab of $n = 3.4$ placed in a liquid of $n = 1.33$, typical of water [63]. The slab was perforated with a lattice of holes in a triangular geometry. Photon localization was achieved by placing a defect or localized state in the slab by omission of one of the holes. This created a rather crude microcavity for light with a Q-factor of 100; this is, of course, several orders of magnitude less than state of the art, but it serves to show the principle of this method for a readily realizable system.

Calculations of the force were performed for a particle in the Rayleigh regime $(\ll l)$ using the coupled dipole method. Computing the forces showed asymmetry in the trapping due to the asymmetric polarization used in the illumination. Potential depths well in excess of kT were achieved, and proposal was put forward for creating more elaborate optical landscapes. Such multiple "nanotraps" may be created with a coupled-cavity systems or by exploiting critical points within a bandgap diagram (a PCS without defects). It is not just photonic crystals that can create such novel trap-

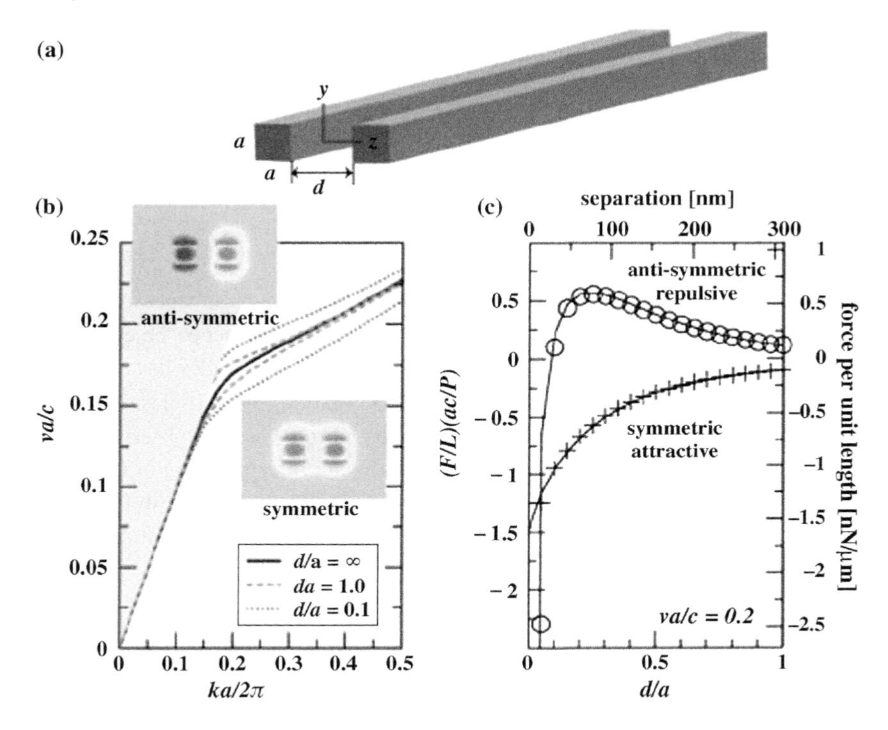

Figure 5.14 Theoretical prediction of optical binding between two freestanding microphotonic waveguides with geometry depicted in (a). The force per unit length is calculated for different guided modes acting on the waveguides at various separations. (d) The symmetric modes produce an attractive force between the waveguides, while antisymmetric modes generate a repulsive force. Reprinted with permission from [64]. Copyright 2005, Optical Society of America. See color insert.

ping systems in the near field. Naturally, one can think of the use of negative refraction that can also achieve this end as well as the use of extraordinary transmission of light through subwavelength apertures. Holes in metallic film have the advantage that the surface plasmon enhancement may significantly aid the nanoparticle trapping.

If one considers the guiding of light along microphotonic waveguides, the very overlap between the guided waves in such waveguide structures may create forces between them [64], as shown in Figure 5.14. This may be termed a form of evanescent wave bonding. The force may be either attractive or repulsive, depending upon the choice of the relative input phase of the fields. Povinelli and colleagues [64] considered forces between two parallel silicon strip waveguides of high refractive index in their study (see Figure 5.14). The forces were sufficient to create experimentally observable displacements, and the authors suggested the use of slow light enhancement as the force scaled inversely with the group velocity of the light. More recent studies have explored such evanescent light wave coupling between two photonic nanowires that permits the detection of motion between the two wires, which results in a change in the light coupling [65].

5.5 CONCLUSIONS

Near-field micromanipulation is coming of age. Significant advances have been made with total internal reflection geometries involving microscope objectives and prisms in the so-termed Kretschmann geometry. Optical waveguides have also been immensely successful, showing fast propulsion and sorting of particles. Some surprises have come to the fore: optical binding and patterning has been observed as well as the onset of nonlinearity in colloidal systems. As our understanding of light propagation in the near field progresses hand in hand with the progression in the field of plasmonics and metamaterials, the future looks very promising for this form of optical micromanipulation.

ACKNOWLEDGMENTS

The authors would like to thank the UK Engineering and Physical Sciences Research Council for funding. Kishan Dholakia would also like to thank Min Gu for useful discussions.

REFERENCES

[1] F. de Fornel, Evanescent waves from Newtonian optics to atomic optics, in: W.T. Rhodes (Ed.), *Springer Series in Optical Sciences*, vol. 73, Springer-Verlag, Berlin/Heidelberg, 2001, p. 31.

[2] J.W. Goodman, Introduction to Fourier Optics, 2nd ed., McGraw–Hill, New York, 1995.

[3] E. Hecht, Optics Fourth Edition, Addison–Wesley, San Francisco, 2002.

[4] S. Kawata, T. Sugiura, Movement of micrometer-sized particles in the evanescent field of a laser-beam, *Opt. Lett. 17* (1992) 772–774.

[5] S. Kawata, T. Tani, Optically driven Mie particles in an evanescent field along a channeled waveguide, *Opt. Lett. 21* (1996) 1768–1770.

[6] R.J. Oetama, J.Y. Walz, Translation of colloidal particles next to a flat plate using evanescent waves, *Colloids Surf. A: Physicochem. Eng. Aspects 211* (2002) 179–195.

[7] L. Novotny, R.X. Bian, X.S. Xie, Theory of nanometric optical tweezers, *Phys. Rev. Lett. 79* (1997) 645–648.

[8] P.C. Chaumet, A. Rahmani, M. Nieto-Vesperinas, Optical trapping and manipulation of nano-objects with an apertureless probe, *Phys. Rev. Lett. 88* (2002) 123601.

[9] K. Okamoto, S. Kawata, Radiation force exerted on subwavelength particles near a nanoaperture, *Phys. Rev. Lett. 83* (1999) 4534–4537.

[10] K. Svoboda, S.M. Block, Optical trapping of metallic rayleigh particles, *Opt. Lett. 19* (1994) 930–932.

[11] A. Ashkin, Forces of a single-beam gradient laser trap on a dielectric sphere in the ray optics regime, *Biophys. J. 61* (1992) 569–582.

[12] P.C. Chaumet, M. Nieto-Vesperinas, Coupled dipole method determination of the electromagnetic force on a particle over a flat dielectric substrate, *Phys. Rev. B 61* (2000) 14119–14127.

[13] M. Lester, J.R. Arias-Gonzalez, M. Nieto-Vesperinas, Fundamentals and model of photonic-force microscopy, *Opt. Lett. 26* (2001) 707–709.

[14] J.D. Jackson, Classical Electrodynamics, third ed., John Wiley and Sons, 1998.

[15] M. Mansuripur, Radiation pressure and the linear momentum of the electromagnetic field, *Opt. Exp. 12* (2004) 5375–5401.

[16] B.A. Kemp, T.M. Grzegorczyk, J.A. Kong, Ab initio study of the radiation pressure on dielectric and magnetic media, *Opt. Exp. 13* (2005) 9280–9291.

[17] E. Almaas, I. Brevik, Radiation forces on a micrometer-sized sphere in an evanescent field, *J. Opt. Soc. Am. B: Opt. Phys. 12* (1995) 2429–2438.

[18] J.Y. Walz, Ray optics calculation of the radiation forces exerted on a dielectric sphere in an evanescent field, *Appl. Opt. 38* (1999) 5319–5330.

[19] I. Brevik, T.A. Sivertsen, E. Almaas, Radiation forces on an absorbing micrometer-sized sphere in an evanescent field, *J. Opt. Soc. Am. B: Opt. Phys. 20* (2003) 1739–1749.

[20] M. Lester, M. Nieto-Vesperinas, Optical forces on microparticles in an evanescent laser field, *Opt. Lett. 24* (1999) 936–938.

[21] H.Y. Jaising, O.G. Helleso, Radiation forces on a Mie particle in the evanescent field of an optical waveguide, *Opt. Commun. 246* (2005) 373–383.

[22] J.R. Arias-Gonzalez, M. Nieto-Vesperinas, Radiation pressure over dielectric and metallic nanocylinders on surfaces: Polarization dependence and plasmon resonance conditions, *Opt. Lett. 27* (2002) 2149–2151.

[23] S. Gaugiran, S. Getin, J.M. Fedeli, J. Derouard, Polarization and particle size dependence of radiative forces on small metallic particles in evanescent optical fields. Evidence for either repulsive of attractive gradient forces, *Opt. Exp. 15* (2007) 8146–8156.

[24] D. Ganic, X.S. Gan, M. Gu, Trapping force and optical lifting under focused evanescent wave illumination, *Opt. Exp. 12* (2004) 5533–5538.

[25] R. Quidant, D. Petrov, G. Badenes, Radiation forces on a Rayleigh dielectric sphere in a patterned optical near field, *Opt. Lett. 30* (2005) 1009–1011.

[26] A. Ashkin, J.M. Dziedzic, J.E. Bjorkholm, S. Chu, Observation of a single-beam gradient force optical trap for dielectric particles, *Opt. Lett. 11* (1986) 288–290.

[27] H. Melville, G.F. Milne, G.C. Spalding, W. Sibbett, K. Dholakia, D. McGloin, Optical trapping of three-dimensional structures using dynamic holograms, *Opt. Exp. 11* (2003) 3562–3567.

[28] A. Constable, J. Kim, J. Mervis, F. Zarinetchi, M. Prentiss, Demonstration of a fiberoptic light-force trap, *Opt. Lett. 18* (1993) 1867–1869.

[29] V. Garces-Chavez, K. Dholakia, G.C. Spalding, Extended-area optically induced organization of microparticles on a surface, *Appl. Phys. Lett. 86* (2005) 031106.

[30] C.D. Mellor, C.D. Bain, Array formation in evanescent waves, *Chemphyschem. 7* (2006) 329–332.

[31] S. Sun, G.J. Leggett, Matching the resolution of electron beam lithography by scanning near-field photolithography, *Nano Lett. 4* (2004) 1381–1384.

[32] S.A. Maier, M.L. Brongersma, P.G. Kik, S. Meltzer, A.A.G. Requicha, H.A. Atwater, Plasmonics—A route to nanoscale optical devices, *Adv. Mater. 13* (2001) 1501.

[33] K. Kneipp, H. Kneipp, I. Itzkan, R.R. Dasari, M.S. Feld, Surface-enhanced Raman scattering and biophysics, *J. Phys.: Condens. Matter 14* (2002) R597–R624.

[34] N. Fang, H. Lee, C. Sun, X. Zhang, Sub-diffraction-limited optical imaging with a silver superlens, *Science 308* (2005) 534–537.

[35] D.R. Smith, J.B. Pendry, M.C.K. Wiltshire, Metamaterials and negative refractive index, *Science 305* (2004) 788–792.

[36] W.L. Barnes, A. Dereux, T.W. Ebbesen, Surface plasmon subwavelength optics, *Nature 424* (2003) 824–830.

[37] T. Esslinger, M. Weidemuller, A. Hemmerich, T.W. Hansch, Surface-plasmon mirror for atoms, *Opt. Lett. 18* (1993) 450–452.

[38] G. Volpe, R. Quidant, G. Badenes, D. Petrov, Surface plasmon radiation forces, *Phys. Rev. Lett. 96* (2006) 238101.

[39] V. Garces-Chavez, R. Quidant, P.J. Reece, G. Badenes, L. Torner, K. Dholakia, Extended organization of colloidal microparticles by surface plasmon polariton excitation, *Phys. Rev. B 73* (2006) 085417.

[40] H.B. Mao, J.R. Arias-Gonzalez, S.B. Smith, I. Tinoco, C. Bustamante, Temperature control methods in a laser tweezers system, *Biophys. J. 89* (2005) 1308–1316.

[41] S. Masuo, H. Yoshikawa, H.G. Nothofer, A.C. Grimsdale, U. Scherf, K. Mullen, H. Masuhara, Assembling and orientation of polyfluorenes in solution controlled by a focused near-infrared laser beam, *J. Phys. Chem. B 109* (2005) 6917–6921.

[42] R.H. Farahi, A. Passian, T.L. Ferrell, T. Thundat, Marangoni forces created by surface plasmon decay, *Opt. Lett. 30* (2005) 616–618.

[43] D. Braun, A. Libchaber, Trapping of DNA by thermophoretic depletion and convection, *Phys. Rev. Lett. 89* (2002) 188103.

[44] M. Righini, A.S. Zelenina, C. Girard, R. Quidant, Parallel and selective trapping in a patterned plasmonic landscape, *Nature Phys. 3* (2007) 477–480.

[45] G. Labeyrie, A. Landragin, J. Von Zanthier, R. Kaiser, N. Vansteenkiste, C. Westbrook, A. Aspect, Detailed study of a high-finesse planar waveguide for evanescent wave atomic mirrors, *Quantum Semiclass. Opt. 8* (1996) 603–627.

[46] P.C. Ke, M. Gu, Effect of the sample condition on the enhanced evanescent wave used for laser-trapping near-field microscopy, *Optik 109* (1998) 104–108.

[47] P.J. Reece, V. Garces-Chavez, K. Dholakia, Near-field optical micromanipulation with cavity enhanced evanescent waves, *Appl. Phys. Lett. 88* (2006) 221116.

[48] M.L. Povinelli, M. Loncar, E.J. Smythe, M. Ibanescu, S.G. Johnson, F. Capasso, J.D. Joannopoulos, Enhancement mechanisms for optical forces in integrated optics, in: K. Dholakia, G.C. Spalding (Eds.), Optical Trapping and Optical Micromanipulation, vol. 6326, 2006, pp. U71–U78.

[49] T.J. Kippenberg, H. Rokhsari, T. Carmon, A. Scherer, K.J. Vahala, Analysis of radiation-pressure induced mechanical oscillation of an optical microcavity, *Phys. Rev. Lett. 95* (2005) 033901.

[50] M.L. Povinelli, S.G. Johnson, M. Loncar, M. Ibanescu, E.J. Smythe, F. Capasso, J.D. Joannopoulos, High-Q enhancement of attractive and repulsive optical forces between coupled whispering-gallery-mode resonators, *Opt. Exp. 13* (2005) 8286–8295.

[51] M.P. MacDonald, G.C. Spalding, K. Dholakia, Microfluidic sorting in an optical lattice, *Nature 426* (2003) 421–424.

[52] T. Cizmar, M. Siler, M. Sery, P. Zemanek, V. Garces-Chavez, K. Dholakia, Optical sorting and detection of submicrometer objects in a motional standing wave, *Phys. Rev. B 74* (2006) 035105.

[53] S.A. Tatarkova, W. Sibbett, K. Dholakia, Brownian particle in an optical potential of the washboard type, *Phys. Rev. Lett. 91* (2003) 038101.

[54] M. Gu, J.B. Haumonte, Y. Micheau, J.W.M. Chon, X.S. Gan, Laser trapping and manipulation under focused evanescent wave illumination, *Appl. Phys. Lett. 84* (2004) 4236–4238.

[55] M. Gu, S. Kuriakose, X.S. Gan, A single beam near-field laser trap for optical stretching, folding and rotation of erythrocytes, *Opt. Exp. 15* (2007) 1369–1375.

[56] J.P. Hole, J.S. Wilkinson, K. Grujic, O.G. Helleso, Velocity distribution of Gold nanoparticles trapped on an optical waveguide, *Opt. Exp. 13* (2005) 3896–3901.

[57] K. Grujic, O.G. Helleso, J.P. Hole, J.S. Wilkinson, Sorting of polystyrene microspheres using a Y-branched optical waveguide, *Opt. Exp. 13* (2005) 1–7.

[58] K. Grujic, O.G. Helleso, Dielectric microsphere manipulation and chain assembly by counter-propagating waves in a channel waveguide, *Opt. Exp. 15* (2007) 6470–6477.

[59] C.D. Mellor, T.A. Fennerty, C.D. Bain, Polarization effects in optically bound particle arrays, *Opt. Exp. 14* (2006) 10079–10088.

[60] M.M. Burns, J.M. Fournier, J.A. Golovchenko, Optical matter—crystallization and binding in intense optical-fields, *Science 249* (1990) 749–754.

[61] P.J. Reece, E.M. Wright, K. Dholakia, Experimental observation of modulation instability and optical spatial soliton arrays in soft condensed matter, *Phys. Rev. Lett. 98* (2007) 203902.

[62] C. Conti, G. Ruocco, S. Trillo, Optical spatial solitons in soft matter, *Phys. Rev. Lett. 95* (2005) 183902.

[63] A. Rahmani, P.C. Chaumet, Optical trapping near a photonic crystal, *Opt. Exp. 14* (2006) 6353–6358.

[64] M.L. Povinelli, M. Loncar, M. Ibanescu, E.J. Smythe, S.G. Johnson, F. Capasso, J.D. Joannopoulos, Evanescent-wave bonding between optical waveguides, *Opt. Lett. 30* (2005) 3042–3044.

[65] I. De Vlaminck, J. Roels, D. Taillaert, D. Van Thourhout, R. Baets, L. Lagae, G. Borghs, Detection of nanomechanical motion by evanescent light wave coupling, *Appl. Phys. Lett. 90* (2007) 233116.

Chapter 6

Holographic Optical Tweezers

Gabriel C. Spalding[1], *Johannes Courtial*[2], *and Roberto Di Leonardo*[3]

[1]*Illinois Wesleyan University, Bloomington, IL, USA*
[2]*University of Glasgow, United Kingdom*
[3]*Università di Roma, Italy*

6.1 BACKGROUND

All around the world, hobbyists and collectors take great pride in the workmanship put into miniaturizations, producing small train sets with clear windows and seats in the cars, and tiny buildings with working doors and detailed interiors. Clearly though, as the size is further reduced, model building becomes more and more a painstaking task, and we are all the more in awe of what has been produced. Ultimately, as we move toward the microscopic scale, we find that for the creation and assembly of machines—or even simple static structures—we can no longer use the same approaches that we would naturally use on the macroscopic scale. *Entirely new techniques are needed.*

Luckily, Newton's second law dictates that as we reduce the mass of an object (as we shrink the size of our components), then a correspondingly weaker force will suffice for a given desired acceleration. So, while we would not think to use *light* to assemble the engine of a train, it turns out that optical forces are not only sufficient for configuring basic micromachines, but that the optical fields required can be arrayed in ways that offer the possibility of complex system integration (with much greater ease than, say, magnetic fields). Indeed, within clear constraints, optics provides a natural interface with the microscopic world, allowing for imaging, interrogation, and control.

There is no doubt that these are powerful technologies. Optical forces have convincingly been calibrated [1] down to 25 fN, and recent efforts have similarly extended

STRUCTURED LIGHT AND ITS APPLICATIONS

the range of optical torque calibration [2]. Careful studies of the optical torques exerted upon microscale components [3–6] provided a much clearer understanding of wave-based spin and orbital angular momentum (both optical [7,8] and, by extension and analogy, the quantum mechanical angular momentum of electrons in atoms). This work clearly deserves a place in the standard canon of the physics curriculum. As for the science that has resulted from the use of optical *forces*, consider the experiments done by Steven Block's group at Stanford. Their apparatus (using 1064 nm light) now has a resolution of order, the Bohr radius (!!) and they have used this exquisitely developed technique to directly observe the details of error correction in RNA transcription of DNA [9]—a true *tour de force*. Another prime example might be the results from the Carlos Bustamante group at Berkeley, whose experimental confirmations [10,11] of new *fluctuation theorems* have allowed recovery of RNA folding free energies, a key feat, given that biomolecular transitions of this sort occur under nonequilibrium conditions and involve significant hysteresis effects that had previously been taken to preclude any possibility of extracting such equilibrium information from experimental data. In fact, the work on fluctuation theorems now being done (both theoretically and experimentally) is among the most significant work in statistical mechanics done in the past two decades. These theorems have great general importance and include extension of the Second Law of Thermodynamics into the realm of micromachines and biomolecules. Clearly such work opens vast new intellectual opportunities.

6.2 *EXAMPLE* RATIONALE FOR CONSTRUCTING EXTENDED ARRAYS OF TRAPS

Basic optical techniques used to manipulate one or two microscopic objects at a time have been available for some time, but the focus of this review is upon recently emerging technical means for creating *arrays* of optical traps. The statistical nature of the experiments described in the previous section—which involve samples that are in the diffusive limit, where Brownian motion is significant—suggests that there might be a significant benefit to simultaneously conducting an array of optical trap experiments by creating independent sets of traps across the experimental field of view.

At the same time, the very issue of whether or not such an array of experiments may be treated *independently* points also toward an altogether different class of studies aimed at either probing or exploiting the wide array of physical mechanisms that might serve to couple spatially separated components. For example, biological studies of cell–cell signaling change character when, instead of dealing with a pair of cells isolated from all others, one deals with an ensemble, as this changes the conditions required for quorum sensing. In such cases, the use of an array of optical traps can ensure well-controlled, reproducible ensembles for systematic studies. As a specific instance, the early stages of biofilm formation are studied using various mutant strains of bacteria, with different biofilm-related genes deleted [12]; here, optical trap arrays are used to ensure geometric consistency from ensemble to ensemble. In fact, quite a wide variety of many-body problems are amenable to study using trap arrays [13–19].

It should be added that optical trap arrays determine not only the equilibrium structures that assemble, but also the dynamics of particles passing through the optical lattices [20–23]: the flow of those particles that are influenced *most* by optical forces tends to be channelled along crystallographic directions in a periodic trap array (often referred to as an *optical lattice*). In such a lattice, the magnitude of the optical forces is an oscillatory function of particle size [24], which means that it is possible to tune the lattice constant so as to make any given particle type either strongly interacting with the light or essentially noninteracting [25]. Therefore, it is possible to perform all-optical sorting of biological/colloidal suspensions and emulsions [24], and strong claims have been made as to the size selectivity associated with this type of separation technology [26]. In any case, this approach does allow massively parallel processing of the particles to be sorted, and so the throughput can be much higher than with *active* microfluidic sorting technologies, as shown by Applegate and colleagues [27], where particles are analyzed one by one and then a deflection control decision is based upon feedback from that analysis. Moreover, while the laser power delivered to the optical lattice may be, in total, significant, the intensity integrated over each biological cell can be a fraction of what is used in conventional optical tweezers, so all-optical sorting does relatively little to stress the extracted cells [28]. Many opportunities remain for studies of colloidal traffic through both static and dynamic optical lattices [29,30], and there is a real potential for the use of optical trap arrays in microfluidic lab-on-a-chip technologies.

While trap arrays have been put to a number of good uses in microfluidics (e.g., for multipoint microscale velocimetry [31]), it bears repeating that optical forces are, in the end, still relatively weak, and so for some purposes it would be reasonable to combine the use of optical forces (to allow sophisticated, integrated assembly of components) with the use of other forces for actuation. That said, the use of optical forces does allow for construction of simple micromachines: components may be separated or brought together, even allowing ship-in-a-bottle-type assembly (e.g., of axel-less microcogs in a microfluidic chamber [32]), oriented (e.g., for lock-and-key assembly of parts), and even actuated.

Here, we have mentioned only a few of the many reasons one might be interested in constructing extended arrays of traps. Clearly, this is a field where a number of new applications are expected to emerge over the coming years. Our next task, then, is to describe in detail some of the techniques used for generation—and dynamic reconfiguration, in three dimensions—of optical trap arrays.

Before we begin, one must note that the holographic techniques we will describe are not limited to array formation. They also enhance the control one can exert over a *single* trap by the ability to shape optical fields in three dimensions, e.g., to create an optical bottle beam [33,34] or even arrays of bottle beams [35]. A bottle beam is akin to a bubble of light (a dark region of space completely enclosed by a skin of high-intensity fields) and is intended for trapping cold atom clouds ("The Atom Motel ... where atoms check in but they do not check out!").

6.3 EXPERIMENTAL DETAILS

6.3.1 The Standard Optical Train

The majority of experiments involving optical traps use the single-beam gradient trap geometry [36] (referred to as *optical tweezers*), so we will focus upon this particular sort of setup. The most important details of experimental realizations of optical tweezers are discussed elsewhere, for example by Neuman and Block [37], but for our purposes it is sufficient to note that in optical tweezers the laser is very tightly focused, so it produces strong gradients in the optical fields in the region surrounding the focal spot, and an associated dipole force on polarizable media in that region. This dipole force, which is usually called a *gradient force*, dominates over radiation pressure for the sorts of samples usually studied with this technique.

There are alternative trapping geometries that may be desired. For gold nanoparticles and for transmissive particles with an index of refraction *much* higher than that of the surrounding medium, radiation pressure is significant. A counterpropagating beam trap is strongly preferred over optical tweezers in these instances [38,39]. While a significant radiation pressure would tend to knock particles out of optical tweezers, radiation pressure actually plays a helpful role in both counterpropagating beam traps and in levitation traps [40], each of which is seeing a resurgence in the literature.

Figure 6.1 shows a schematic representation of a standard *holographic optical tweezers* (HOT) setup. Beginning at the laser, we first have a beam-expanding telescope, so that the beam fills the holographic element, which is shown here as a reflective programmable *spatial light modulator* (SLM) but may be replaced with any (reflective or transmissive) *diffractive optical element* (DOE). The size of the illuminating Gaussian beam is a compromise between power efficiency and resolution: On the one hand, the larger the beam, the better use it makes of the area of the SLM,

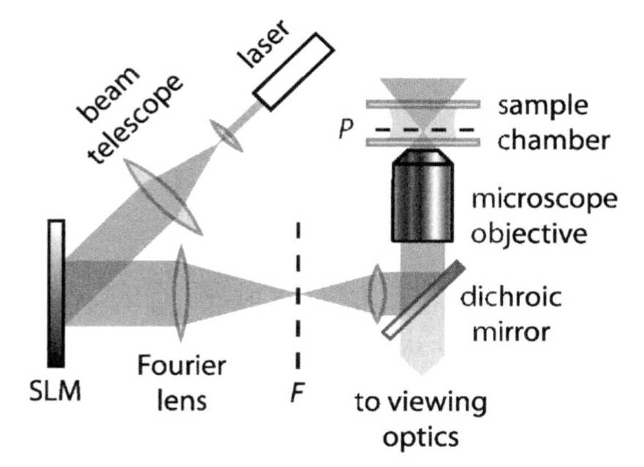

Figure 6.1 Simplified standard holographic optical tweezers (HOT) setup. The beam from a laser is widened in a beam telescope and illuminates a spatial light modulator (SLM). The first-order diffracted beam is collected by the Fourier lens; as the SLM is positioned in the front focal plane of the Fourier lens, the complex amplitude in the back focal plane, F, is the Fourier transform of the complex amplitude in the SLM plane. The remaining combination of lenses, usually including a microscope objective, images the beam in the Fourier plane into the central trapping plane, P, which is usually chosen to be in a liquid-filled sample chamber.

which in turn means higher resolution of the resulting light pattern in the trapping plane. On the other hand, if the Gaussian beam illuminating the SLM is *too* large, a fraction of the incident power misses the active area of the SLM and is therefore either lost or goes into the zeroth order. The standard compromise is a beam diameter that is roughly matched to that of the SLM. But it is not only for the sake of improved efficiency that the beam diameter is matched to that of the holographic element: when an SLM is used, any significant incident beam power *must* be distributed, so as to avoid boiling the liquid–crystal active element, permanently destroying the device. Following the SLM, there is a second telescope that ensures that the beam diameter is appropriate given the diameter of the back aperture of the lens to be used for tweezing (usually a standard microscope objective lens).

In practice, the two lenses comprising this second telescope are separated by the sum of their focal lengths; the SLM is separated from the first lens in that telescope, the Fourier lens, by the focal length of that lens, and the second lens in the telescope is separated by its own focal length from the back aperture of the objective. In this way, this second telescope actually serves four separate roles. First, it adjusts the beam diameter so as to fill the entrance pupil (i.e., the back aperture) of the objective lens. Second, as the placement of the SLM is equivalent to that of the steering mirror in conventional optical tweezers, the telescope projects the hologram onto the back aperture of the objective lens, thereby ensuring that any beam deflections created by the hologram do not cause the beam to walk off of the entrance pupil of the objective lens. Third, the fact that the hologram and the back aperture of the objective are conjugate image planes also creates a simple relationship between the beam at the output of the hologram and the beam in the focal plane, P: the beam's complex amplitude in the trapping plane is the Fourier transform of that in the SLM plane. Fourth, because the telescope used is Keplerian, rather than Galilean, the amount of real estate taken up on the optical table is somewhat greater, but the benefit that comes with this cost is that it allows for spatial filtering to be done in the plane labeled F in Figure 6.1. Plane F is conjugate to the focal plane P, so an enlarged image of the trapping pattern is there, where it may be manipulated [41]. (Commonly, a spot block is introduced at plane F to remove the undeflected, zeroth-order spot. An alternative is to add a blazing to the hologram, shifting the desired output pattern array away from the zeroth order, with all subsequent optics aligned along the path of the centroid of the first-order beamlets.)

In optical tweezers, the strong field gradient along the direction of propagation is produced by the peripheral rays in the tightly focused beam and not by the rays along the optic axis. For this reason, it is essential to use a final tweezing lens with a high numerical aperture (NA) and to ensure that the input beam delivers significant power to those peripheral rays. Therefore, a simple telescope is usually used to match the beam diameter to that of the entrance pupil (back aperture) of a high-NA microscope objective or, in the case of a Gaussian beam, where the intensity of peripheral rays is weak, to slightly overfill the back aperture of the objective. (Excessive overfilling simply throws away light that would otherwise go toward the desired mechanical effect, and also produces an undesired Airy pattern around each trap site created.)

It should be noted that the original work using holographic methods for optical trapping did not require that the DOE be projected onto the back focal plane of the tweezing lens. Fournier and colleagues [42,43] simply illuminated a binary phase hologram with a quasi-plane wave that, through Fresnel diffraction, generates self-images of the grating in planes that are periodically positioned along the direction of propagation (a phenomenon known as the *Talbot effect*). Already in 1995, scientist observed the very strong trapping of 3-micron spheres in these Talbot planes, and they discussed the means of creating a variety of lattices. In their 1998 paper, Fournier and colleagues also proposed the use of programmable "spatial light modulators to obtain a time-dependent optical potential that can be easily monitored." Fresnel holographic optical trapping is now seeing a revival for use in holographic optical trapping [44] for the generation of *large* periodic array structures. The first demonstration of sorting on an optical lattice [24] used a DOE that was conjugate to the image plane rather than the Fourier plane. Because of this conjugacy condition, the use of a fan-out DOE results in the convergence of multiple beams in the trapping volume, and the associated formation of an interference pattern. This setup can produce high-quality 3D arrays of traps over a large region of space, which can be tuned in ways that include the lattice constant, the lattice envelope, and the degree of connectivity between trap sites [45]. That said, compared to Fresnel holographic optical trapping, the Fourier-plane holography that we describe here offers a number of benefits, e.g., the ability to fully utilize such trade-offs as, for example, restricting the area in the trapping-plane over which the beam is shaped in order to gain resolution [46]. Moreover, generating a 3D array of traps intended to deflect the flow of particles passing through (i.e., optical sorting) is less of a challenge than filling a 3D array of traps so as to create a static structure: in the latter case, particles filling traps in each layer perturb the optical fields in all subsequent layers. Direct manipulation of the beam's Fourier-space properties can allow the creation of optical fields in the trapping volume that are self-healing (or self-reconstructing) in a number of arbitrarily chosen directions [47], a significant feature for the generation of some types of 3D filled arrays. Primarily, though, the benefits of Fourier-plane HOTs are the ability to generate generalized arrays without any special requirements regarding symmetry, and the ability to provide flexible, highly precise individual trap positioning [48].

Also in 1995, the group of Heckenberg and Rubinsztein-Dunlop at the University of Queensland produced phase-modulating holograms for mode conversion of conventional optical tweezers into traps capable of transmitting orbital angular momentum [3]. Because the Laguerre–Gaussian modes utilized are structurally stable solutions of the Helmholtz equation, the DOE/SLM need not be positioned in any particular plane. Therefore, this work in Australia, which is compatible with Fourier-plane holographic trapping, established (along with work in Scotland [5]) the techniques that are still in use today for the generation of traps carrying optical angular momentum.

The basic optical train shown in Figure 6.1, projecting the hologram onto the back focal plane of a microscope objective, was first demonstrated (using a prefabricated, commercial DOE) in 1998 [49], with a description of the most common algorithm for creating tailored holograms for optical trapping following in 2001 [50] and again in 2002 [51]. In this work, and in much of what has followed, it *is* assumed that the phase-modulating hologram will be positioned as we have described, so that the complex amplitude in the DOE/SLM plane and the complex amplitude in the central trapping plane form a Fourier-transform pair. With this relationship established, the complex amplitude distribution in one plane can be calculated very efficiently from the other one using a Fast Fourier Transform [52].

Given this simple relationship, it may seem surprising that there is any need to discuss algorithms at all: given a desired intensity distribution in the focal plane of the microscope objective, one need only take the inverse Fourier transform to determine the appropriate hologram. However, the result of that simple operation would be a hologram that modulates *both the phase and amplitude* of the input beam. The required amplitude modulation would remove power from the beam, with catastrophic consequences for efficiency in many cases. If, then, for the sake of efficiency, we constrain the hologram to *phase-only* modulation, there is usually no analytical solution that will yield the desired intensity distribution in the trapping plane.

Some of the required phase modulations are obvious. The simplest DOE would be a blazed diffraction grating, which is the DOE equivalent of a prism, and the Fresnel lens, which is the DOE equivalent of a lens. With these basic elements, we can move the optical trap sideways (with the blazing), but also in and out of the focal plane (with the Fresnel lens), and by superposition of such gratings and lenses we can create *multiple* foci, which can be moved and, through the addition of further phase modulation, can even be *shaped* individually [51].

Clearly we are not limited to a superposition of gratings and lenses. Because a standard HOT setup has holographic control over the field in the trapping plane, it can shape not only the intensity of the light field but also its phase. For example, there are very simple holograms that turn individual traps into optical vortices—that is, bright rings with a phase gradient around the ring. Because light will *repel* particles with a lower index of refraction than the surrounding medium, holographically produced arrays of vortex beams can be used to trap and create ordered arrays of low-index particles [53]. Other simple holograms can create nondiffracting and self-healing (over a finite range) Bessel beams [54,55] (though holographic generation of Bessel beams is *not* compatible with Fourier-plane placement of the DOE). Still, in all of the cases described previously, the phase profile required for the DOE is something you could guess, with good results.

However, while it is possible to intuit a phase modulation that will form extended periodic structures [56], matters become more complex when considering arbitrary arrangements of particles. As shown in the next section, the simplest guesses do not always yield the best results. Luckily, as we will describe in detail, the quality of the trap array can be greatly improved through the selection of an optimized iterative algorithm, though what might be optimal within the context of a particular experiment may involve trade-offs in terms of efficiency and computational speed.

6.3.1.1 The diffractive optical element or spatial light modulator

In this section, we will discuss the physical specifications that are required of the diffractive optical element to be used for the generation of optical traps. This DOE can either be static (etched in glass or, inexpensively, stamped into plastic), but is often in the reconfigurable form of a phase-only spatial light modulator (SLM)—a phase hologram under real-time computer control.

Dufresne and colleagues discuss the issues involved in lithographically manufactured DOE, such as the influence of the number of phase levels created and of surface roughness upon the resulting trap array [50]. One can gain a sense of these issues noting that while the light intensity directed to each spot in a 10×10 array was predicted to vary by $\pm 10\%$ for a binary DOE (given the specific algorithm used to calculate them), in an actual experiment it varied by $\pm 23\%$, with the additional variation attributed to manufacturing defects associated with the process of etching the phase modulation profile into glass [57]. As more phase levels are added, the lithographic challenges increase (though, with careful alignment, 2^N phase levels can be created by N etch steps). The greater the extent of the array, the more sensitive the uniformity of trap intensity becomes to fabrication errors.

Although holograms etched into glass clearly produce static images, one can rapidly raster the laser between a tiled array of such holograms, thereby producing a dynamic hologram with a refresh rate limited only by the speed of rastering. However, because of the heightened opportunity for real-time reconfiguration and optimization, we will emphasize here the special case where the DOE is a programmable SLM.

There are various types of SLMs [58]: they can, for example, be divided according to which aspect of light they modulate, which mechanism they use to do this, or whether they work in reflection or in transmission. Most HOTs use phase-only, liquid–crystal (LC), reflective SLMs: phase-only because of the efficiency advantage mentioned in the previous section; liquid crystal because of the cheap availability and maturity of this technology; and reflective because of a speed (and efficiency) advantage (the switching time depends upon the thickness of the liquid–crystal layer, which can be halved for reflective devices).

The LC character of most SLMs used in HOTs determines many of their properties. We will discuss here a few of these; more detailed discussions can be found, for example, in [59]. They work by applying locally defined voltages to an array of areas spread across an LC layer. The liquid–crystal molecules realign themselves in response to the voltage, thereby changing their optical properties. In parallel-aligned nematic liquid–crystals, for example, they tilt relative to the substrate [59]. LC SLMs were first used as holograms after correction of thickness nonuniformities [60]. Nowadays a number of different models are commercially available, for example from Boulder Nonlinear Systems [61], Holoeye Photonics AG [62], and Hamamatsu [63].

Liquid–crystal-based SLMs use either nematic or ferroelectric liquid crystals. Nematic SLMs offer a large number of phase levels (typically 256, though the sigmoidal grayscale-to-phase-level function of some of these systems compresses the number of useful levels to a number well below the nominal rating), but are slow (in practice, an update rate for a Near-IR nematic SLM would typically be 20 Hz or less). Ferroelectric SLMs have only two phase levels (phase shift 0 and π), which limits the choice of algorithms for the calculation of the hologram patterns and the diffraction efficiency (by always creating a symmetric -1st order of the same brightness as the $+1$st order) [64]. On the other hand, with update rates of typically tens of kilohertz, ferroelectric SLMs are significantly faster. While nematic SLMs are the usual choice, ferroelectric SLMs [65,66] have also been used in HOTs [64,67].

The maximum phase delay an LC SLM can introduce increases with the thickness of its LC layer. Depending on the LC-layer thickness, an LC SLM has a maximum wavelength for which it can achieve a full 2π phase delay; this is often taken as the upper limit of an SLM's specified wavelength working range. On the other hand, the response time is proportional to the square of the LC layer's thickness [68]. Again, in reflective SLMs, light passes through the LC layer twice, halving the required thickness.

Due to lack of surface flatness, reflective SLMs usually aberrate the light beam more than transmissive SLMs. Like the spherical aberration introduced by the objective, this can deteriorate the quality of the traps and limit the trapping range [37]. Aberrations in the flatness of the SLM can (together with other aberrations in the optics) be corrected by displaying a suitable phase hologram on the SLM [69–72]. This is perhaps not surprising, as SLMs are also used in other adaptive-optics systems [73,69].

Commercial LC SLMs are either optically (Hamamatsu) or electrically (Boulder, Holoeye) addressed. Electronically addressed SLMs contain (square or rectangular) pixels, with a small gap—an associated dead area—between pixels. The convolution theorem dictates that the field in the SLM's Fourier plane (the trapping plane) is the Fourier transform of the field of coherent point light sources centered in each pixel (the phase and intensity is that of the pixel) multiplied by the Fourier transform of the field of a single centered pixel. The first term (the Fourier transform of the point light sources) is periodic in x and y. The copies of the central field are essentially diffraction orders of the grating formed by the point light sources. The second term (the Fourier transform of a single pixel) is usually close to a sinc function in x and y (whereby the sinc function gets wider when the pixel gets narrower, and vice versa), which means that the intensity of the field in the Fourier plane falls off to zero at the nodes of the sinc function. This limits the power that is wasted into higher diffraction orders. By optimizing the pixel shape, the power diffracted into higher grating orders can be further reduced. This is essentially what is happening in the optically addressed SLMs by Hamamatsu, where a "write light" pattern produced with the help of a pixelated liquid-crystal display is projected onto a photoconductive layer. The projected pattern is slightly out of focus, resulting in an image consisting of pixels with smoothed edges. The photoconductive layer controls the voltage across the LC layer, which modulates the phase of "readout light" passing through it. The resulting phase modulation is such that diffraction into the higher grating orders is almost completely suppressed.

It should again be noted that a light beam with too much power can destroy an LC SLM by boiling the liquid crystal. This limits the number of optical traps that has been achieved with LC-SLM-based Fourier-plane HOTs to $\approx$200 [74], a limit that can perhaps be overcome by using cooled SLMs (SLMs with a larger area over which the beam's power can be distributed) or other types of SLMs such as deformable mirrors [75].

6.4 ALGORITHMS FOR HOLOGRAPHIC OPTICAL TRAPS

Holographic optical tweezers are usually employed to individually trap and manipulate an ensemble of small objects in three-dimensional space. The desired light intensity distribution for such applications consists of a set of high-intensity, diffraction-limited spots on a dark background. Many pre-existing algorithmic strategies developed for computer-generated holograms have been adapted and optimized for this special task. Other approaches, such as genetic algorithms, have been applied to holographic lithography [76] and would complement the work described here. Some of the algorithms that we explicitly examine are valued primarily because they are fast and are therefore particularly suited to interactive use (e.g., in reference [77]). When a higher degree of control in trap intensities is required, one has to resort to calculating the holograms *prior* to the experiment, using the more iterative algorithms required for high-quality traps. This is the case when, for example, one requires a set of traps with *uniform* (or precisely controlled) potential well depths, or when ghost traps cannot be tolerated (for example when they prevent filling of the intended trapping sites). Following and extending a recent review of the literature contained in [78], we will review the algorithms that are presently available for optical trapping using the standard optical train described in the previous section. We will then discuss both why and how they work.

We assume a uniform plane-wave illumination of the pixellated SLM and call $u_j = |u| \exp(i\phi_j)$ the complex amplitude of the electric field reflecting off of the jth pixel, where ϕ_j is the corresponding phase shift. The total, time-averaged energy flux through the SLM is given by $W_0 = c\epsilon_0 N |u|^2 d^2/2$, where N is the total number of pixels and d^2 is the surface area of a pixel. We can use scalar diffraction theory to propagate the complex electric field from the jth pixel's surface to the location of the mth trap in the image space [79]. Summing up the contributions from all of the N

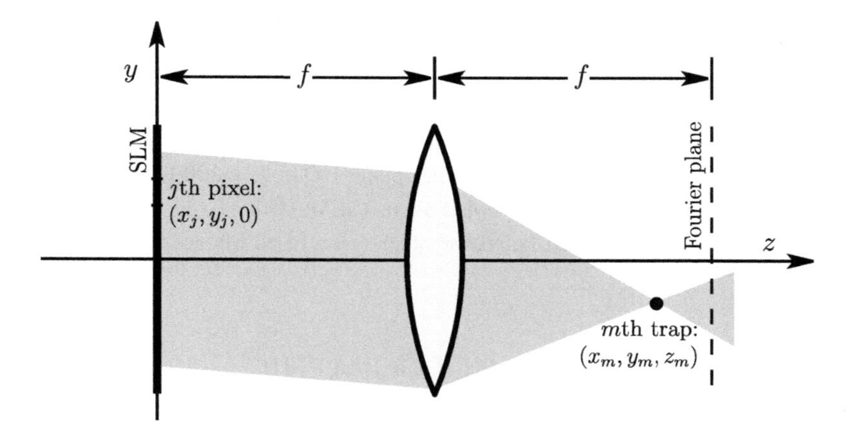

Figure 6.2 Geometry of pixel and trap positions relative to the (effective) Fourier lens's focal planes. The transverse position of the jth pixel in the SLM plane is (x_j, y_j); the position of the mth trap relative to the center of the Fourier plane is (x_m, y_m, z_m), x and y are the two transverse coordinates; z is the longitudinal coordinate.

pixels, we obtain the complex amplitude v_m of electric field at the position of trap m:

$$v_m = \frac{e^{i2\pi(2f+z_m)/\lambda}}{i} \frac{d^2}{\lambda f} \sum_{j=1,N} |u| e^{i(\phi_j - \Delta_j^m)}, \tag{1}$$

where

$$\Delta_j^m = \frac{\pi z_m}{\lambda f^2}\left(x_j^2 + y_j^2\right) + \frac{2\pi}{\lambda f}(x_j x_m + y_j y_m). \tag{2}$$

x_j, y_j are the jth SLM pixel's coordinates (projected into the back focal plane of the objective), and x_m, y_m, z_m are the mth trap's coordinates referred to the Fourier plane, as shown in Figure 6.2. We can easily generalize the Δ_j^m to add orbital angular momentum to trapping beams [51]. To make the notation more compact, we will introduce the dimensionless variable V_m:

$$V_m = \sum_{j=1,N} \frac{1}{N} e^{i(\phi_j - \Delta_j^m)}, \tag{3}$$

whose physical meaning can be understood by noting that $I_m = |V_m|^2$ measures the energy flux (in units of W_0) flowing through an area $f^2\lambda^2/(Nd)$ (the area of a diffraction-limited spot) centered at the mth trap site. For $z_m = 0$, V_m corresponds to the discrete Fourier transform of $e^{i\phi_j}$ evaluated at the spatial frequencies $(x_m/\lambda f, y_m/\lambda f)$.

Our task here is to search, given a set of Δ_j^m, for the best choice of ϕ_js to maximize the modulus of V_m on all traps. We will use as a benchmark the task of computing an $N = 768 \times 768$, 8-bit hologram for the generation, of $M = 100$ traps arranged on a 10×10 square lattice located in the Fourier ($z_m = 0$) plane. The performance of different strategies is quantified by three parameters: efficiency (e), uniformity (u), and percent standard deviation (σ)

$$e = \sum_m I_m, \qquad u = 1 - \frac{\max[I_m] - \min[I_m]}{\max[I_m] + \min[I_m]}, \qquad \sigma = \sqrt{\langle (I - \langle I \rangle)^2 \rangle} / \langle I \rangle. \quad (4)$$

In the above equation, $\langle \cdots \rangle$ denotes the average over all trap indices m.

In the trivial case of one trap, $M = 1$, this best choice is easily found by setting $\phi_j = \Delta_j^1$, which makes all terms in the sum (3) real and equal to $1/N$ and $|V_1|^2 = 1$. In the following section, we will refer to the array Δ_j^m as the single mth trap hologram.

6.4.1 Random Mask Encoding

For multiple traps, $M > 1$, we must seek a compromise between the M different choices $\phi_j = \Delta_j^m$ (one choice for each m value) that would divert all energy onto trap m. One of the fastest routes is the *random mask encoding technique* (RM) [80].

The compromise is obtained by setting:

$$\phi_j = \Delta_j^{m_j}, \quad (5)$$

where m_j is a number between 1 and M randomly chosen for each j. The technique is very fast, and performs remarkably well as far as uniformity is concerned. However, the overall efficiency can be very low when M is large. In fact, on average, for each m, only N/M pixels will interfere constructively, and all others will give a vanishing contribution. Therefore, $|V_m|^2 \simeq 1/M^2$ and $e \simeq 1/M$, which can be significantly smaller than one when M is large. In the present case, where $M = 100$, we numerically obtained $u = 0.58$ but $e = 0.01 = 1/M$. However, the RM algorithm is particularly useful to quickly generate either one or a few additional traps on top of a complex light structure obtained via a precalculated hologram. Such "helper tweezers" are useful for filling in the precalculated array of traps, allowing the user to interactively trap, drag, and drop initially free particles into the desired locations. For such purposes, a fraction of the SLM pixels can be randomly chosen and temporarily used to display the "service-trap" hologram.

6.4.2 Superposition Algorithms

In superposition algorithms, the phase on each pixel is chosen as the argument of the complex sum of separate, single-trap holograms. Though we are completely disregarding the amplitude of the complex sum, the resulting hologram produces usable trap arrays. In terms of the quantities V_m, a superposition hologram maximizes the sum of the projections of all of the V_m on a fixed axis in the complex plane.

To illustrate this, we will now try to maximize the real part of $\sum_m V_m$ with respect to ϕ_j. The stationary points are easily obtained by imposing the condition of a vanishing gradient:

$$\frac{\partial}{\partial \phi_j} \sum_m \mathrm{Re}\{V_m\} = \mathrm{Re}\left\{ \frac{i e^{i\phi_j}}{N} \sum_m e^{-i\Delta_j^m} \right\} = 0 \tag{6}$$

whose solutions are given by

$$\bar{\phi}_j = \arg\left[\sum_m e^{i\Delta_j^m} \right] + n_j \pi, \quad n_j = 0, 1. \tag{7}$$

For the stationary point to be a local maximum, the corresponding Hessian matrix has to be negative definite. In the present case, the Hessian is purely diagonal, and when evaluated on the stationary points reads

$$\left. \frac{\partial^2}{\partial \phi_j \partial \phi_k} \sum_m \mathrm{Re}\{V_m\} \right|_{\phi_j = \bar{\phi}_j} = -\delta_{jk} (-1)^{n_j} \left| \frac{1}{N} \sum_m e^{-i\Delta_j^m} \right|. \tag{8}$$

The maximum condition is obtained when all of the n_j are set to zero, and, therefore,

$$\phi_j = \arg\left[\sum_m e^{i\Delta_j^m} \right], \tag{9}$$

which can be read as the phase of the sum of single-trap holograms. We call this algorithm "superposition of prisms and lenses" (S) [81,82]. The S algorithm, though slower than RM (due to the extra N arg function evaluations), gives efficiencies of order 1, but gives very poor uniformities. In fact, in our benchmark case seven though the uniformity is only $u = 0.01$, the efficiency has risen to $e = 0.29$. Moreover, when highly symmetrical trap geometries are sought, as for the square lattice, a consistent part of the energy is diverted to unwanted ghost traps [83].

A better compromise can be obtained if we only try to maximize the sum of the amplitudes of V_m projected on randomly chosen directions in the complex plane. In other words, we can seek a maximum of $\sum_m \mathrm{Re}\{V_m \exp(-i\theta_m)\}$, where θ_m are random numbers uniformly distributed in $[0, 2\pi]$. In this case we obtain

$$\phi_j = \arg\left[\sum_m e^{i(\Delta_j^m + \theta_m)}\right], \tag{10}$$

which is the phase of the linear superposition of single-trap holograms with coefficients of unit modulus and random phase. This last choice, usually called *Random Superposition* (SR) [84], has the same computational cost as S, but results in a much better efficiency than S though trap intensities will still vary significantly (e.g., for our benchmark case, $e = 0.69$, $u = 0.01$).

We want to stress that, when dealing with low-symmetry geometries, SR holograms can also produce good uniformity levels, and no further refinement is needed. If precise trap positioning is not an issue, one can deliberately reduce the pattern symmetry by adding a small amount of random displacement to trap locations as demonstrated by Curtis and colleagues [83].

Though slower than RM, the computational speed of SR still allows for interactive manipulation. Since the entire hologram is continuously recalculated, SR is preferred to RM for interactive applications requiring dynamic deformation of the entire trapping pattern.

6.4.3 Gerchberg–Saxton Algorithms

The GS algorithm [84–89] was developed by the crystallographers Ralph Gerchberg and Owen Saxton to infer an electron beams phase distribution in a transverse plane, given the intensity distributions in two planes. It can also be applied to light, specifically to find a phase distribution that turns a given input intensity distribution arriving at the hologram plane (SLM) into a desired intensity distribution in the trapping plane. In GS, the complex amplitude is propagated back and forth between the two planes taking care at each step to replace the intensity on the trapping plane with the target intensity and that on the SLM plane with the laser's actual intensity profile.

The algorithm can be extended to 3D trap geometries where multiple planes are considered for forward propagation. The back-propagated field is then obtained as the complex sum of the corrected and back-propagated fields from the target planes. Generalization to *full* 3D shaping [90,91] is currently far too slow for interactive use (taking days to calculate the desired phase modulation).

In these implementations of GS, forward and back propagations were performed with FFT transforms. However when the target intensity is an array of point traps, it is rather pointless to calculate the field complex amplitude in points whose amplitude will be replaced by zero before back propagation. FFT has also the drawback of discretizing trap transverse coordinates in units of the Nyquist spatial frequency. A much faster and more versatile implementation of GS for HOT only computes the field at the trap locations.

In terms of the V_m, it can be shown that GS converges on a phase hologram that maximizes the sum of the amplitudes of V_m without projecting them onto any special directions in the complex plane (real axis for S or randomly chosen directions for SR). Again, differentiating with respect to ϕ_j, we obtain the stationary points

$$\frac{\partial}{\partial \phi_j} \sum_m |V_m| = \mathrm{Re}\left\{ \frac{i e^{i\phi_j}}{N} \sum_m e^{-i\Delta_j^m} \frac{V_m^*}{|V_m|} \right\} = 0. \tag{11}$$

$$\bar{\phi}_j = \arg\left[\sum_m e^{i\Delta_j^m} V_m / |V_m| \right] + n_j \pi, \quad n_j = 0, 1. \tag{12}$$

This time the Hessian computed on the stationary point is not purely diagonal

$$\frac{\partial^2}{\partial \phi_j \partial \phi_k} \sum_m |V_m| \bigg|_{\phi_j = \bar{\phi}_j} = -\delta_{jk} (-1)^{n_j} \left| \frac{1}{N} \sum_m e^{-i\Delta_j^m} \frac{V_m^*}{|V_m|} \right| + O\left(\frac{1}{N^2} \right). \tag{13}$$

However the nondiagonal terms are $1/N$ smaller than the diagonal ones. It can be shown that such a perturbation will only affect the sign of one eigenvalue at most [92]. When N is very large, we can neglect this eventuality and call the stationary point

$$\phi_j = \arg\left[\sum_m e^{i\Delta_j^m} \frac{V_m}{|V_m|} \right] \tag{14}$$

a maximum. In this case, ϕ_j are obtained as the phase of the linear superposition of single-trap holograms with coefficients of unit modulus and a phase given by the phase of V_m, that is, the field produced by the ϕ_j themselves on trap site m. It is now impossible to write the ϕ_j in an explicit form given the implicit dependence of V_m on ϕ_j. One possible approach to a solution is simply to start with a guess for ϕ_j (i.e., the one obtained from SR) and use equation (14) in an iterative procedure. We refer to this iterative procedure as Gerchberg–Saxton (GS) [93,94] and note that it converges after a few tens of iterations. In particular, after thirty iterations, we obtained $e = 0.94$ and $u = 0.60$ for our benchmark case.

The reason the algorithms discussed so far usually result in poor uniformities can be understood if one observes that these algorithms were *only* designed to maximize the sum of the amplitudes of V_m, having no bias toward uniformity. There are a number of optimization criteria one could use. For example, a bias toward uniformity is included if we instead seek a maximum in a quantity like $\prod_m |V_m|$ or equivalently

$\sum_m \log|V_m|$. By differentiating the biased function with respect to ϕ_j we obtain

$$\frac{\partial}{\partial\phi_j}\left[(1-\xi)\sum_m |V_m| + \xi\sum_m \log|V_m|\right]$$
$$= \mathrm{Re}\left\{\frac{ie^{i\phi_j}}{N}\sum_m e^{-i\Delta_j^m}\frac{V_m^*}{|V_m|}\left(1-\xi+\frac{\xi}{|V_m|}\right)\right\} = 0. \quad (15)$$

It is easy to show that in this case, the Hessian matrix is once again diagonal (in the limit of large N) and negative definite at the stationary point

$$\phi_j = \arg\left[\sum_m e^{i\Delta_j^m}\frac{V_m}{|V_m|}\left(1-\xi+\frac{\xi}{|V_m|}\right)\right]. \quad (16)$$

If we seek the solution of equation (16) by an iterative procedure, we obtain the Generalized Adaptive Additive algorithm (GAA) [50,51]. With the choice $\xi = 0.5$, GAA produces a uniformity improvement $u = 0.79$ with the same efficiency $e = 0.93$ as GS.

We may wonder if an improvement in uniformity can be obtained by aiming at a slightly modified target intensity distribution. To do this, we can introduce the M extra degrees of freedom w_m that maximize the weighted sum $\sum_m w_m|V_m|$, with the constraint that $|V_m|$ are all equal. By differentiating with respect to ϕ_j, we obtain the maximum condition

$$\phi_j = \arg\left[\sum_m e^{i\Delta_j^m}\frac{w_m V_m}{|V_m|}\right]. \quad (17)$$

Again, the above formula expresses ϕ_j in an implicit form, this time containing also the unknown weights w_m. Starting from a SR guess for ϕ_j and setting $w_m = 1$, the iteration proceeds as follows

$$\text{0th step:}\quad w_m^0 = 1, \qquad \phi_j^0 = \phi_j^{\mathrm{SR}},$$
$$k\text{th step:}\quad w_m^k = w_m^{k-1}\frac{\langle|V_m^{k-1}|\rangle}{|V_m^{k-1}|}, \qquad \phi_j^k = \arg\left[\sum_m e^{i\Delta_j^m}\frac{w_m^k V_m^{k-1}}{|V_m^{k-1}|}\right].$$

In other words, at each step we adjust the weight w_m in such a way to reduce $|V_m|$ deviations from the average $\langle|V|\rangle$. The above procedure converges, with a speed typical of GS and GAA, to a hologram having the almost optimal performance of $e = 0.93$, $u = 0.99$. This is referred to as the Weighted Gerchberg–Saxton algorithm or GSW [78].

6.4.4 Direct-Search Algorithm and Simulated Annealing

One can also achieve hologram refinement through a direct search for a maximum in a "gain function" defined over the space of phase levels. There is some freedom in how the gain function is defined, but for simplicity we might choose a linear combination of our efficiency and uniformity metrics, such as

$$e/M - f\sigma. \tag{18}$$

Starting from a good initial guess for the hologram, such as what one obtains from SR, we pick 1 pixel at random and cycle through all the $P = 256$ gray levels while looking for an improvement (increase) in the gain function. The Direct Search algorithm (DS) involves continuation of this process [7,95,96]. As suggested in [96], when starting from an SR hologram and setting $f = 0.5$, the algorithm achieves a perfect uniformity ($u = 1.00$) after $1.3N$ steps, with the computational cost scaling as $M \times P$, though the overall efficiency is diminished to $e = 0.68$. Better holograms can be obtained by giving more bias to efficiency ($f = 0.25$) and waiting for a substantially longer time ($\sim 10N$ steps—that is, about a hundred times longer than GS). However, we have observed that reducing the number of gray-levels P to just 8 can significatively reduce the computational cost (by a factor of 32) without affecting performance too much (see [95] for a systematic exploration of parameter space). With eight "grayscale" phase levels and all other parameters set as before, we obtained $e = 0.84$ and $u = 1.00$ after $7N$ steps (that is, about three times longer than GS). At this point, the whole hologram has been reduced to 3-bit depth, and a comparison with other algorithms working at full, 8-bit depth is not necessarily a fair comparison. For a better exploration of the gain function landscape, we could define more complex acceptance rules, allowing moves that temporarily decrease the gain (as in a Metropolis algorithm). This sort of "simulated annealing" of the gain function was investigated for 2D light distributions by Yoshikawa and Yatagai [97]. Though applications to 3D trap arrays could lead to better holograms than simple DS (which might be referred to as a "quenched" simulated annealing), longer computational times would be needed.

6.4.5 Summary

Table 6.1 summarizes the results of the performed benchmark test on a 10×10 square grid. The conclusion is that when aiming at such a symmetric pattern, GSW gives the best performance in terms of both quality and computational time. However, it is worth reiterating that when working with lower-symmetry patterns, superposition algorithms can provide reasonably good traps in a much reduced time [83]. When

Table 6.1

Summary of *theoretical* performances of the investigated algorithms

Algorithm	Detail	e	u	σ (%)	K	Scaling
RM		0.01	0.58	16	–	N
S		0.29	0.01	257	–	$N \times M$
SR		0.69	0.01	89	–	$N \times M$
GS		0.94	0.60	17	30	$K \times M \times N$
GAA		0.93	0.79	9	30	$K \times N \times M$
DS		0.68	1.00	0	7.5×10^5	$K \times P \times M$
GSW		0.93	0.99	1	30	$K \times N \times M$

The target trap structure is a 10×10 square grid. Column 2 contains a 100×100 detail of the total 768×768 hologram. Performance parameters after K (column 6) iterations are reported in columns 3, 4, and 5. Computational cost scaling is reported in column 7, where: M = number of traps, N = number of pixels in hologram, K = number of iterations, and P = number of gray levels (256 here).

speed is an important issue, the time needed to compute an SR hologram, or equivalently to perform a single step in GS-based algorithms, can be efficiently reduced by relying on the Graphic Processing Unit (GPU) of a graphic board for both computational and rendering tasks [98]. Being concerned with algorithms for calculating fields only on the trap sites, the overall performance ratings in Table 6.1 are basically unchanged when moving to 3D geometries. As an example, Table 6.2 shows the results of a similar test performed on a three-dimensional target consisting of 18 traps arranged in the conventional unit cell of the diamond lattice.

6.4.6 Alternative Means of Creating Extended Optical Potential Energy Landscapes

Large arrays of optical traps can be generated in several ways. While holographic optical tweezers are very flexible, there are situations in which alternative approaches might be considered.

Table 6.2

Predicted algorithm performances on a 3D target

Algorithm	e	u	σ (%)	K
RM	0.07	0.79	13	–
S	0.69	0.52	40	–
SR	0.72	0.57	28	–
GS	0.92	0.75	14	30
GAA	0.92	0.88	6	30
DS	0.67	1.00	0	1.7×10^5
GSW	0.93	0.99	1	30

There are 18 traps arranged on the sites of the conventional unit cell of the diamond lattice. Performance parameters after K (column 5) iterations are reported in columns 2, 3, and 4.

For example, multibeam interference is simple, produces high-quality optical lattices over extended 3D volumes, and can tolerate high beam powers. Extensive work on laser-induced freezing and other novel phase transitions has utilized this approach [13,19]. Burns and colleagues used the standing optical field resulting from the interference of several beams to trap polystyrene spheres, thereby producing a 2D colloidal crystal, and to propose the existence of optically mediated particle–particle interactions in this system (optical binding) [99,100]. However, such approaches are limited to symmetric patterns.

Alternatively, galvan mirrors [101] or piezoelectrics [37,102] have served as the basis for designs involving scanning laser tweezers. These approaches are briefly summarized in the next section. Another (SLM-based) strategy that allows flexible generation of trap arrays uses the Generalized Phase Contrast (GPC) method. In addition, arrays covering large areas have now been produced using evanescent waves. In the remainder of the section, we will discuss these promising alternative techniques.

6.4.6.1 Time-sharing of traps

For generating a *simple*, smooth potential, such as a ring trap, the use of analog galvo-driven mirrors might be preferred over either HOTs or acousto-optic deflectors (AODs). Coupled with an electro-optical modulator, this can yield smoothly varying intensity modulations in a continuous optical potential [103,104]. Galvo systems provide much higher throughput of the incident light than either AODs or Fourier-plane HOTs, and have been put to good use in many experiments (e.g., the Bechinger group uses galvo-driven mirrors to create simple optical "corrals" that controllably adjust the density of 2D colloidal ensembles). However, inertia limits the scan speed of any macroscopic mirror to a fraction of what is available via AODs.

Acousto-optic deflectors *are* simply another class of reconfigurable diffractive optic element—a class that is limited to simple blazings but has a much higher refresh rate: AODs can be scanned at *hundreds* of kilohertz, repositioning the laser on such a short time scale that, under some circumstances, the trapped particles experience only a time-averaged potential. This short time constant allows for the creation of multiple "time-shared" traps using AOD deflection of the same (first-order) diffraction spot [105,106].

In order for each trapped particle to feel only the time-averaged potential, the maximal time that the laser can spend away from any one trap would be something like a tenth of the time scale set by the corner frequency in the power spectrum of particle displacements. The inverse of this corner frequency indicates, essentially, the time it takes the particle to diffuse across the trap. The smaller the particle, the shorter this time scale becomes. Less viscous environments also present challenging time scales: for aerosols, the corner frequency can be 2 kHz even for an 8-micron sphere [107]. Because of these high corner frequencies, it could be a challenge to use AODs to generate large arrays in such samples and, indeed, HOT-based array generation has been preferred for such work [108]. It should be emphasized that, for *any* type of sample, "while the laser's away, the beads will stray!" That is, over time interval t when the laser is elsewhere (at other trap sites or traveling between traps), a microbead will diffuse away from its nominal trap site a distance $d = \sqrt{2Dt}$. For a 1-micron-diameter bead in water, the diffusion coefficient is $D = k_B T / 6\pi \eta r = 0.4\ \mu\text{m}^2/\text{s}$. So if the laser is absent for 25 *micro*seconds, the bead is expected to diffuse 5 nm. Clearly, smaller spheres would diffuse further during the same time, and (even for micron-scale spheres) as the number of trapping sites increases, the demand for speed from all components of the control system (both hardware and software) becomes significant. This (coupled with the requirement that the laser spend sufficient time at each trap site to produce the time-averaged power required for the desired trap strength, and the fact that while trap strength depends only on the time-average power, sample damage due to two-photon absorption and local heating contain a dependence upon the peak power [109]) represents a maximum limit to both the type of arrays that can be constructed and the accuracy to which the spheres can be positioned when using time-shared trapping.

That said, impressive results have been obtained. A 20×20, two-dimensional array of traps was constructed and (mostly) filled with 1.4 μm silica spheres by the van Blaaderen group [110]. Moreover, by physically splitting the beam and creating two optical paths with offset image planes, the group was able to make small arrays of traps in two separate planes. Because nearly all of the colloidal particles in their sample were index-matched to the surrounding medium, they were able to tweeze a dilute concentration of nonindex-matched tracer particles so as to controllably seed nucleation of 3D order in a concentrated colloidal sample [110], and then use fluorescent confocal microscopy to image the results. Primarily, though, work utilizing AODs has been limited to the generation of 2D arrays of traps.

For experiments involving just one or two spheres, the positioning resolution of AODs driven by low-jitter digital frequency synthesizers cannot be beaten. It is not unheard of for such systems to claim a positioning resolution of less than one-tenth of a nanometer, though, at this level, the accuracy of positioning is not only affected by diffusion of the particle in the optical trap, it is also affected by the pointing stability of the trapping laser and by hysteresis and drift of the sample stage and of the objective lens. There are a number of technical points one must be aware of. For analog AOD systems, there can remain problems with "ghost traps" (as the beam is often sequentially repositioned in x and then in y, so the generation of two traps along the diagonal of a square yields an unintended spot at one of the other corners). Moreover, because the AOD efficiency falls off as a function of the deflection angle, for applications requiring uniform arrays, one must compensate, either by spending more time at peripheral traps or by increasing the power sent to those traps. If such steps are taken, AODs offer good trap uniformity, precise positioning in 2D, and fast updates. Also, in some sense, AOD-generated arrays can be thought of as being made of incoherent light (different beams do not interfere, being present only one at a time).

More details, including references, can be found in [37]. For generating *arrays* of traps using AODs, pages 2964–2965 of [110] clearly and concisely describe many of the relevant parameters that must be considered.

Unlike the SLM-based techniques, AOD-based systems cannot normally do mode conversion or aberration correction, and they cannot generate arrays of traps dispersed in 3D.

6.4.6.2 Generalized phase-contrast method

Frits Zernike won the 1953 Nobel Prize for developing an imaging technique that could turn a phase modulation caused by a transparent sample (e.g., a biological cell) into an intensity modulation at the plane of a camera or the eye. While Zernike's approach was valid for *weak* phase modulation, workers at Risø National Laboratory in Denmark have created a Generalized Phase-Contrast technique (GPC), which provides a simple method for using the SLM to produce arbitrary user-defined arrays of traps in 2D [111], and can also be made to work in 3D [112,113].

In the GPC approach, the SLM is conjugate to the trapping plane, and no computations are required in order to convert the phase-only modulation of the SLM into an intensity modulation in the image plane; instead there is a direct, one-to-one correspondence between the phase pattern displayed on the SLM and the intensity pattern created in the trapping plane. In the Fourier plane, a small π-phase filter shifts focused light coming from the SLM so that at the image plane it will interfere with a plane-wave component. The result is a system that only requires the user to write the desired 2D patterns on the SLM. The downside to this is that xy positions are limited to pixel positions, meaning that ultra-high-precision trap positioning is not possible to the degree it is with the other techniques we have described. Also, beam-shaping Fourier-holography tricks are not applicable to the GPC approach.

Extension of the GPC method to 3D requires the use of counterpropagating beam traps, rather than optical tweezers, and so three-dimensional control is, in some sense, more involved. For this reason, the Risø team has developed an automated alignment protocol [114] for users interested in 3D control. It is not possible to controllably place traps behind each other with this method, but there are now many impressive demonstrations of 3D manipulation using the GPC technique.

Notably, in this "imaging mode," the SLM efficiency is much higher, providing a throughput of up to 90% of the light, as there is no speckle noise and no diffraction losses (i.e., there are no ghost orders; there is no zero-order beam) [115]. The version of GPC using counterpropagating beam traps can also use low-NA optics, which can have a large field of view, and a large Rayleigh range. So, while Fourier-plane holographic optical traps can provide only a small range of axial displacements, limited by spherical aberration, the low-NA GPC trap arrays are sometimes called *optical elevators* because of the large range over which the traps can be displaced along the optic axis. Moreover, the working distance can be up to 1 cm, which is 100 times that of a conventional optical tweezers setup. This long working distance even makes it possible to image the trapped structure from the side [116]. Because no immersion fluid is required for low-NA imaging optics, one could imagine performing experiments using this approach in extreme environments, such as vacuum or zero gravity. That said, the use of a high-NA, oil-immersion trapping lens is necessary in order to provide high trap stiffness along the axial direction.

6.4.6.3 Evanescent-wave optical trap arrays

Evanescent fields hold promise for future generation of trap arrays, primarily for two reasons. One is not subject to the free-space diffraction limit and can therefore create significantly subwavelength structures in the optical fields. Also, it has been shown that patterned evanescent fields can create large numbers of traps spanning macroscopically large areas [117]. Interestingly, in an *unpatterned*, but resonantly enhanced, evanescent field, arrays of trapped particles have recently been observed to self-organize, due to the onset of nonlinear optical phenomena (optical solitons) [118]. Several schemes have been proposed for holographic control of evanescent fields, and a tailored algorithm for doing this sort of light shaping has recently appeared [119].

The disadvantages of evanescent fields are that one *must* obviously work very close to the surface, create patterns using only the range of incident angles beyond the critical angle for total internal reflection, and allow for the strong variation in penetration depth as a function of incident angle (which, on the other hand, allows 3D shaping of the evanescent field [119]). Taken together, these necessarily constrain the shapes that can be holographically achieved in the near field, and clearly require the development of specialized algorithms.

6.5 THE FUTURE OF HOLOGRAPHIC OPTICAL TWEEZERS

Good, fast algorithms currently exist for Fourier-plane holographic optical tweezers consisting of arbitrary 3D trap configurations. Already, today, HOTs can be bought commercially [74]. A number of researchers have begun to use HOTs as the centerpiece of integrated biophotonic workstations [120–122]. With the capability of doing holographic work established, it is reasonable to combine HOTs with digital holographic microscopy [123–125]. Moreover, as the traps are already under computer control, it is relatively easy to combine HOTs with pattern recognition to automate particle capture and sorting [126], and to trigger key events, localized in space and time, in systems near instabilities [127]. Related techniques are also undergoing significant development. So, while a great deal of science has been accomplished with one or two pointlike traps, there clearly remains enormous potential associated with extended arrays of optical traps.

ACKNOWLEDGMENTS

G.C.S. was supported by an award from the Research Corporation and by the Donors of the Petroleum Research Fund of the American Chemical Society. J.C. acknowledges the support of the Royal Society.

REFERENCES

[1] A. Rohrbach, Switching and measuring a force of 25 femtonewtons with an optical trap, *Opt. Exp. 13* (2005) 9695–9701.

[2] G. Volpe, D. Petrov, Torque detection using Brownian fluctuations, *Phys. Rev. Lett. 97* (2006) 210603.

[3] H. He, M.E.J. Friese, N.R. Heckenberg, H. Rubinsztein-Dunlop, Direct observation of transfer of angular momentum to absorptive particles from a laser beam with a phase singularity, *Phys. Rev. Lett. 75* (1995) 826–829.

[4] M.E.J. Friese, J. Enger, H. Rubinsztein-Dunlop, N.R. Heckenberg, Optical angular-momentum transfer to trapped absorbing particles, *Phys. Rev. A 54* (1996) 1593–1596.

[5] N.B. Simpson, K. Dholakia, L. Allen, M.J. Padgett, Mechanical equivalence of spin and orbital angular momentum of light: An optical spanner, *Opt. Lett. 22* (1997) 52–54.

[6] M.E.J. Friese, T.A. Nieminen, N.R. Heckenberg, H. Rubinsztein-Dunlop, Optical alignment and spinning of laser-trapped microscopic particles, *Nature 394* (6691) (1998) 348–350.

[7] J. Leach, M.J. Padgett, S.M. Barnett, S. Franke-Arnold, J. Courtial, Measuring the orbital angular momentum of a single photon, *Phys. Rev. Lett. 88* (2002) 257901.

[8] S. Franke-Arnold, S. Barnett, E. Yao, J. Leach, J. Courtial, M. Padgett, Uncertainty principle for angular position and angular momentum, *New J. Phys. 6* (2004) 103.

[9] E.A. Abbondanzieri, W.J. Greenleaf, J.W. Shaevitz, R. Landick, S.M. Block, Direct observation of base-pair stepping by RNA polymerase, *Nature 438* (7067) (2005) 460–465.

[10] J. Liphardt, S. Dumont, S.B. Smith, I. Tinoco, C. Bustamante, Equilibrium information from nonequilibrium measurements in an experimental test of Jarzynski's equality, *Science 296* (2002) 1832–1835.

[11] D. Collin, F. Ritort, C. Jarzynski, S.B. Smith, I. Tinoco, C. Bustamante, Verification of the Crooks fluctuation theorem and recovery of RNA folding free energies, *Nature 437* (2005) 231–234.

[12] J.C. Butler, I. Smalyukh, J. Manuel, G.C. Spalding, M.J. Parsek, G.C.L. Wong, Generating biofilms with optical tweezers: The influence of quorum sensing and motility upon pseudomonas aeruginosa aggregate formation, 2007, in preparation.

[13] A. Chowdhury, B.J. Ackerson, N.A. Clark, Laser-induced freezing, *Phys. Rev. Lett. 55* (1985) 833–836.

[14] C. Bechinger, M. Brunner, P. Leiderer, Phase behavior of two-dimensional colloidal systems in the presence of periodic light fields, *Phys. Rev. Lett. 86* (2001) 930–933.

[15] M. Brunner, C. Bechinger, Phase behavior of colloidal molecular crystals on triangular light lattices, *Phys. Rev. Lett. 88* (2002) 248302.

[16] C. Reichhardt, C.J. Olson, Novel colloidal crystalline states on two-dimensional periodic substrates, *Phys. Rev. Lett. 88* (2002) 248301.

[17] K. Mangold, P. Leiderer, C. Bechinger, Phase transitions of colloidal monolayers in periodic pinning arrays, *Phys. Rev. Lett. 90* (2003) 158302.

[18] C.J.O. Reichhardt, C. Reichhardt, Frustration and melting of colloidal molecular crystals, *J. Phys. A: Math. Gen. 36* (2003) 5841–5845.

[19] J. Baumgartl, M. Brunner, C. Bechinger, Locked-floating-solid to locked-smectic transition in colloidal systems, *Phys. Rev. Lett. 93* (2004) 168301.

[20] P.T. Korda, G.C. Spalding, D.G. Grier, Evolution of a colloidal critical state in an optical pinning potential landscape, *Phys. Rev. B 66* (2002) 024504.

[21] P.T. Korda, M.B. Taylor, D.G. Grier, Kinetically locked-in colloidal transport in an array of optical tweezers, *Phys. Rev. Lett. 89* (2002) 128301.

[22] C. Reichhardt, C.J.O. Reichhardt, Directional locking effects and dynamics for particles driven through a colloidal lattice, *Phys. Rev. E 69* (2004) 041405.

[23] C. Reichhardt, C.J.O. Reichhardt, Cooperative behavior and pattern formation in mixtures of driven and nondriven colloidal assemblies, *Phys. Rev. E 74* (2006) 011403.

[24] M.P. MacDonald, G.C. Spalding, K. Dholakia, Microfluidic sorting in an optical lattice, *Nature 426* (2003) 421–424.

[25] W. Mu, Z. Li, L. Luan, G.C. Spalding, G. Wang, J.B. Ketterson, Measurements of the force on polystyrene microspheres resulting from interferometric optical standing wave, *Opt. Exp.* (2007), submitted for publication.

[26] M. Pelton, K. Ladavac, D.G. Grier, Transport and fractionation in periodic potential-energy landscapes, *Phys. Rev. E 70* (2004) 031108.

[27] R. Applegate, J. Squier, T. Vestad, J. Oakey, D. Marr, Optical trapping, manipulation, and sorting of cells and colloids in microfluidic systems with diode laser bars, *Opt. Exp. 12* (2004) 4390–4398.

[28] M.P. MacDonald, S. Neale, L. Paterson, A. Richies, K. Dholakia, G.C. Spalding, Cell cytometry with a light touch: Sorting microscopic matter with an optical lattice, *J. Biol. Regul. Homeost. Agents 18* (2004) 200–205.

[29] R.L. Smith, G.C. Spalding, S.L. Neale, K. Dholakia, M.P. MacDonald, Colloidal traffic in static and dynamic optical lattices, *Proc. Soc. Photo. Opt. Instrum. Eng. 6326* (2006) 6326N.

[30] R.L. Smith, G.C. Spalding, K. Dholakia, M.P. MacDonald, Colloidal sorting in dynamic optical lattices, *J. Opt. A: Pure Appl. Opt. 9* (2007) S1–S5.

[31] R. Di Leonardo, J. Leach, H. Mushfique, J.M. Cooper, G. Ruocco, M.J. Padgett, Multipoint holographic optical velocimetry in microfluidic systems, *Phys. Rev. Lett. 96* (2006) 134502.

[32] A. Terray, J. Oakey, D.W.M. Marr, Microfluidic control using colloidal devices, *Science 296* (2002) 1841–1844.

[33] J. Arlt, M.J. Padgett, Generation of a beam with a dark focus surrounded by regions of higher intensity: An optical bottle beam, *Opt. Lett. 25* (2000) 191–193.

[34] D. McGloin, G.C. Spalding, H. Melville, W. Sibbett, K. Dholakia, Applications of spatial light modulators in atom optics, *Opt. Exp. 11* (2003) 158–166.

[35] D. McGloin, G.C. Spalding, H. Melville, W. Sibbett, K. Dholakia, Three-dimensional arrays of optical bottle beams, *Opt. Commun. 225* (2003) 215–222.

[36] A. Ashkin, J.M. Dziedzic, J.E. Bjorkholm, S. Chu, Observation of a single-beam gradient force optical trap for dielectric particles, *Opt. Lett. 11* (1986) 288–290.

[37] K.C. Neuman, S.M. Block, Optical trapping, *Rev. Sci. Instrum. 75* (2004) 2787–2809.

[38] A. van der Horst, High-refractive index particles in counter-propagating optical tweezers—manipulation and forces, PhD thesis, Utrecht University, 2006.

[39] A. van der Horst A. Moroz, A. van Blaaderen, M. Dogterom, High trapping forces for high-refractive index particles trapped in dynamic arrays of counter-propagating optical tweezers, 2007, in preparation.

[40] A. Ashkin, Acceleration and trapping of particles by radiation pressure, *Phys. Rev. Lett. 24* (1970) 156–159.

[41] P. Korda, G.C. Spalding, E.R. Dufresne, D.G. Grier, Nanofabrication with holographic optical tweezers, *Rev. Sci. Instrum. 73* (2002) 1956–1957.

[42] J.M. Fournier, M.M. Burns, J.A. Golovchenko, Writing diffractive structures by optical trapping, *Proc. Soc. Photo. Instrum. Eng. 2406* (1995) 101–111.

[43] C. Mennerat-Robilliard, D. Boiron, J.M. Fournier, A. Aradian, P. Horak, G. Grynberg, Cooling cesium atoms in a Talbot lattice, *Europhys. Lett. 44* (1998) 442–448.

[44] E. Schonbrun, R. Piestun, P. Jordan, J. Cooper, K.D. Wulff, J. Courtial, M. Padgett, 3D interferometric optical tweezers using a single spatial light modulator, *Opt. Exp. 13* (2005) 3777–3786.

[45] M.P. MacDonald, S.L. Neale, R.L. Smith, G.C. Spalding, K. Dholakia, Sorting in an optical lattice, *Proc. Soc. Photo. Instrum. Eng. 5907* (2005) 5907E.

[46] L.C. Thomson, Y. Boissel, G. Whyte, E. Yao, J. Courtial, Superresolution holography for optical tweezers, 2007, in preparation.

[47] L.C. Thomson, J. Courtial, Holographic shaping of generalized self-reconstructing light beams, 2007, submitted for publication.

[48] C.H.J. Schmitz, J.P. Spatz, J.E. Curtis, High-precision steering of multiple holographic optical traps, *Opt. Exp. 13* (2005) 8678–8685.

[49] E.R. Dufresne, D.G. Grier, Optical tweezer arrays and optical substrates created with diffractive optics, *Rev. Sci. Instrum. 69* (1998) 1974–1977.

[50] E.R. Dufresne, G.C. Spalding, M.T. Dearing, S.A. Sheets, D.G. Grier, Computer-generated holographic optical tweezer arrays, *Rev. Sci. Instrum. 72* (2001) 1810–1816.

[51] J.E. Curtis, B.A. Koss, D.G. Grier, Dynamic holographic optical tweezers, *Opt. Commun. 207* (2002) 169–175.

[52] J.W. Goodman, Introduction to Fourier Optics, 2nd ed., McGraw–Hill, New York, 1996.

[53] P.A. Prentice, M.P. MacDonald, T.G. Frank, A. Cuschieri, G.C. Spalding, W. Sibbett, P.A. Campbell, K. Dholakia, Manipulation and filtration of low index particles with holographic Laguerre–Gaussian optical trap arrays, *Opt. Exp. 12* (2004) 593–600.

[54] J. Arlt, V. Garcés-Chávez, W. Sibbett, K. Dholakia, Optical micromanipulation using a Bessel light beam, *Opt. Commun. 197* (2001) 239–245.

[55] V. Garcés-Chávez, D. McGloin, H. Melville, W. Sibbett, K. Dholakia, Simultaneous micromanipulation in multiple planes using a self-reconstructing light beam, *Nature 419* (2002) 145–147.

[56] L.Z. Cai, X.L. Yang, Y.R. Wang, All fourteen Bravais lattices can be formed by interference of four noncoplanar beams, *Opt. Lett. 27* (2002) 900–902.

[57] P.T. Korda, Kinetics of Brownian particles driven through periodic potential energy landscapes, PhD thesis, University of Chicago, 2002.

[58] J.A. Neff, R.A. Athale, S.H. Lee, 2-dimensional spatial light modulators—a tutorial, *Proc. IEEE 78* (1990) 826–855.

[59] Y. Igasaki, F. Li, N. Yoshida, H. Toyoda, T. Inoue, N. Mukohzaka, Y. Kobayashi, T. Hara, High efficiency electrically-addressable phase-only spatial light modulator, *Opt. Rev. 6* (1999) 339–344.

[60] F. Mok, J. Diep, H.K. Liu, D. Psaltis, Real-time computer-generated hologram by means of liquid–crystal television spatial light-modulator, *Opt. Lett. 11* (1986) 748–750.

[61] Boulder Nonlinear Systems, Spatial Light Modulators, retrieved on 2007 at http://www.bouldernonlinear.com/products/XYphaseFlat/XYphaseFlat.htm.

[62] HOLOEYE Photonics AG, Spatial Light Modulator, retrieved on 2007 at http://www.holoeye.com/spatial_light_modulators-technology.html.

[63] Hamamatsu Corporation, Programmable Phase Modulator (Spatial Light Modulator), retrieved on 2007 at http://sales.hamamatsu.com/en/products/electron-tube-division/detectors/spatial-light-modulator.php.

[64] W.J. Hossack, E. Theofanidou, J. Crain, K. Heggarty, M. Birch, High-speed holographic optical tweezers using a ferroelectric liquid crystal microdisplay, *Opt. Exp. 11* (2003) 2053–2059.

[65] G. Moddel, K.M. Johnson, W. Li, R.A. Rice, L.A. Paganostauffer, M.A. Handschy, High-speed binary optically addressed spatial light-modulator, *Appl. Phys. Lett. 55* (1989) 537–539.

[66] L.K. Cotter, T.J. Drabik, R.J. Dillon, M.A. Handschy, Ferroelectric–liquid–crystal silicon-integrated-circuit spatial light-modulator, *Opt. Lett. 15* (1990) 291–293.

[67] A. Lafong, W.J. Hossack, J. Arlt, T.J. Nowakowski, N.D. Read, Time-multiplexed Laguerre–Gaussian holographic optical tweezers for biological applications, *Opt. Exp. 14* (2006) 3065–3072.

[68] J. Amako, T. Sonehara, Kinoform using an electrically controlled birefringent liquid–crystal spatial light-modulator, *Appl. Opt. 30* (1991) 4622–4628.

[69] G.D. Love, Wave-front correction and production of Zernike modes with a liquid–crystal spatial light modulator, *Appl. Opt. 36* (1997) 1517–1524.

[70] Y. Roichman, A. Waldron, E. Gardel, D.G. Grier, Optical traps with geometric aberrations, *Appl. Opt. 45* (2006) 3425–3429.

[71] K.D. Wulff, D.G. Cole, R.L. Clark, R. Di Leonardo, J. Leach, J. Cooper, G. Gibson, M.J. Padgett, Aberration correction in holographic optical tweezers, *Opt. Exp. 14* (2006) 4170–4175.

[72] A. Jesacher, A. Schwaighofer, S. Furhapter, C. Maurer, S. Bernet, M. Ritsch-Marte, Wavefront correction of spatial light modulators using an optical vortex image, *Opt. Exp. 15* (2007) 5801–5808.

[73] T.R. O'Meara, P.V. Mitchell, Continuously operated spatial light modulator apparatus and method for adaptive optics, *U.S. Patent 5* (396, 364) (1995).

[74] Arryx, Inc., retrieved on 2007 at http://www.arryx.com/.

[75] T. Ota, S. Kawata, T. Sugiura, M.J. Booth, M.A.A. Neil, R. Juškaitis, T. Wilson, Dynamic axial-position control of a laser-trapped particle by wave-front modification, *Opt. Lett. 28* (2003) 465–467.

[76] J.W. Rinne, P. Wiltzius, Design of holographic structures using genetic algorithms, *Opt. Exp. 14* (2006) 9909–9916.

[77] J. Leach, K. Wulff, G. Sinclair, P. Jordan, J. Courtial, L. Thomson, G. Gibson, K. Karunwi, J. Cooper, Z.J. Laczik, M. Padgett, Interactive approach to optical tweezers control, *Appl. Opt. 45* (2006) 897–903.

[78] R. Di Leonardo, F. Ianni, G. Ruocco, Computer generation of optimal holograms for optical trap arrays, *Opt. Exp. 15* (2007) 1913–1922.

[79] J.W. Goodman, Introduction to Fourier Optics, McGraw–Hill, New York, 1996.

[80] M. Montes-Usategui, E. Pleguezuelos, J. Andilla, E. Martin-Badosa, Fast generation of holographic optical tweezers by random mask encoding of Fourier components, *Opt. Exp. 14* (2006) 2101–2107.

[81] M. Reicherter, T. Haist, E.U. Wagemann, H.J. Tiziani, Optical particle trapping with computer-generated holograms written on a liquid–crystal display, *Opt. Lett. 24* (1999) 608–610.

[82] J. Liesener, M. Reicherter, T. Haist, H.J. Tiziani, Multi-functional optical tweezers using computer-generated holograms, *Opt. Commun. 185* (2000) 77–82.

[83] J.E. Curtis, C.H.J. Schmitz, J.P. Spatz, Symmetry dependence of holograms for optical trapping, *Opt. Lett. 30* (2005) 2086–2088.

[84] L.B. Lesem, P.M. Hirsch, J.A. Jordon Jr., The kinoform: A new wavefront reconstruction device, *IBM J. Res. Develop. 13* (1969) 150–155.

[85] J.N. Mait, Diffractive beauty, *Opt. Photon. News 9* (1998) 21–25, 52.

[86] R.W. Gerchberg, W.O. Saxton, A practical algorithm for the determination of the phase from image and diffraction plane pictures, *Optik 35* (1972) 237–246.

[87] N.C. Gallagher, B. Liu, Method for computing kinoforms that reduces image reconstruction error, *Appl. Opt. 12* (1973) 2328–2335.

[88] R.W. Gerchberg, Super-resolution through error energy reduction, *Optica Acta 21* (1974) 709–720.

[89] B. Liu, N.C. Gallagher, Convergence of a spectrum shaping algorithm, *Appl. Opt. 13* (1974) 2470–2471.

[90] G. Shabtay, Three-dimensional beam forming and Ewald's surfaces, *Opt. Commun. 226* (2003) 33–37.

[91] G. Whyte, J. Courtial, Experimental demonstration of holographic three-dimensional light shaping using a Gerchberg–Saxton algorithm, *New J. Phys. 7* (2005) 117.

[92] L. Angelani, L. Casetti, M. Pettini, G. Ruocco, F. Zamponi, Topological signature of first-order phase transitions in a mean-field model, *Europhys. Lett. 62* (2003) 775–781.

[93] T. Haist, M. Schönleber, H.J. Tiziani, Computer-generated holograms from 3D-objects written on twisted-nematic liquid crystal displays, *Opt. Commun. 140* (1997) 299–308.

[94] G. Sinclair, J. Leach, P. Jordan, G. Gibson, E. Yao, Z.J. Laczik, M.J. Padgett, J. Courtial, Interactive application in holographic optical tweezers of a multi-plane Gerchberg–Saxton algorithm for three-dimensional light shaping, *Opt. Exp. 12* (2004) 1665–1670.

[95] M. Meister, R.J. Winfield, Novel approaches to direct search algorithms for the design of diffractive optical elements, *Opt. Commun. 203* (2002) 39–49.

[96] M. Polin, K. Ladavac, S. Lee, Y. Roichman, D. Grier, Optimized holographic optical traps, *Opt. Exp. 13* (2005) 5831–5845.

[97] N. Yoshikawa, T. Yatagai, Phase optimization of a kinoform by simulated annealing, *Appl. Opt. 33* (1994) 863–868.

[98] T. Haist, M. Reicherter, M. Wu, L. Seifert, Using graphics boards to compute holograms, *Comput. Sci. Eng. 8* (2006) 8–13.

[99] M.M. Burns, J.M. Fournier, J.A. Golovchenko, Optical binding, *Phys. Rev. Lett. 63* (1989) 1233–1236.

[100] M.M. Burns, J.M. Fournier, J.A. Golovchenko, Optical matter—crystallization and binding in intense optical-fields, *Science 249* (1990) 749–754.

[101] K. Sasaki, M. Koshioka, H. Misawa, N. Kitamura, H. Masuhara, Laser-scanning micromanipulation and spatial patterning of fine particles, *JJAP Part 2-Letters 30* (1991) L907–L909.

[102] C. Mio, T. Gong, A. Terray, D.W.M. Marr, Design of a scanning laser optical trap for multiparticle manipulation, *Rev. Sci. Instrum. 71* (2000) 2196–2200.

[103] V. Blickle, T. Speck, U. Seifert, C. Bechinger, Characterizing potentials by a generalized Boltzmann factor, *Phys. Rev. E* (2007) 060101.

[104] V. Blickle, T. Speck, C. Lutz, U. Seifert, C. Bechinger, The Einstein relation generalized to non-equilibrium, *Phys. Rev. Lett. 98* (2007) 210601.

[105] K. Visscher, G.J. Brakenhoff, J.J. Kroll, Micromanipulation by multiple optical traps created by a single fast scanning trap integrated with the bilateral confocal scanning laser microscope, *Cytometry 14* (1993) 105–114.

[106] K. Visscher, S.P. Gross, S.M. Block, Construction of multiple-beam optical traps with nanometer-resolution position sensing, *IEEE J. Selected Topics Quantum Electron. 2* (1996) 1066–1076.

[107] R. Di Leonardo, G. Ruocco, J. Leach, M.J. Padgett, A.J. Wright, J.M. Girkin, D.R. Burnham, D. McGloin, Parametric resonance of optically trapped aerosols, *Phys. Rev. Lett. 99* (2007) 029902.

[108] D.R. Burnham, D. McGloin, Holographic optical trapping of aerosol droplets, *Opt. Exp. 14* (2006) 4175–4181.

[109] B. Agate, C.T.A. Brown, W. Sibbett, K. Dholakia, Femtosecond optical tweezers for in-situ control of two-photon fluorescence, *Opt. Exp. 12* (2004) 3011–3017.

[110] D.L.J. Vossen, A. van der Horst, M. Dogterom, A. van Blaaderen, Optical tweezers and confocal microscopy for simultaneous three-dimensional manipulation and imaging in concentrated colloidal dispersions, *Rev. Sci. Instrum. 75* (2004) 2960–2970.

[111] P.C. Mogensen, J. Glückstad, Dynamic array generation and pattern formation for optical tweezers, *Opt. Commun. 175* (2000) 75–81.

[112] P.J. Rodrigo, V.R. Daria, J. Gluckstad, Dynamically reconfigurable optical lattices, *Opt. Exp. 13* (2005) 1384–1394.

[113] P.J. Rodrigo, I.R. Perch-Nielsen, J. Gluckstad, Three-dimensional forces in GPC-based counterpropagating-beam traps, *Opt. Exp. 14* (2006) 5812–5822.

[114] J.S. Dam, P.J. Rodrigo, I.R. Perch-Nielsen, C.A. Alonzo, J. Gluckstad, Computerized drag-and-drop alignment of gpc-based optical micromanipulation system, *Opt. Exp. 15* (2007) 1923–1931.

[115] D. Palima, V.R. Daria, Effect of spurious diffraction orders in arbitrary multifoci patterns produced via phase-only holograms, *Appl. Opt. 45* (2006) 6689–6693.

[116] I.R. Perch-Nielsen, P.J. Rodrigo, J. Gluckstad, Real-time interactive 3D manipulation of particles viewed in two orthogonal observation planes, *Opt. Exp. 13* (2005) 2852–2857.

[117] V. Garcés-Chávez, K. Dholakia, G.C. Spalding, Extended-area optically induced organization of microparticies on a surface, *Appl. Phys. Lett. 86* (2005) 031106.

[118] P.J. Reece, E.M. Wright, K. Dholakia, Experimental observation of modulation instability and optical spatial soliton arrays in soft condensed matter, *Phys. Rev. Lett. 98* (2007) 203902.

[119] L.C. Thomson, J. Courtial, G. Whyte, M. Mazilu, Algorithm for 3D intensity shaping of evanescent wave fields, 2007, in preparation.

[120] M. Kyoung, K. Karunwi, E.D. Sheets, A versatile multimode microscope to probe and manipulate nanoparticles and biomolecules, *J. Microsc. Oxford 225* (2007) 137–146.

[121] V. Emiliani, D. Cojoc, E. Ferrari, V. Garbin, C. Durieux, M. Coppey-Moisan, E. Di Fabrizio, Wave front engineering for microscopy of living cells, *Opt. Exp. 13* (2005) 1395–1405.

[122] D. Stevenson, B. Agate, X. Tsampoula, P. Fischer, C.T.A. Brown, W. Sibbett, A. Riches, F. Gunn-Moore, K. Dholakia, Femtosecond optical transfection of cells: Viability and efficiency, *Opt. Exp. 14* (2006) 7125–7133.

[123] E. Cuche, F. Bevilacqua, C. Depeursinge, Digital holography for quantitative phase-contrast imaging, *Opt. Lett. 24* (1999) 291–293.

[124] T.M. Kreis, Frequency analysis of digital holography with reconstruction by convolution, *Opt. Eng. 41* (2002) 1829–1839.

[125] S.H. Lee, D.G. Grier, Holographic microscopy of holographically trapped three-dimensional structures, *Opt. Exp. 15* (2007) 1505–1512.

[126] S.C. Chapin, V. Germain, E.R. Dufresne, Automated trapping, assembly, and sorting with holographic optical tweezers, *Opt. Exp. 14* (2006) 13095–13100.

[127] A.J. Pons, A. Karma, S. Akamatsu, M. Newey, A. Pomerance, H. Singer, W. Losert, Feedback control of unstable cellular solidification fronts, *Phys. Rev. E 75* (2007) 021602.

Chapter 7

Atomic and Molecular Manipulation Using Structured Light

Mohamed Babiker[1] *and David L. Andrews*[2]

[1]*University of York, UK*
[2]*University of East Anglia, UK*

7.1 INTRODUCTION

Optical manipulation is a field that encompasses a wide range of well-attested mechanisms and methods. The mechanisms that are most prominent in any specific system are primarily dictated by the size of the target particles. The particle size, in turn, determines the nature of the physical system in which such effects can be observed; the scale of size for optical tweezers and allied methods runs up to a significant fraction of the beam width. Microscopic particles such as cells and polymer beads represent optically controllable particles at this higher limit of size [1–3]. To offset gravitational forces, such materials are most conveniently studied in liquid suspension, and in such cases the particle position and motion are controllable by various means, including intensity gradients (optical tweezing of individual particles at a laser focus, or for large numbers of particles in holographically generated traps) and multiple-scattering (optical binding). Together, such methods represent a branch of optical technology that already has extensive applications in the fields of medicine, sensors, and micromechanical devices.

At the opposite extreme, the lower end of the size scale, most such methods are clearly unusable. Individual atoms and small molecules do not present sufficient cross-section to respond differentially across their own dimensions to wavelength-scale vari-

STRUCTURED LIGHT AND ITS APPLICATIONS

ations in intensity, nor are they so readily localizable. In the condensed phase, the optical manipulation of particles smaller than 100 nm becomes problematic because of Brownian motion. In the gas phase, laser cooling schemes such as the configuration known as optical molasses, based on momentum exchange and exploiting Doppler shift to its own ultimate demise, allow the optical generation of traps within which further optical manipulation can be achieved. This science of cold atoms [4–8] has, of course, recently developed into another burgeoning area of study, the generation and control of Bose–Einstein condensates [9]. In such systems, the responsive motion of individual atoms or molecules, or of the whole assembly in the Bose–Einstein case, is determined by an optically generated potential well.

Against this background, the theoretical and technological developments that have recently led to the production of structured laser light introduce another tier of possibilities associated with orbital angular momentum (OAM) content [10–14]. The intricate wavefront structures of Laguerre–Gaussian, Bessel, and Mathieu light beams, for example, allow the production of force fields and torques that have no counterpart in conventional optical beams. Already it has been shown that the exploitation of such beams for atomic and molecular manipulation can lead to a variety of lattice structures, clusters, and rings [15–17]. In this chapter, we will describe the general principles and give the key equations. We will also exhibit some of the results that have emerged from studies of optically trapped atoms, and molecules in the quasi-static environment of a liquid crystal [18]. First, we will begin with a brief overview of the context of the engagement of light possessing orbital angular momentum with atoms and molecules.

7.2 A BRIEF OVERVIEW

The literature dealing with study of the engagement of light endowed with orbital angular momentum with atoms and molecules is relatively sparse in comparison with that involving optical manipulation of the larger particles [5] mentioned earlier, such as biological cells and polymer beads. Most published works on atoms and molecules are concerned with theoretical studies, but there are also a few experimental studies.

The possibility that orbital angular momentum effects can influence matter at the atomic and molecular level was first mentioned as a speculation in pioneering work by Allen and colleagues [19]. This was followed by a number of theoretical investigations [20–24] that led to the prediction of the light-induced torque [20], the azimuthal Doppler shift [21], and a number of studies on the motion of atoms and ions in Laguerre–Gaussian beams, including optical molasses in one, two, and three dimensions [22–24]. The role of photon spin when considered in the same context as OAM was clarified [23]. This led to the identification of a spin–orbit term and a contribution involving l–s coupling in the azimuthal force due to circularly polarized Laguerre–Gaussian light. More recent work concentrated on trapping in dark regions of the beam profile, indicating that under such circumstances the trapped atoms would experience diminished heating effects [25]. Studies dealing with the selection rules governing the interaction of light with the internal and external degrees of freedom were undertaken by van Enk [26] and Babiker and colleagues [27], while Juzeliūnas and colleagues identified novel features in the interactions of Bose–Einstein condensates [28,29]. As interest in the subject has grown, many other groups have also engaged in issues surrounding the effects of OAM on atoms and molecules [30–35].

Tabosa and Petrov conducted the first experimental study involving atoms interacting with orbital angular momentum of light [36]. They demonstrated the transfer of OAM from the beam to cold cesium atoms. Other studies have dealt with the channeling of atoms in material structures possessing cylindrical symmetry, where the optical modes are distinguished by orbital angular momentum features. Theoretical studies [32,37] have shown that, in such structures, the channeling of atoms involves light torques similar to those produced by free space Laguerre–Gaussian beams—which have also been employed as atom guides [37–42]. In the molecular context, particular interest has focused on liquid crystals, despite their complexity, since their distinctive combination of anisotropic local structure and relatively labile orientational motion is directly amenable to optical interrogation. The effects of OAM on liquid crystals have been studied recently by Piccirillo and colleagues [43,44], and a subsequent analysis employing the dielectric model of the nematic liquid crystal has been reported by Carter and colleagues [18].

7.3 TRANSFER OF OAM TO ATOMS AND MOLECULES

For both structured and unstructured light, most optical processes involve electric dipole interactions with the radiation fields, this type of interaction generally being the strongest form of coupling. Depending upon the specific process, the dipolar interactions entailed may invoke either static or transition moments.

If, as shown by Berry [45], orbital angular momentum is an intrinsic property of all types of azimuthal phase-bearing light, then it could be argued that orbital angular momentum should be exchanged in an optical transition, just as spin angular momentum is exchanged in a radiative transition, leading to modified selection rules for electronic transitions. These matters have been explored by an explicit analysis [27], which concluded that the exchange of OAM occurs in the electric dipole approximation and couples only the center of mass to the light beam. The internal degrees of freedom associated with the "electronic" motion are not involved in any OAM exchange with the light beam to this leading order. It is only in the next order, namely in an electric quadrupole transition, that an exchange involving the light, the center of mass, and the internal degrees of freedom can be realized. One unit of OAM is exchanged between the light and the internal dynamics, so that the light beam possesses $(l \pm 1)\hbar$ units of OAM and the center of mass motion gains ± 1 units. These conclusions suggest that no experiments can detect OAM exchange between Laguerre–Gaussian light and molecular systems through changes involving electric dipole transitions. The analysis confirms the fact that OAM effects are manifest primarily in the center mass motion by the imposition of additional forces and associated torques. The study of these additional forces is best carried out by a extending the formalism of Doppler cooling and trapping to the case of light possessing OAM, as we will discuss next.

7.4 DOPPLER FORCES AND TORQUES

It has long been known that the Doppler effect responsible for broadening atomic transitions can be exploited for laser cooling. On irradiating an atomic gas with a laser beam detuned to the red of an absorption frequency, only a subset of the atoms—those that experience a compensating (blue) Doppler shift due to motion toward the light source—can absorb the light. The decay of the resulting excited state releases a photon in a random direction. Due to the extremely short lifetime usually associated with the excited state, this is a process that can recur with great rapidity, and the net effect over a series of absorption and emission cycles is that such atoms experience a net imparted momentum against their direction of travel, slowing them down. For the self-selected group of atoms within the laser beam profile, this loss of translational energy signifies cooling, to the extent that such a term can meaningfully be applied to a nonequilibrium system. With two counterpropagating beams, the velocities of the fastest atoms in each direction can be reduced; and as the laser frequency is gradually increased, the breadth of the initially Maxwellian velocity distribution becomes increasingly narrow. Transverse motions can be controlled by the addition of further counterpropagating beams, with each pair of sources in a mutually orthogonal configuration; this is the essence of optical molasses.

The significant features introduced by the use of structured light possessing orbital angular momentum are: (1) there is, in addition to translational effects, a light-induced torque that causes a rotational motion of the atoms about the beam axis and (2) there are regions of maximum and minimum intensities in the beam cross-section. The forces and torque are, in general, time-dependent as well as position-dependent. As we will discuss, the full space- and time-dependence of the motion is, in general, characterized by a transient regime, followed by a steady state regime after a sufficiently large time has elapsed from the instant in which the beam is switched on (typically for elapsed times much larger than the characteristic timescale of the problem).

7.4.1 Essential Formalism

Consider an atom or a molecule for which the gross motion is that of the center of mass and the internal dynamics is modeled in terms of a two-level atom. In the presence of a laser field, the total Hamiltonian for the whole system is

$$H = \hbar\omega a^{\dagger}a + \frac{\mathbf{P}^2}{2M} + \hbar\omega_0\pi^{\dagger}\pi - i\hbar\big[\tilde{\pi}^{\dagger}f(\mathbf{R}) - \text{h.c.}\big], \tag{1}$$

where $\tilde{\pi}$ and $f(\mathbf{R})$ are given by

$$\tilde{\pi} = \pi e^{i\omega t}, \qquad f(\mathbf{R}) = (\boldsymbol{\mu}_{12}\cdot\hat{\boldsymbol{\epsilon}})\alpha\mathcal{F}_{klp}(\mathbf{R})e^{i\Theta_{klp}(\mathbf{R})}/\hbar. \tag{2}$$

Here π and $\pi^{\dagger}$ are the ladder operators for the two-level system; $\mathbf{P}$ is the center-of-mass momentum operator with M the total mass and ω_0 the dipole transition frequency. The operators a and $a^{\dagger}$ are the annihilation and creation operators of the laser light, and ω is its frequency. In the classical limit, appropriate for the case of a coherent beam, the a and $a^{\dagger}$ operators become c-numbers involving the parameter α such that

$$a(t) \to \alpha e^{-i\omega t}, \qquad a^{\dagger}(t) \to \alpha^{*}e^{i\omega t}. \tag{3}$$

The last term in equation (1) is the interaction Hamiltonian coupling the laser light to the two-level system in the electric dipole and rotating wave approximations, evaluated at the center of mass position vector $\mathbf{R}$. The coupling function $f(\mathbf{R})$ in equation (2) involves $\boldsymbol{\mu}_{12}$, the transition dipole matrix element of the atom interacting with a Laguerre–Gaussian light mode characterized by $\hat{\boldsymbol{\epsilon}}$, the mode polarization vector, the mode amplitude function, $\mathcal{F}_{klp}(\mathbf{R})$ and phase $\Theta_{klp}(\mathbf{R})$, given by

$$\mathcal{F}_{klp}(\mathbf{R}) = \mathcal{F}_{k00}\frac{N_{lp}}{(1+z^2/z_R^2)^{1/2}}\left(\frac{\sqrt{2}r}{w(z)}\right)^{|l|}L_p^{|l|}\left(\frac{2r^2}{w^2(z)}\right)e^{-r^2/w^2(z)}, \tag{4}$$

$$\Theta_{klp}(\mathbf{R}) = \frac{kr^2z}{2(z^2+z_R^2)} + l\phi + (2p+l+1)\tan^{-1}(z/z_R) + kz. \tag{5}$$

Here $\mathcal{F}_{k00}$ may be identified as the amplitude for a plane wave propagating along the z-axis with wave vector k; the coefficient $N_{lp} = \sqrt{p!/(|l|+p!)}$ is a normalization constant; $w(z)$ is a characteristic width of the beam at axial coordinate z and is explicitly given by $w^2(z) = 2(z^2+z_R^2)/kz_R$, where z_R is the Rayleigh range. The LG mode indices l and p determine the field intensity distribution and are such that $l\hbar$ is the orbital angular momentum content carried by each quantum.

We will now assume that the position $\mathbf{R}$ and the momentum operator $\mathbf{P}$ of the atomic center of mass take their average values $\mathbf{r}$ and $\mathbf{P}_0 = M\mathbf{V}$, where $\mathbf{V}$ is the center of mass velocity. Thus, we are treating the atom gross motion classically, while its internal motion continues to be treated quantum mechanically. This treatment is justified provided that the spread in the atomic wave packet is much smaller than the wavelength of the light, and that the recoil energy is much smaller than the linewidth. The system density matrix can then be written as

$$\rho_S = \delta(\mathbf{R} - \mathbf{r})\delta(\mathbf{P} - M\mathbf{V})\rho(t), \tag{6}$$

where $\rho(t)$ is the internal density matrix, which follows the time evolution

$$\frac{d\rho}{dt} = -\frac{i}{\hbar}[H, \rho] + \mathcal{R}\rho, \tag{7}$$

and where the term $\mathcal{R}\rho$ represents the relaxation processes in the two-level system. The optical Bloch equations governing the evolution of the density matrix elements can be written as follows:

$$\begin{bmatrix} \dot{\hat{\rho}}_{21}(t) \\ \dot{\hat{\rho}}_{12}(t) \\ \dot{\rho}_{22}(t) \end{bmatrix} = \begin{bmatrix} -(\Gamma_2 - i\Delta) & 0 & 2f(\mathbf{r}) \\ 0 & -(\Gamma_2 + i\Delta) & 2f^*(\mathbf{r}) \\ -f^*(\mathbf{r}) & -f(\mathbf{r}) & -\Gamma_1 \end{bmatrix} \begin{bmatrix} \hat{\rho}_{21}(t) \\ \hat{\rho}_{12}(t) \\ \rho_{22}(t) \end{bmatrix} + \begin{bmatrix} -f(\mathbf{r}) \\ -f^*(\mathbf{r}) \\ 0 \end{bmatrix}. \tag{8}$$

Here the relaxation processes are assumed to be characterized by an inelastic collision rate, Γ_1, and an elastic collision one, Γ_2. The effective, velocity-dependent, detuning Δ is given by $\Delta = \Delta_0 - \nabla\Theta \,.\, \mathbf{V}$ and we have set $\hat{\rho} = \tilde{\rho}\exp(-it\mathbf{V}\cdot\nabla\Theta)$. We have also made use of the relation $\rho_{11}(t) + \rho_{22}(t) = 1$.

The average force due to the light acting on the center of mass is the expectation value of the trace of $-\rho\nabla H$,

$$\langle\mathbf{F}\rangle = -\big\langle \mathrm{tr}(\rho\nabla H)\big\rangle. \tag{9}$$

The total force can be written as the sum of two types of force, a dissipative force $\langle\mathbf{F}_{\mathrm{diss}}\rangle$ and a dipole force $\langle\mathbf{F}_{\mathrm{dipole}}\rangle$, and these are related to the density matrix elements as follows:

$$\big\langle\mathbf{F}_{\mathrm{diss}}(\mathbf{R}, t)\big\rangle = -\hbar\nabla\Theta\big(\hat{\rho}_{21}^* f(\mathbf{r}) + \hat{\rho}_{21} f^*(\mathbf{r})\big) \tag{10}$$

$$\big\langle\mathbf{F}_{\mathrm{dipole}}(\mathbf{R}, t)\big\rangle = i\hbar\frac{\nabla\Omega}{\Omega}\big(\hat{\rho}_{21}^* f(\mathbf{r}) - \hat{\rho}_{21} f^*(\mathbf{r})\big). \tag{11}$$

Here we have introduced the position dependent Rabi frequency $\Omega(\mathbf{R})$, defined as

$$\hbar\Omega(\mathbf{R}) = \left|(\boldsymbol{\mu}_{12}\cdot\hat{\boldsymbol{\epsilon}})\alpha\mathcal{F}(\mathbf{R})\right|, \qquad f(\mathbf{R}) = \Omega(\mathbf{R})e^{i\Theta(\mathbf{R})}. \tag{12}$$

Clearly all quantities depend upon the mode type and implicitly carry the labels klp.

The center of mass dynamics is determined by Newton's second law, written in the form

$$M\frac{d^2\mathbf{R}}{dt^2} = \langle\mathbf{F}(t)\rangle, \tag{13}$$

where $\langle\mathbf{F}(t)\rangle$ is the total average force. Since this differential equation is second order in time, values of the position vector components $\mathbf{R}(0)$ and initial velocity vector components $\mathbf{V}(0)$ should be stated as initial conditions. The main outcome of solving equation (13) is a complete determination of the trajectory function $\mathbf{R}(t)$, along with $\mathbf{V}(t) = \dot{\mathbf{R}}(t)$. Furthermore, as will become apparent, the development furnishes important information about the evolution of the light-induced torque.

7.4.2 Transient Dynamics

The transient effects are most prominent for transitions with a long excited state lifetime. Rare-earth ions provide such a context. Consider an Eu^{3+} ion that has $M = 25.17\times10^{-26}$ kg, and for its ${}^5D_0 \to {}^7D_1$ transition $\lambda = 614$ nm and $\Gamma = 1111$ Hz. We focus on the $l = 1$, $p = 0$ Laguerre–Gaussian mode and assume the laser intensity to be $I = 10^5$ W cm^{-2}, and the beam waist $w_0 = 35\lambda$. The transient regime can be explored for three special cases, namely (a) exact resonance; (b) strong collisions and (c) intense field. For the latter case, we will assume the higher intensity $I = 10^8$ W cm^{-2}. Evaluations have been carried out for a period $t_{\max} \approx 5\Gamma^{-1} \approx 4.5$ ms, which is sufficiently long to exhibit effects both for the transient regime and the steady state.

The results are shown in Figure 7.1 for the cases of strong collisions. The atom follows a characteristic path with an axial motion superimposed on an in-plane motion. The in-plane motion is seen to be in the form of loops in the shape of petals. It is in this characteristic motion that the effects of the optical torque are evident. A similar trajectory arises in the case of an intense external field, as shown in Figure 7.2, but here it is seen that there are many more loops due to the atom gaining kinetic energy with a larger force and torque. Figure 7.3 explores the initial stages of the trajectory before the second petal is formed. In the case of exact resonance, as shown in Figure 7.4, there is no dipole force acting upon the atom and therefore no radial force, due to the zero-detuning. The atomic radial position is constant.

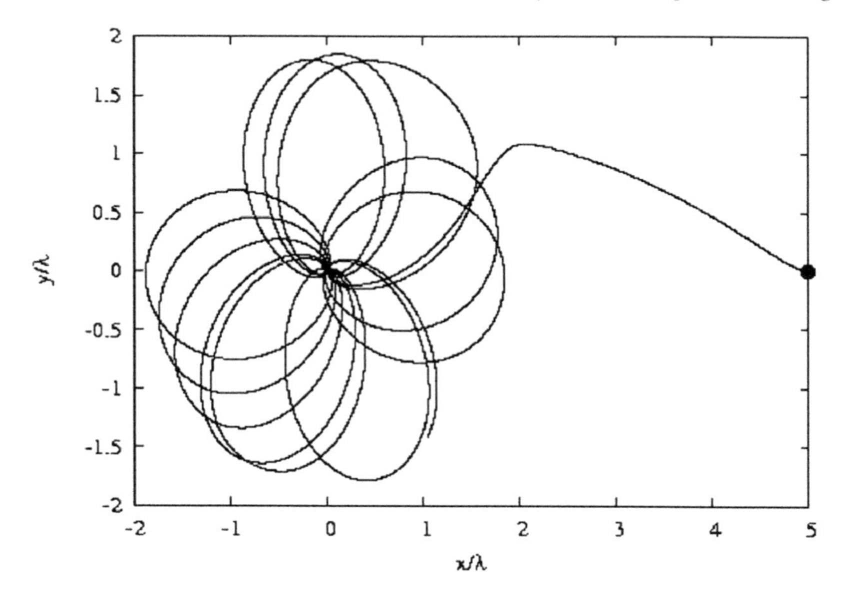

Figure 7.1 The path in the x–y-plane of an Eu^{3+} ion subject to an $LG_{1,0}$ mode in the case of strong collisions. The initial position is represented by a dot. Other parameters used for the generation of this figure are described in text.

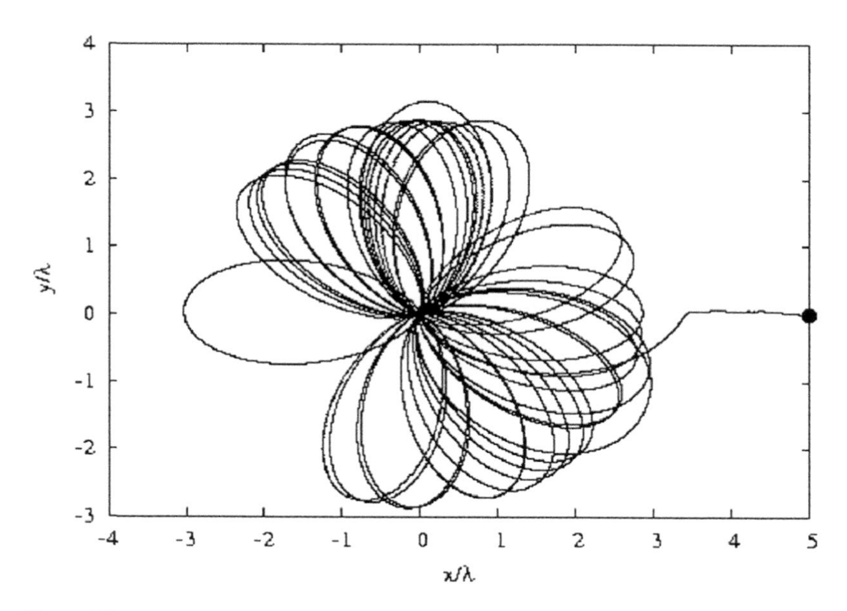

Figure 7.2 As in Figure 7.1, but for the case of a strong external field, as described in the text.

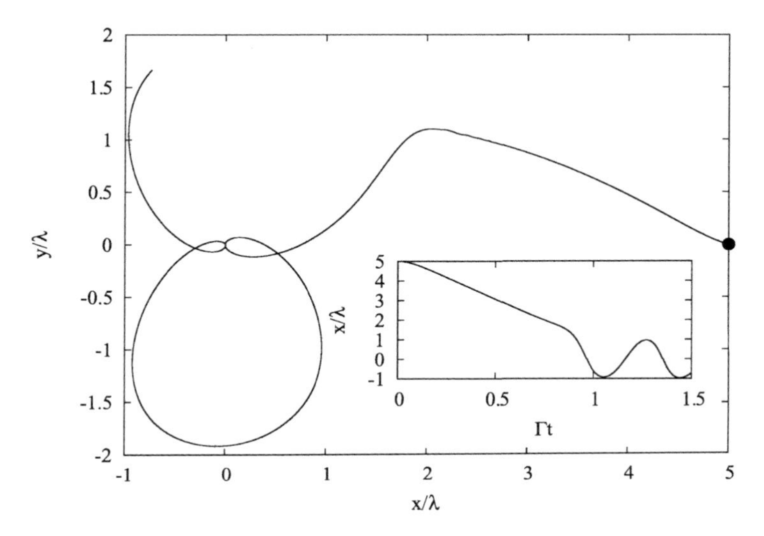

Figure 7.3 The initial stages of the trajectories shown in Figures 7.1 and 7.2 exhibiting the initiation of the first petal-like loop. Inset: variation of the x-component of the position with time.

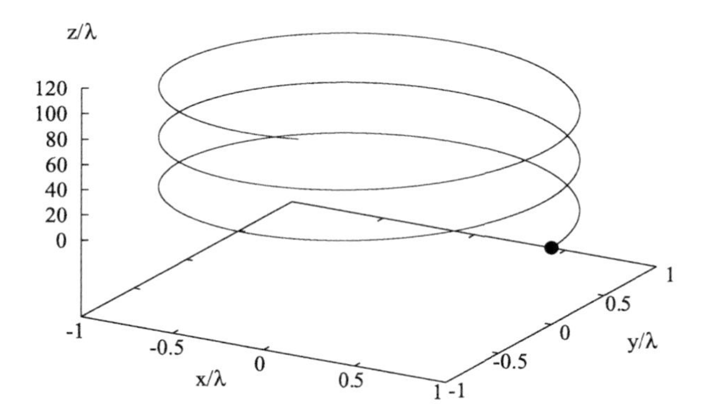

Figure 7.4 The path of an Eu^{3+} ion subject to an $LG_{1,0}$ in the case of exact resonance. Other parameters are the same as those shown in Figures 7.1 and 7.2.

The time-dependent torque is defined as

$$\mathcal{T}(t) = \mathbf{r}(t) \times \langle \mathbf{F}(t) \rangle. \tag{14}$$

The evolution of this torque can be determined along with the corresponding trajectories. The torque experienced by the Eu^{3+} ion for the case of strong collisions is shown in Figure 7.5. It is evident from the figure that once the beam is switched on, there is

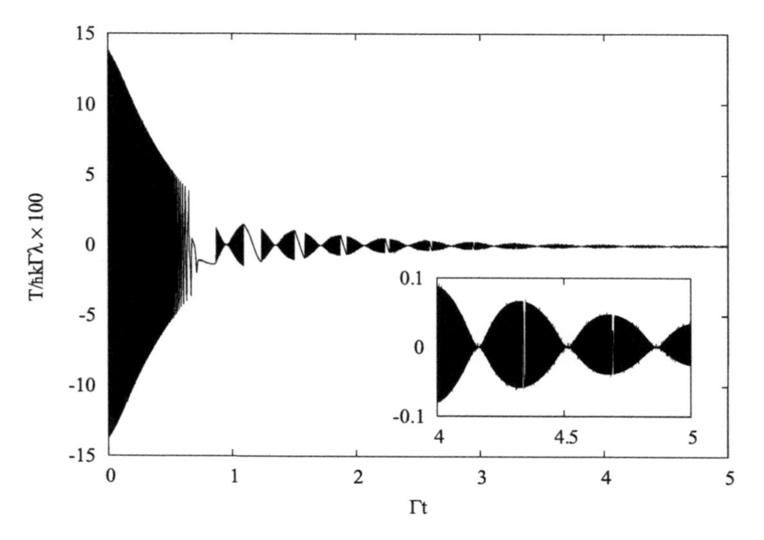

Figure 7.5 The variation of the (axial) light-induced torque acting on the Eu^{3+} ion subject to an $LG_{1,0}$ in the case of strong collisions. The detuning $\Delta_0 = 500\Gamma$ and the initial radial position is $r = 5\lambda$.

an abrupt increase in the magnitude of the torque, which then oscillates and rapidly decays toward a steady state value. Furthermore, the evolution exhibits a collapse and revival pattern, with each cycle corresponding to a loop in the trajectory. The peak of the torque corresponds to the outer tip of the loop, and the collapse corresponds to the points near the beam axis. The sudden jumps in the torque are real events arising from the change in the direction of motion as the atom is repelled from regions of extremum field intensity values.

7.4.3 Steady State Dynamics

The formal expressions for the steady state forces can be deduced by taking the limit $t \to \infty$, or the time derivatives in the optical Bloch equations set equal to zero. In the steady state, where $\Gamma t \gg 1$ (where Γ is the de-excitation rate of the upper state of the atomic transition), the total force on a two-level atom exhibits position-dependence and is naturally divisible into two terms. Restoring the explicit reference to a specific Laguerre–Gaussian mode, the steady state force on a moving atom due to a single beam propagating along the positive z axis is written

$$\langle \mathbf{F} \rangle_{klp} = \langle \mathbf{F}_{\text{diss}} \rangle_{klp} + \langle \mathbf{F}_{\text{dipole}} \rangle_{klp}, \tag{15}$$

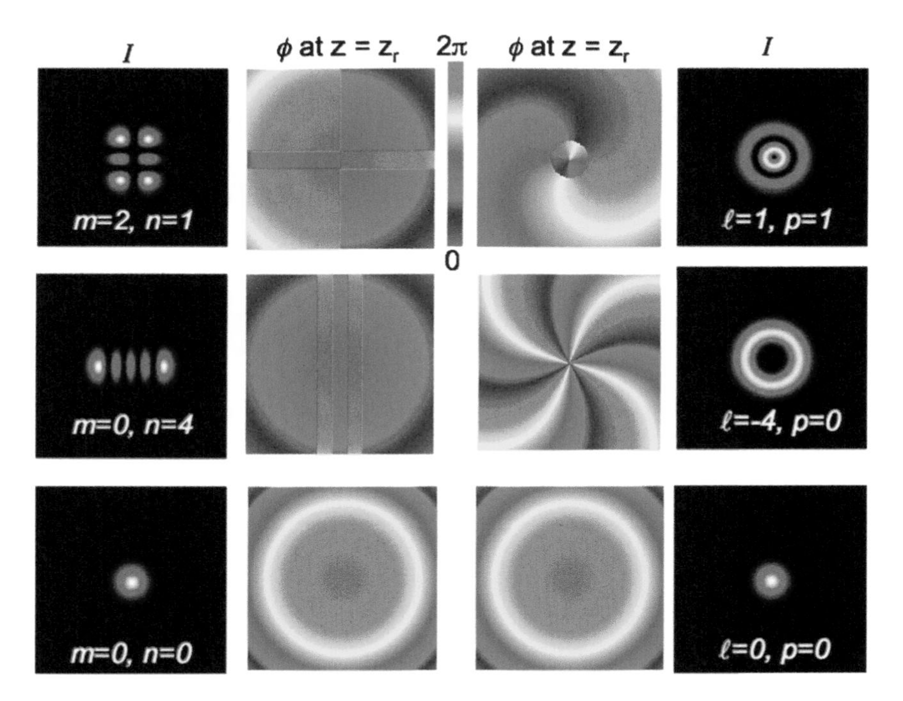

Figure 1.1 Examples of the intensity and phase structures of Hermite–Gaussian modes (*left*) and Laguerre–Gaussian modes (*right*), plotted at a distance from the beam waist equal to the Rayleigh range.

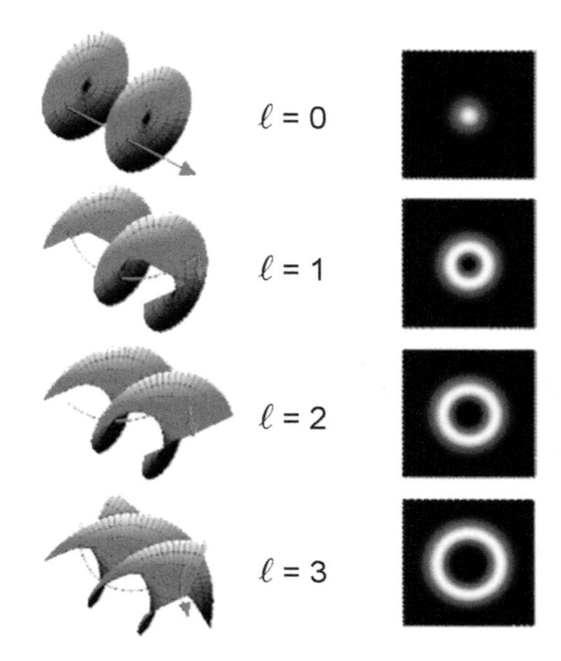

Figure 1.2 Helical phase fronts corresponding to the $\exp(i\ell\phi)$ phase structure, and corresponding intensity profiles of Laguerre–Gaussian modes.

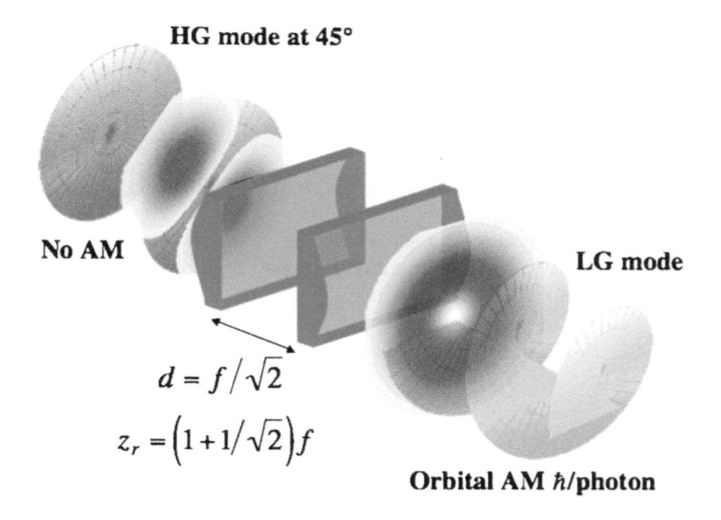

Figure 1.3 Cylindrical lens mode converter used to transform Hermite–Gaussian modes into Laguerre–Gaussian modes.

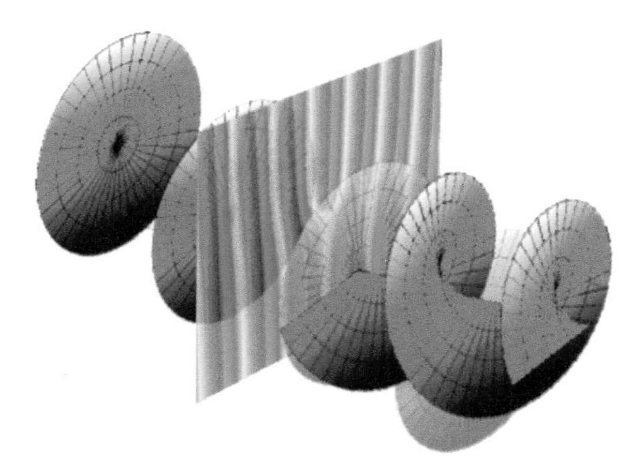

Figure 1.4 The "classic" forked hologram for the production of helically phased beams.

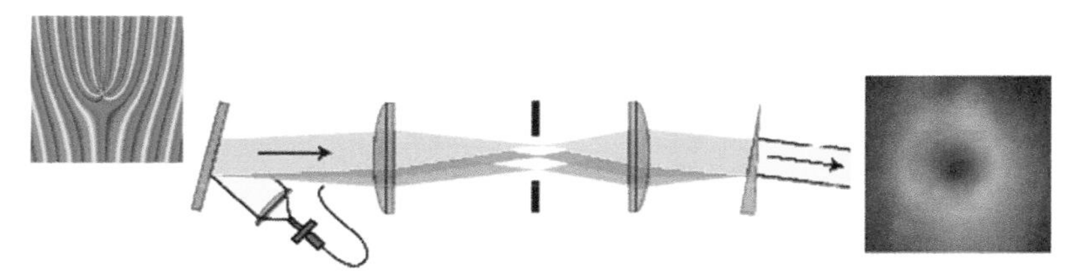

Figure 1.5 Use of a prism to compensate for the chromatic dispersion of the forked hologram allows the creation of white-light optical vortices.

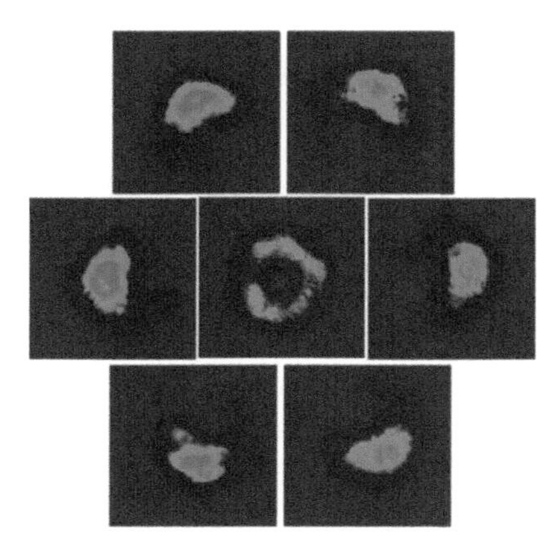

Figure 3.1 False-color image showing the interference of two collinear beams: a Laguerre–Gaussian beam with $l = 1$ (at the center) with a zero-order beam. The surrounding frames correspond to the interference patterns for phase differences in steps of $\pi/3$.

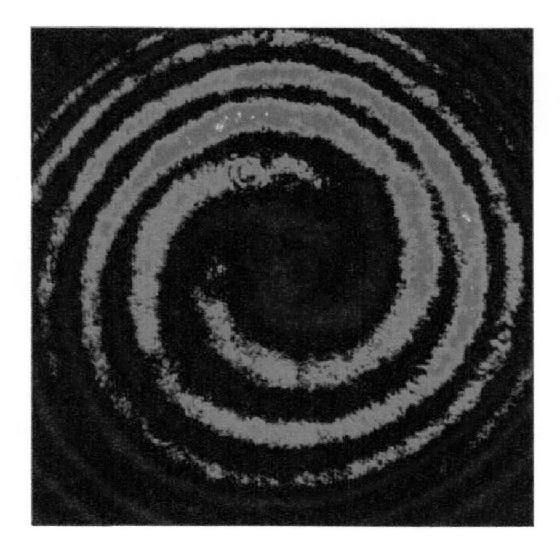

Figure 3.3 False color image of the double spiral interference pattern produced by the interference of a beam with $l = 2$ and an expanded beam in a zero-order (i.e., $l = 0$) mode.

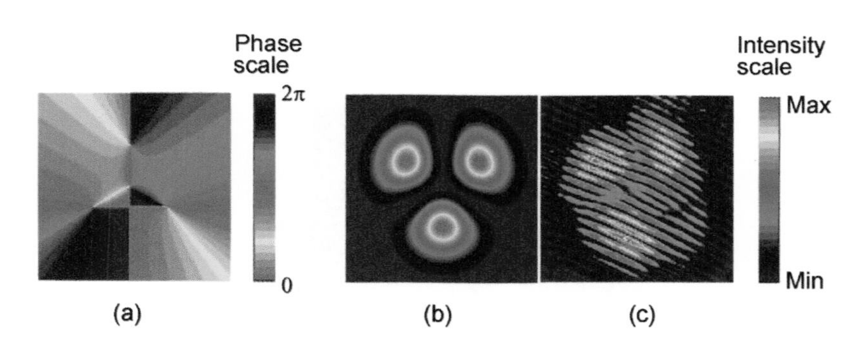

Figure 3.5 Composite vortex beam created by superposition of Laguerre–Gaussian beams with $l = +2$ with $l = -1$ at a ratio of amplitudes $A_{l=+2}/A_{l=-1} = 0.84$. Frames (a) and (b) show the computed phase and intensity of the composite beam. Frame (c) shows a measured interference pattern of the composite beam with a reference plane-wave.

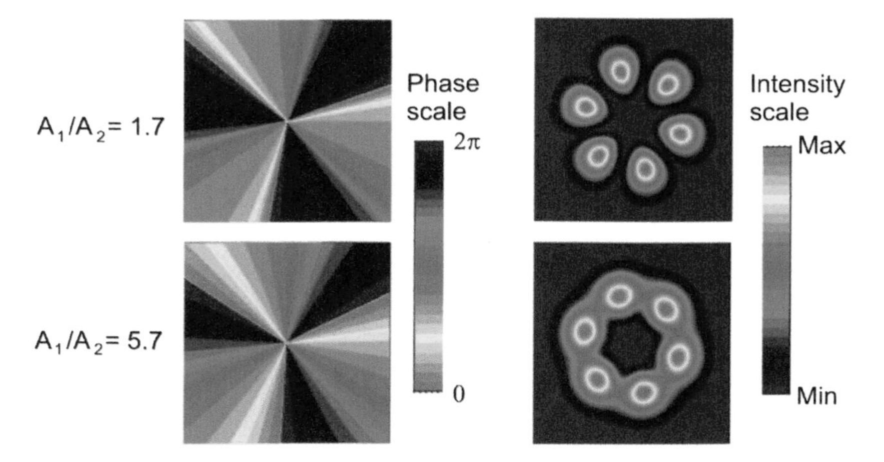

Figure 3.7 Composite beam when $l_1 = +3$ and $l_2 = -3$ for different relative amplitudes A_1 and A_2.

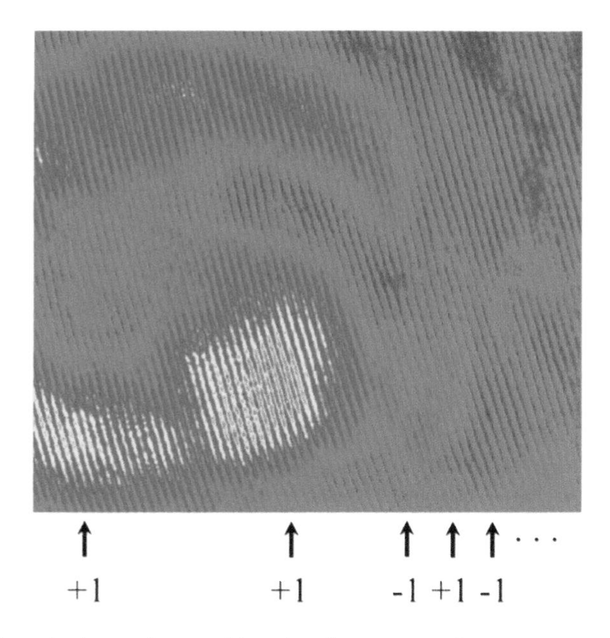

Figure 3.8 False-color image of a noncollinear interference pattern between an expanded fundamental mode and the field of a half-integer vortex beam with $q = 1.5$. The arrows indicate the horizontal positions of easily identified singly charged vortices, denoted by forks in the interference pattern. The direction of the tines denotes the sign of the charge.

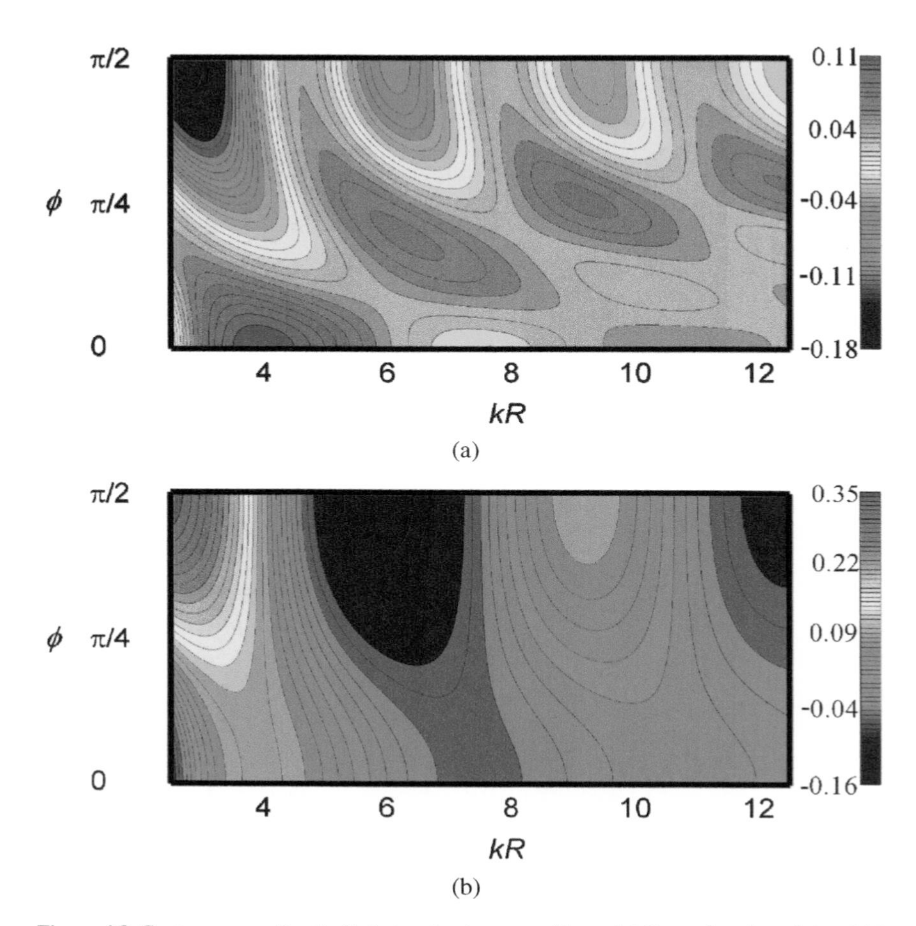

Figure 4.8 Contour maps of optically induced pair energy. Plots of ΔE as a function of ϕ and kR: (a) $\zeta = 0$; (b) $\zeta = \pi/2$. The variation of ΔE with kR along the abscissa, $\phi = 0$, shows its first two maxima at $kR \sim 4.0$, 10.5, and the first (nonproximal) minimum, at $kR \sim 7.5$ (compare to Figure 8.1). The horizontal scale typically spans distances R of several hundred nanometers, depending upon the value of k (see text). The units of the color scale are $\alpha_0^{(A)}\alpha_0^{(B)}2Ik^3/(4\pi\varepsilon_0^2c)$. Adapted from references [38,39].

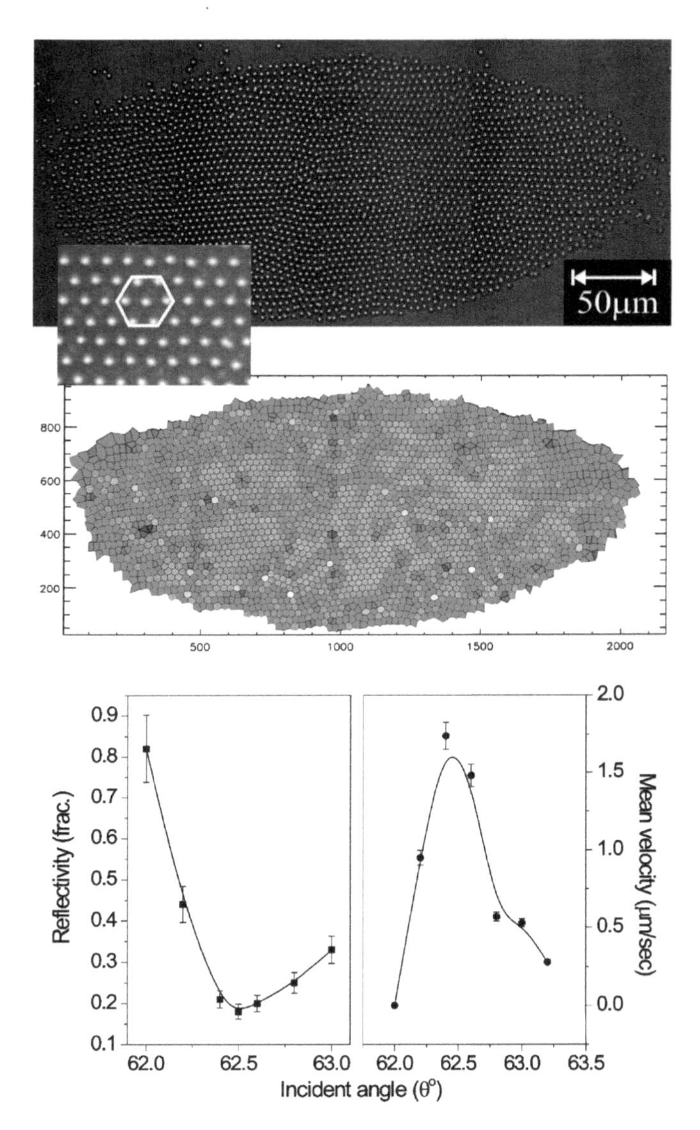

Figure 5.4 *Top*: Experimental observation of an array of colloids (approximately 2800 particles) accumulated in the presence of a surface plasmon polariton excited by attenuated total reflection. *Middle*: Voronoi plot showing the Wigner–Seitz cell for each colloid in the array indicates that at the center there is a predominantly close-packed crystalline arrangement (hexagonal). At the periphery, particles are more fluidlike with no preferred nearest neighbor arrangement. *Bottom*: Reflectivity of the incident light and mean velocity (curve is a guide to the eye) of the particles as a function of incident angle. Copyright 2006 by the American Physical Society [39].

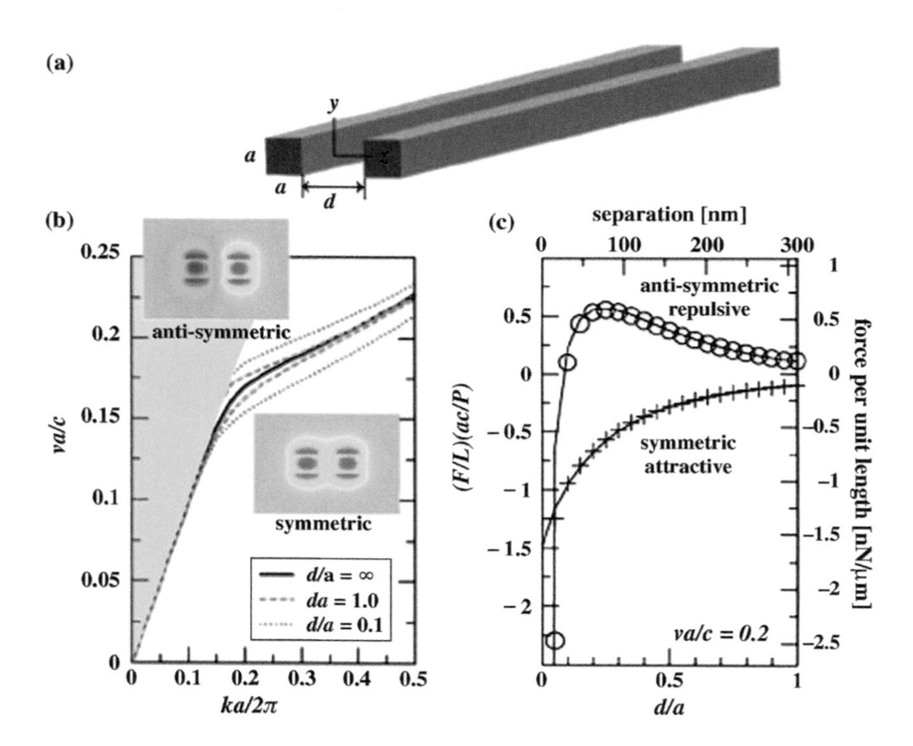

Figure 5.14 Theoretical prediction of optical binding between two freestanding microphotonic waveguides with geometry depicted in (a). The force per unit length is calculated for different guided modes acting on the waveguides at various separations. (d) The symmetric modes produce an attractive force between the waveguides, while antisymmetric modes generate a repulsive force. Reprinted with permission from [64]. Copyright 2005, Optical Society of America.

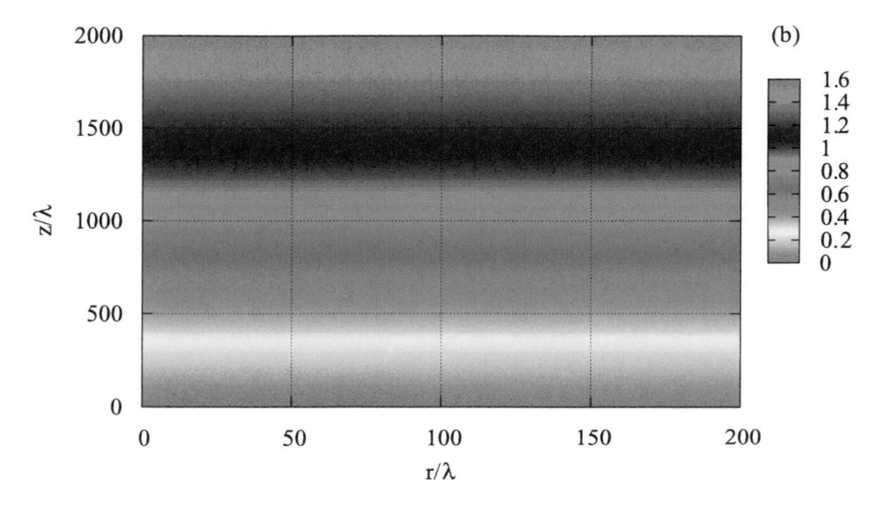

Figure 7.11 Color-coded contour representation of the same data as shown in Figure 7.10. The key represents a scale of the local director orientation angle Ψ relative to the r-axis. It spans the angular range $\Psi = 0$ at the bottom (*red*) to $\Psi = \pi/2$ at the top (*magenta*).

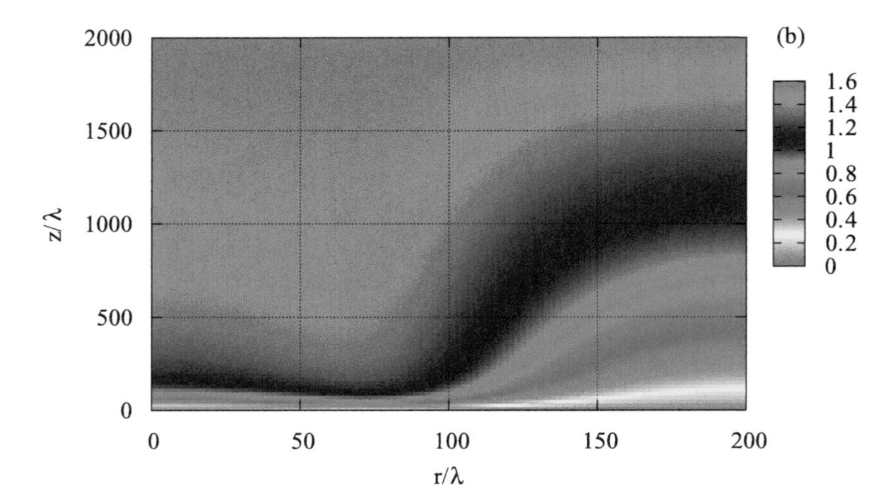

Figure 7.13 Color-coded contour representation of the same data as shown in Figure 7.12. The key represents a scale of the local director orientation angle Ψ relative to the r-axis. It spans the angular range $\Psi = 0$ at the bottom (*red*) to $\Psi = \pi/2$ at the top (*magenta*).

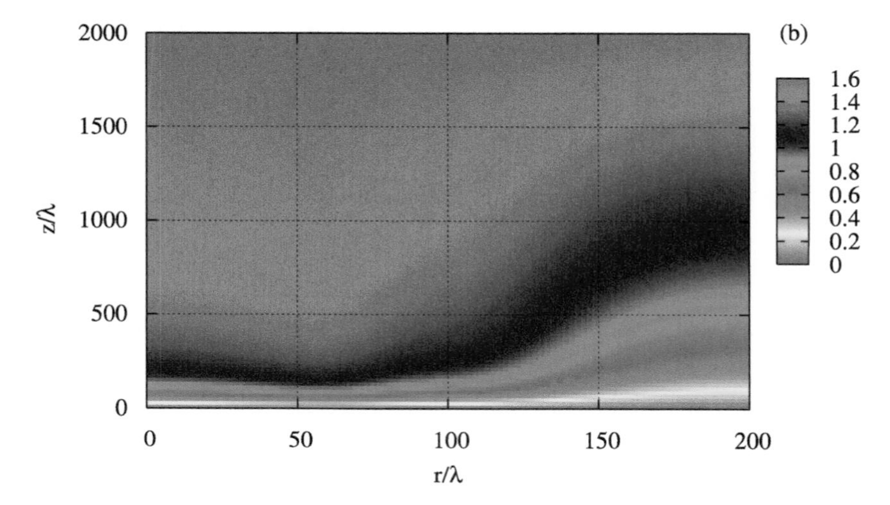

Figure 7.15 Color-coded contour representation of the same data as shown in Figure 7.14. The key represents a scale of the local director orientation angle Ψ relative to the r-axis. It spans the angular range $\Psi = 0$ at the bottom (*red*) to $\Psi = \pi/2$ at the top (*magenta*).

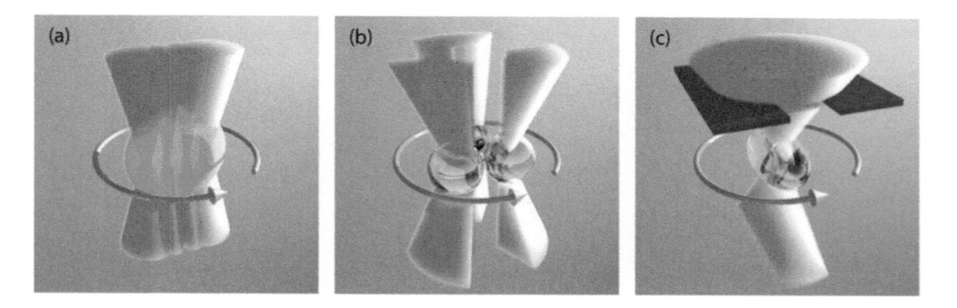

Figure 9.1 Rotation in optical tweezers using the rotation of intensity shaped beams. Sato and colleagues (a) used a Hermite–Gaussian laser mode to trap and rotate a red blood cell, Paterson and colleagues (b) used a spiral interference pattern, and O'Neil and colleagues (c) used a rotating mechanical aperture to shape a Gaussian beam.

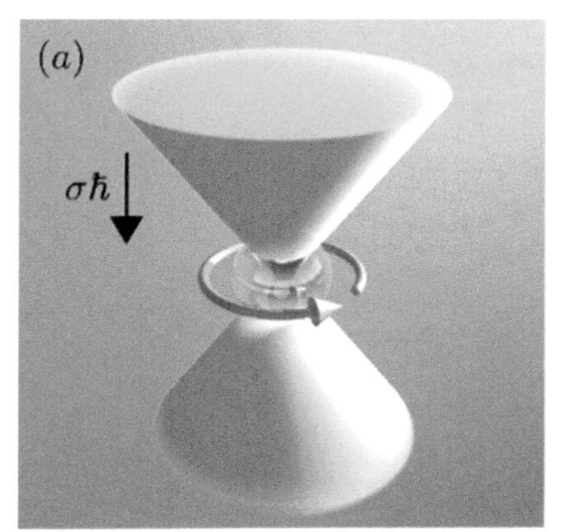

Figure 9.2 Angular momentum transfer in optical tweezers. Rotation of a trapped object can be induced by (a), the transfer of spin angular momentum using a circularly polarized beam or (b), the transfer of orbital angular momentum using a beam such as a high-order Laguerre–Gaussian or Bessel beam.

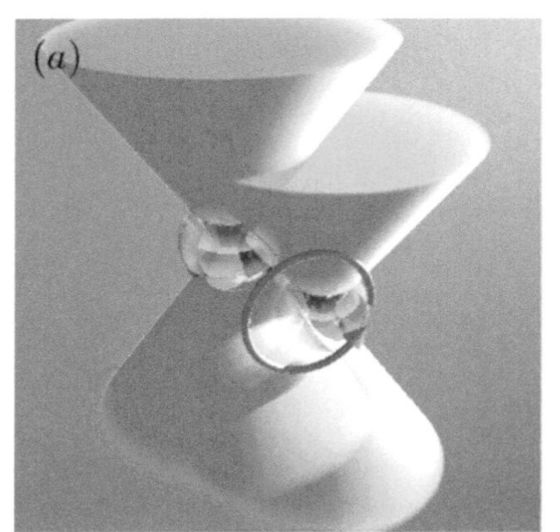

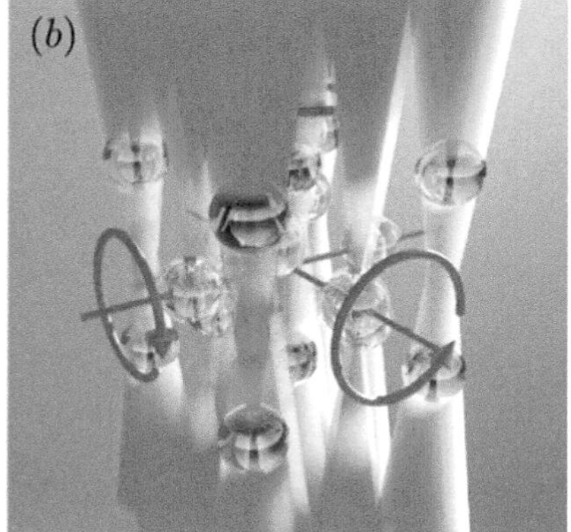

Figure 9.3 Out of plane rotation in optical tweezers. (a) Bingelyte and colleagues used a spatial light modulator to generated two independent optical traps that were controlled independently. (b) Sinclair and colleagues used a similar setup to generate a diamond unit cell using silica particles. In these examples, the rotation axis can be perpendicular to the optical axis of the trapping beams.

Figure 9.4 A helically shaped object placed in an optical tweezers will introduce orbital angular momentum to the transmitted or scattered light. The corresponding reaction torque will result in rotation of the object.

Figure 9.5 Applications of rotational control in optical tweezers. Leach and colleagues used counter-rotating vaterite spheres placed in a microfluidic channel to generate an optically driven micropump.

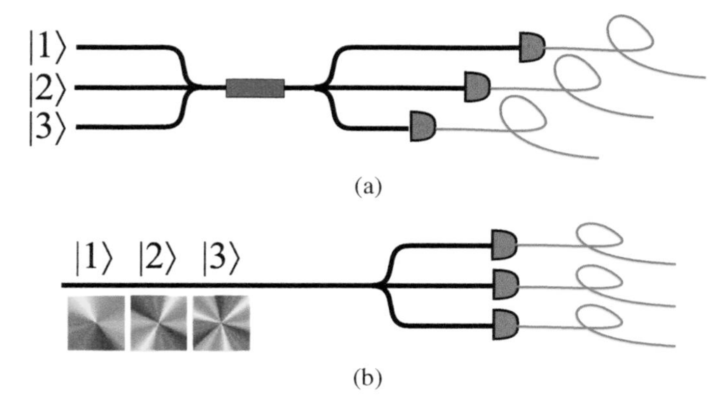

Figure 11.1 The data capacity of quantum communication can be increased by combining many two-bit systems as in time-binning (a), or by using higher-dimensional base system like the OAM of light (b).

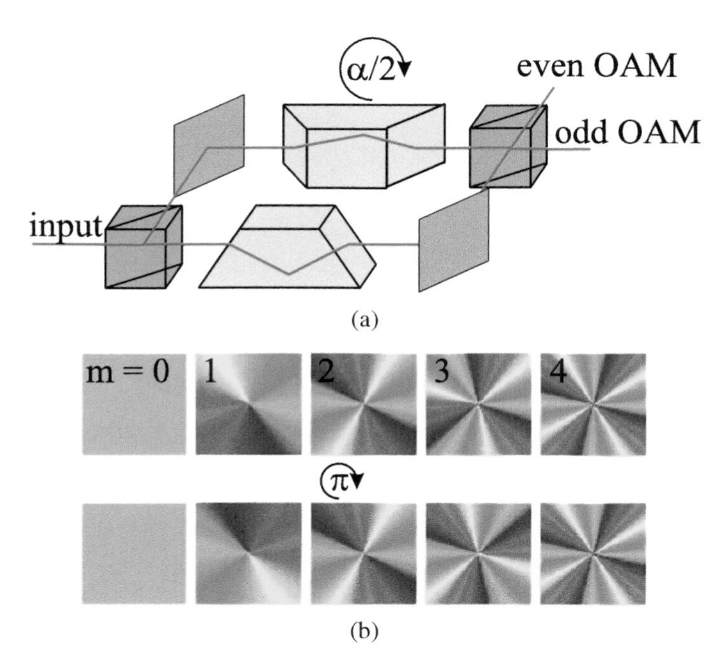

Figure 11.2 (a) The first stage of the OAM channel analyzer [15] is sorting photons in even and odd OAM classes. Two Dove prisms at an angle $\pi/2$ rotate the mode profile by $\alpha = \pi$. (b) Even OAM modes with twofold symmetry remain the same under rotation by π, whereas odd OAM modes become exactly out of phase.

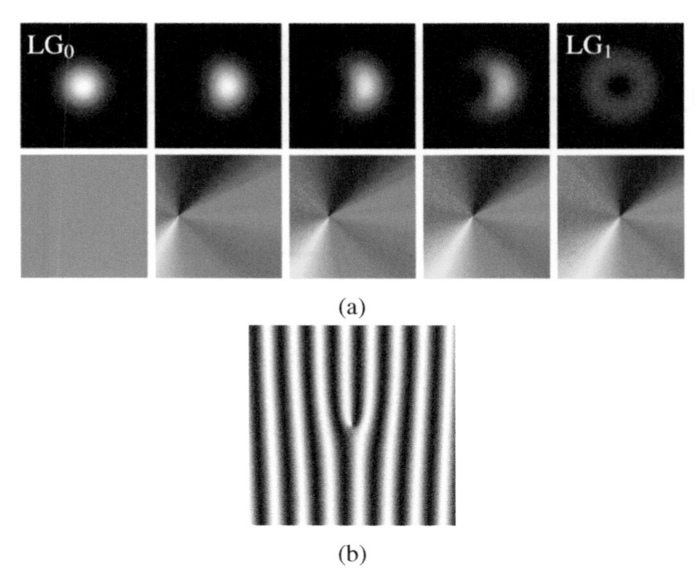

Figure 11.3 (a) Intensities (*top*) and phases (*bottom*) of various superpositions of the fundamental mode, LG_0, and a mode carrying one unit of OAM, LG_1. (b) Hologram generating the LG_1 mode.

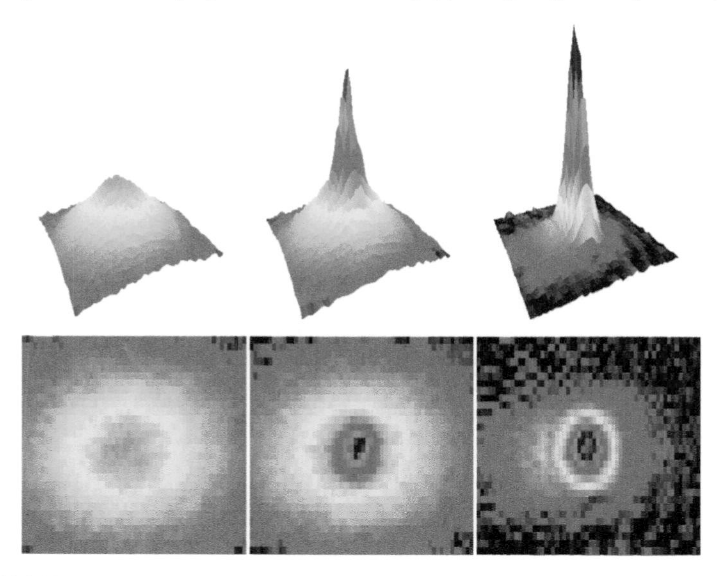

Figure 12.2 The onset of BEC is seen as a sharp peak in the density in the center of the trap. In the figures, the temperature is lowered from left to right. To the far right, we see a pure condensate with a negligible thermal component. The pictures are from the BEC experiment at University of Strathclyde, Glasgow, UK [22].

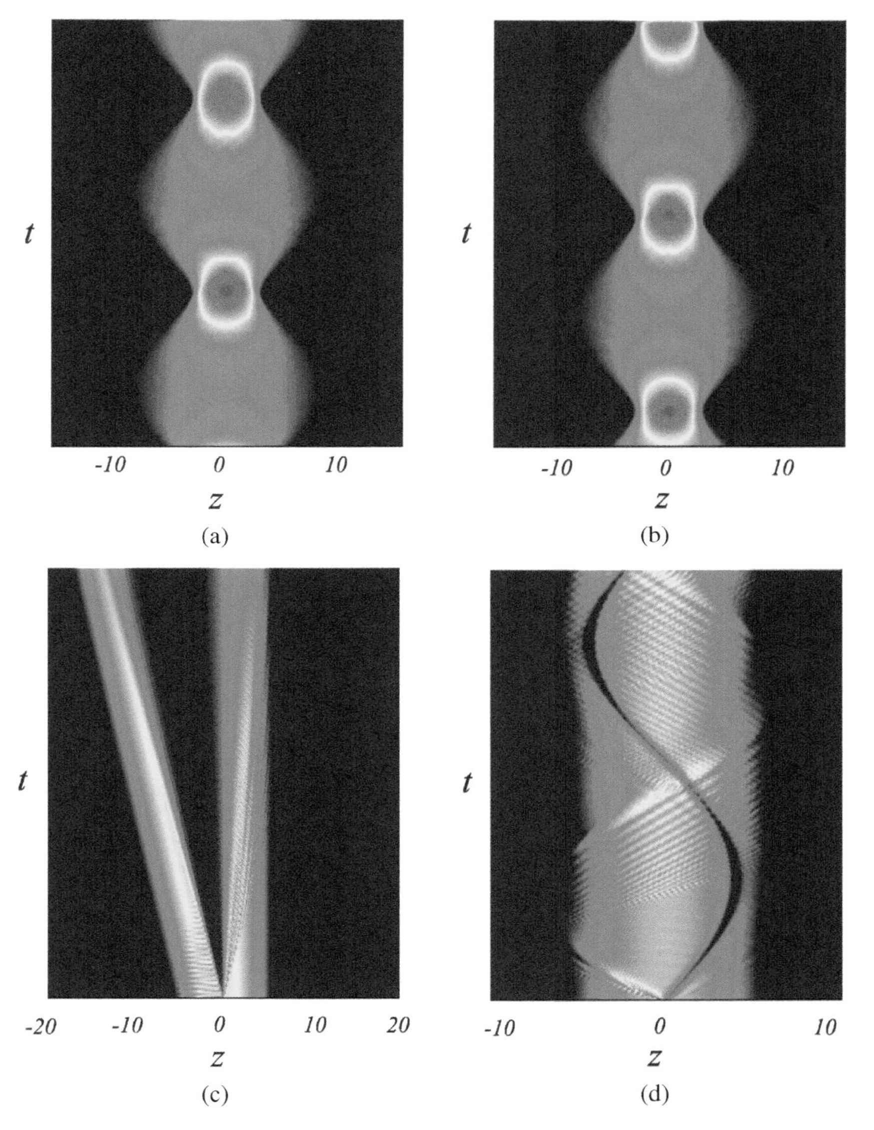

Figure 12.5 The condensate is situated in a harmonic trap. (a) and (b) An imprinted phase that is quadratic in z induces defocusing or focusing, depending upon the sign of the phase gradient. (c) If the phase is chosen such that it is zero for $z > 0$ and linear in z for $z < 0$, the result is a splitting of the cloud where part of the cloud is separated and the remaining part stays stationary during a time much smaller than $1/\omega_z$. (d) With a sharp phase slip imprinted, the result is a dark soliton that oscillates in the cloud (see Figure 12.3).

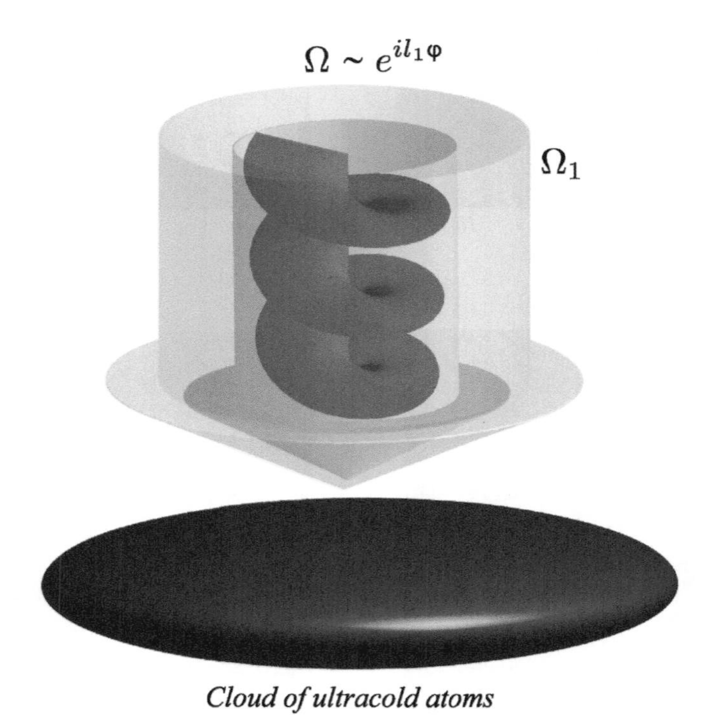

Figure 12.7 At least one of the two coupling beams in the EIT configuration should have an orbital angular momentum.

where $\langle \mathbf{F}_{\text{diss}} \rangle_{klp}$ is the dissipative force

$$\left\langle \mathbf{F}_{\text{diss}}(\mathbf{R}, \mathbf{V}) \right\rangle_{klp} = 2\hbar \Gamma \Omega_{klp}^2(\mathbf{R}) \left(\frac{\nabla \Theta_{klp}(\mathbf{R})}{\Delta_{klp}^2(\mathbf{R}, \mathbf{V}) + 2\Omega_{klp}^2(\mathbf{R}) + \Gamma^2} \right), \tag{16}$$

and $\langle \mathbf{F}_{\text{dipole}}(\mathbf{R}, \mathbf{V}) \rangle_{klp}$ is the dipole force

$$\begin{aligned} &\left\langle \mathbf{F}_{\text{dipole}}(\mathbf{R}, \mathbf{V}) \right\rangle_{klp} \\ &\quad = -2\hbar \Omega_{klp}(\mathbf{R}) \nabla \Omega_{klp} \left(\frac{\Delta_{klp}(\mathbf{R}, \mathbf{V})}{\Delta_{klp}^2(\mathbf{R}, \mathbf{V}) + 2\Omega_{klp}^2(\mathbf{R}) + \Gamma^2} \right). \end{aligned} \tag{17}$$

The effective detuning $\Delta_{klp}(\mathbf{R}, \mathbf{V})$ is now both position- and velocity-dependent.

$$\Delta_{klp}(\mathbf{R}, \mathbf{V}) = \Delta_0 - \mathbf{V} \,.\, \nabla \Theta_{klp}(\mathbf{R}, \mathbf{V}). \tag{18}$$

The dissipative force is due to absorption followed by spontaneous emission of light by the atom, while the dipole force is seen to be proportional to the gradient of the Rabi frequency. Both types of force feature prominently in atom cooling and trapping, with the dissipative force creating a net a frictional force in optical molasses, and the dipole force trapping the atom in regions of extremum field intensity.

The inference that a light-induced torque is automatically created in this context can be confirmed by examining at the velocity-independent forces. For $\mathbf{V} = 0$ and for $z \ll z_R$, we have

$$\left\langle \mathbf{F}_{\text{diss}}^0(\mathbf{R}) \right\rangle_{klp} = \frac{2\hbar \Gamma \Omega_{klp}^2(\mathbf{R})}{\Delta_0^2 + 2\Omega_{klp}^2(\mathbf{R}) + \Gamma^2} \left[k\hat{\mathbf{z}} + \frac{l}{r}\hat{\phi} \right]. \tag{19}$$

There are thus two components of force: an axial component and an azimuthal component. The latter has a nonvanishing moment about the axis, i.e., a torque given by

$$\mathcal{T} = \frac{2\hbar \Gamma \Omega_{klp}^2(\mathbf{R})}{\Delta_0^2 + 2\Omega_{klp}^2(\mathbf{R}) + \Gamma^2} l\hat{\mathbf{z}}. \tag{20}$$

In the saturation limit, where $\Omega \gg \Delta_0$ and $\Omega \gg \Gamma$, we have

$$\mathcal{T} \approx \hbar l \Gamma \hat{\mathbf{z}}. \tag{21}$$

This simple form of the light-induced torque was first pointed out by Babiker and colleagues [20]. In general, the torque is velocity- and position-dependent, and therefore changes along the path of the atom.

7.4.4 Dipole Potential

The velocity-independent dipole force can be derived from the dipole potential

$$\big\langle U(\mathbf{R})\big\rangle_{klp} = \frac{\hbar\Delta_0}{2}\ln\left[1+\frac{2\Omega_{klp}^2(\mathbf{R})}{\Delta_0^2+\Gamma^2}\right], \tag{22}$$

such that $\langle \mathbf{F}^0_{\text{dipole}}\rangle_{klp} = -\nabla\langle U(\mathbf{R})\rangle_{klp}$. This potential would trap atoms in the high intensity regions of the beam for $\Delta_0 < 0$ (red-detuning). For blue detuning, $\Delta_0 > 0$, the trapping would be in the dark regions of the field. For example, consider the LG mode for which $l = 1$, $p = 0$. On the plane of the beam waist $z = 0$, the potential minimum occurs at $r = r_0 = w_0/\sqrt{2}$. For a beam propagating along the z-axis, the locus of the potential minimum is a circle in the xy plane given by

$$x^2 + y^2 = r_0^2. \tag{23}$$

Expanding $\langle U(\mathbf{R})\rangle_{k10}$ about r_0 we have

$$\langle U\rangle_{k10} \approx U_0 + \frac{1}{2}\Lambda_{k10}(r-r_0)^2, \tag{24}$$

where $|U_0|$ is the potential depth given by

$$|U_0| = \frac{1}{2}\hbar|\Delta_0|\ln\left[1+\frac{2\Omega_{k10}^2(r_0)}{\Delta_0^2+\Gamma^2}\right] \tag{25}$$

and Λ_{k10} is an elastic constant given by

$$\Lambda_{k10} = \frac{4\hbar|\Delta_0|}{\Delta_0^2 + 2e^{-1}\Omega_{k00}^2 + \Gamma^2}\left(\frac{e^{-1}\Omega_{k00}^2}{w_0^2}\right). \tag{26}$$

An atom of mass M, trapped if its energy is less than $|U_0|$, will exhibit a vibrational motion about $r = r_0$ of angular frequency approximately equal to $\sqrt{\Lambda_{k10}/M}$.

7.5 THE DOPPLER SHIFT

The force expressions contain the effective detuning Δ_{klp} defined by

$$\Delta_{klp} = \omega - \omega_0 - \nabla\Theta_{klp}\,.\,\mathbf{V}, \tag{27}$$

this can be written as

$$\Delta_{klp} = \omega - \omega_0 - \delta, \tag{28}$$

where δ is an effective Doppler shift associated with the beam. In the limit $z \ll z_R$, we have

$$\delta = kV_z + \frac{l}{r}V_\phi, \tag{29}$$

where V_z and V_ϕ are the components of the atomic velocity in cylindrical polar coordinates. The first term is the Doppler shift associated with the axial component, as one would expect from a plane wave traveling along the z-axis. The second term is entirely new, arising from the orbital angular momentum content of the beam. It is important to note that this azimuthal Doppler shift increases with the angular momentum quantum number l.

7.5.1 Trajectories

Newton's second law determines the form of the atom dynamics, subject to initial conditions. The solutions lead to the trajectory $\mathbf{R}(t)$, and they also determine the evolution of other variables of the system. Unfortunately $\mathbf{R}(t)$ cannot, in general, be determined analytically, and it is necessary to proceed using numerical analysis. It is easy to check that the trajectories for two cases in which an Mg^+ ion is subject to single separate beams differing only in the sign of l will only display a reversal of the direction of rotation. This is consistent with the existence of the light-induced torque.

7.5.2 Multiple Beams

Doppler cooling manifests itself in the so-called optical molasses configurations in one, two, and three dimensions. For beams endowed with OAM, a description of optical molasses requires the specification of individual field distributions to be referred to the laboratory coordinate system. We therefore need to apply multiple coordinate transformations with reference to the original Cartesian axes. The total force acting on the atom is the vector sum of individual forces in a given configuration of light beams.

For a light beam of frequency ω, axial wave vector k, and quantum numbers l and p coupled to an atom or an ion at a general position vector $\mathbf{R} = (r, \phi, z)$ in cylindrical coordinates, the phase $\Theta_{klp}(\mathbf{R})$ and the Rabi frequency $\Omega_{klp}(\mathbf{R})$ can be taken as

$$\Theta_{klp} = l\phi + kz \tag{30}$$

and

$$\Omega_{klp}(\mathbf{R}) \approx \Omega_0\left(\frac{r\sqrt{2}}{w_0}\right)^{|l|} \exp(-r^2/w_0^2) L_p^{|l|}\left(\frac{2r^2}{w_0^2}\right). \tag{31}$$

These expressions are applicable for a Laguerre–Gaussian beam in the limit $z \ll z_R$, where z_R is the Rayleigh range, and $w(z) = w_0$, i.e., we ignore all beam curvature effects.

The total forces acting upon the atomic center of mass moving with velocity $\mathbf{V} = \dot{\mathbf{R}}$ are given above in the steady state, but with the approximate phase $\Theta_{klp}(\mathbf{R})$ and Rabi frequency $\Omega_{klp}(\mathbf{R})$, as in equations (30) and (31). These are given in cylindrical polar coordinates, with the beams propagation parallel to the z-axis. However, in order to consider multiple beams, it is convenient to begin by expressing the position dependence in the Rabi frequency and phase in Cartesian coordinates $\mathbf{R} = (x, y, z)$, simply by substituting $r = \sqrt{x^2 + y^2}$ and $\phi = \arctan(y/x)$. A beam propagating in an arbitrary direction is determined by applying two successive transformations. The first transformation is a rotation of the beam about the y-axis by an angle θ, and the second is a subsequent rotation about the x-axis by an angle ψ. This signifies to the following overall coordinate transformation:

$$x \to x' = \cos(\theta)x + \sin(\theta)z, \tag{32}$$

$$y \to y' = -\sin(\theta)\sin(\psi)x + \cos(\psi)y + \cos(\theta)\sin(\psi)z, \tag{33}$$

$$z \to z' = -\sin(\theta)\cos(\psi)x - \sin(\psi)y + \cos(\theta)\cos(\psi)z. \tag{34}$$

By suitable choice of the angles θ and ψ, we obtain the force distribution due to a twisted light beam propagating in any direction. In this manner, we are able to consider geometrical arrangements involving counterpropagating beams (especially those corresponding to one-, two-, and three-dimensional optical molasses configurations) for beams possessing OAM.

We will concentrate on the case of optical molasses of magnesium ions Mg^+ with a transition of frequency ω_0 corresponding to the transition wavelength $\lambda = 280.1$ nm and transition rate $\Gamma = 2.7 \times 10^8$ s^{-1}. The Mg^+ mass $M = 4.0 \times 10^{-26}$ kg. We consider red-detuned light to induce trapping in areas of high intensity, $\Delta_0 = -\Gamma$ and $w_0 = 35\lambda$. The equation of motion for the Mg^+ ion is now

$$M\frac{d^2}{dt^2}\mathbf{R}(t) = \sum_i \langle \mathbf{F}_i \rangle, \tag{35}$$

where the sum is taken over individual (total) force contributions from each beam present. In the one-dimensional molasses configuration, a pair of counterpropagating beams is set up along the z-axis. The specification of the force due to the beam propagating in the negative z-direction is shown in equations (32), (33), and (34) with $\theta = \pi$ and $\psi = 0$. Figure 7.6 shows the trajectory of the Mg^+ ion with $l_1 = -l_2 = 1$ and $p_1 = p_2 = 0$. The initial radial position is $r = 10\lambda$, and the initial velocity is $v = 5\hat{z}$ ms^{-1}. The motion is for a time duration equal to 2×10^5 Γ^{-1}. It is clear that

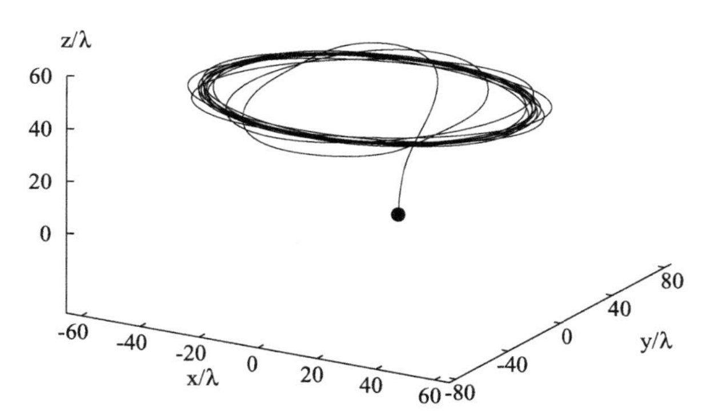

Figure 7.6 Path of an Mg^+ ion in the one-dimensional twisted optical molasses created by two counterpropagating Laguerre–Gaussian beams, with $l_1 = -l_2 = 1$ and $p_1 = p_2 = 0$ propagating along the z-axis. The initial velocity is $v = 5\hat{z}$ ms^{-1}.

the atom is slowed down to a halt in the z-direction, while in its motion in the x–y plane it is attracted to the region of high-beam intensity at approximately $r_0 > w_0/\sqrt{2}$. The long-time motion is a uniform circular motion, as can be deduced from Figure 7.7, which exhibits the corresponding evolution of the velocity components. Once the Mg^+ ion is trapped axially, it continues to rotate clockwise about the axis, subject to a torque which, in the saturation limit, is given by $|\langle \boldsymbol{T} \rangle| \approx l_1 \hbar \Gamma - l_2 \hbar \Gamma = 2\hbar \Gamma$. The motion of the ion gives rise to an electric current equal to $e/\tau \equiv e v_s / 2\pi r_0$. With v_s of about 2 ms^{-1} and $r_0 \approx w_0 = 35\lambda$, we have an ionic current of the order of a femtoAmp if a single ion is involved. It is significant that the current scales with the number of trapped ions; obviously, 1 million or so ions can produce a current on the nA scale.

7.5.3 Two- and Three-Dimensional Molasses

We will now introduce a second pair of counterpropagating beams along the x-axis, and one pair could be characterized by a different width w_0'. The total force is now the vector sum of individual forces from the four beams. The specifications of three of the beams is made with the help of transformation equations (32) to (34). The trajectories of two Mg^+ ions positioned at different initial points, each having an initial velocity of $v_z = 5$ ms^{-1}, are shown in Figure 7.8, where each of the four beams has an azimuthal index $l = 1$, and radial index $p = 0$. Since by choice of l-values in this case, the total torque arising from either pair of beams is zero, each ion ends up at a specific fixed point where it remains essentially motionless. To understand this, note that the deepest potential well is four times as deep as that of a single beam, with the potential minima situated along the locus of spatial points defined simultaneously by two equations: $x^2 + y^2 = w_0^2/2$ and $y^2 + z^2 = w_0'^2/2$. For

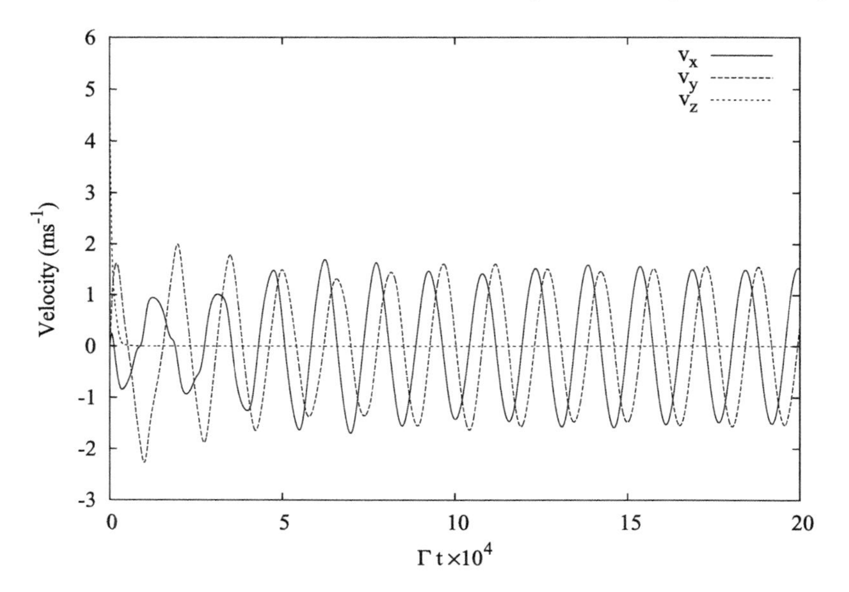

Figure 7.7 Evolution of the three velocity components of the Mg^+ ion in the one-dimensional optical molasses shown in Figure 7.6. The axial component of the velocity rapidly approaches zero, consistent with Doppler cooling, while the in-plane components v_x and v_y show a convergence toward uniform circular motion.

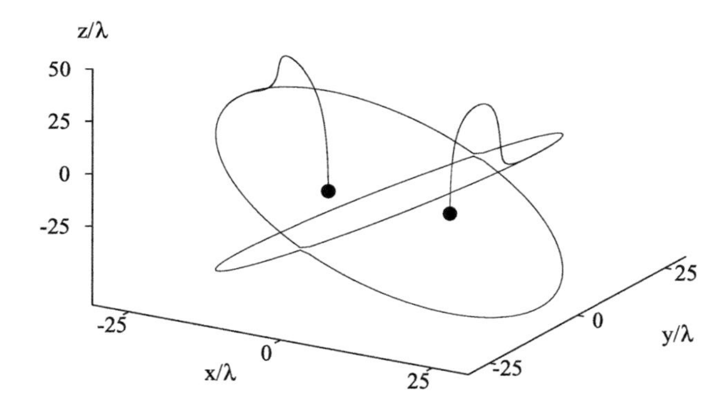

Figure 7.8 Trajectories of two Mg^+ ions with different initial locations subject to a two-dimensional optical molasses formed by two pairs of counterpropagating twisted beams, with $l_i = 1$ and $p_i = 0$ for $i = 1 - 4$. Each ion ends up motionless on the locus of lowest potential energy minima corresponding to two oblique orthogonal circles, as explained in text.

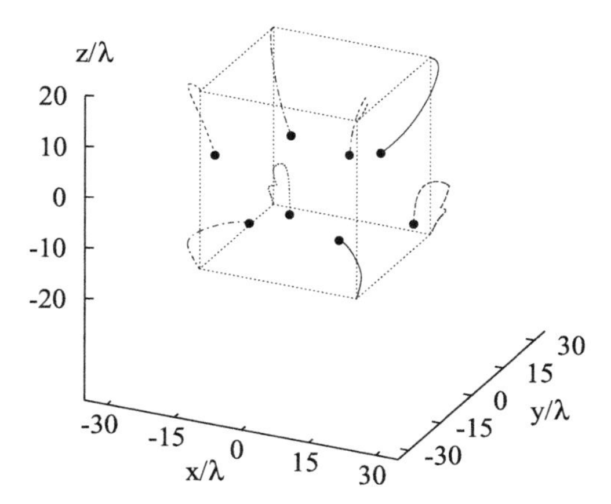

Figure 7.9 Trajectories of eight Mg^+ ions in a three-dimensional twisted optical molasses formed by three pairs of counterpropagating Laguerre–Gaussian beams with $l_i = 1$ and $p_i = 0$ for $i = 1 - 8$. The initial velocity of each ion is $v_z = 5\ \mathrm{ms}^{-1}$. The ions end up motionless at the corners of a cube of side w_0.

$w_0' = w_0$, these equations describe two orthogonal oblique circles representing the intersection curves of two cylinders of radii $w_0/\sqrt{2}$. Solving for x and y, we have $x = \pm z$ and $y = \pm\sqrt{w_0^2/2 - z^2}$. The locus of spatial points where the dipole potential is minimum can be described by the parametric equations $x(u) = (w_0'/\sqrt{2})\cos u$; $y(u) = (w_0'/\sqrt{2})\sin u$; $z(u) = \pm\sqrt{w_0^2/2 - (w_0'^2/2)\sin^2 u}$. All Mg^+ ions in the two-dimensional configuration of orthogonal counterpropagating pairs of twisted beams will be trapped at points lying on one of the two oblique circles, as determined by the initial conditions. An ensemble of Mg^+ ions with a distribution of initial positions and velocities will populate the two circles, producing two orthogonal essentially static Mg^+ ion loops. Associated with this system of charges would be a Coulomb field whose spatial distribution, for example, for ions uniformly distributed in the ring can easily be evaluated. When the values of l are such that each pair of beams generates a torque, the motion becomes more complicated, but the ions will seek to congregate in the region of potential minima, while responding to the combined effects of two orthogonal torques and orthogonal axial cooling forces.

When a third pair of counterpropagating beams is added to the two-dimensional configuration, orthogonal to the plane containing the original beams, we have a three-dimensional configuration. In this case, the deepest potential minima are located at eight discrete points defined by the coordinates $x = \pm\frac{w_0}{2}$, $y = \pm\frac{w_0}{2}$, and $z = \pm\frac{w_0}{2}$. These coincide with the eight corners of a cube of side w_0, centered at the origin of coordinates (see Figure 7.9).

7.6 ROTATIONAL EFFECTS ON LIQUID CRYSTALS

As observed earlier, a liquid crystal is another physical system where new physical effects should arise when subject to twisted light. The most prominent effect in this case can be expected to be an optical influence on the angular distribution of the director $\hat{\mathbf{n}}(\mathbf{r})$ in the illuminated region. To focus on a case of direct and practical relevance, let us consider a liquid crystal film of thickness d occupying the region $0 \leqslant z \leqslant d$, with the light incident in such a manner that the beam waist coincides with the plane $z = 0$.

To begin with, we first note that in the absence of the light, the free energy density of the system is given by [46]

$$\mathcal{F}_0(\mathbf{r}) = \frac{1}{2}\mathcal{K}_1\big[\nabla \,.\, \hat{\mathbf{n}}(\mathbf{r})\big]^2 + \frac{1}{2}\mathcal{K}_2\big[\hat{\mathbf{n}} \,.\, \nabla \times \hat{\mathbf{n}}(\mathbf{r})\big]^2 + \frac{1}{2}\mathcal{K}_3\big[\hat{\mathbf{n}} \times \nabla \times \hat{\mathbf{n}}(\mathbf{r})\big]^2, \quad (36)$$

where $\mathcal{K}_{1,2,3}$ are the Frank elastic constants and the terms represent splay, twist, and bend contributions to the free energy density, respectively. As an approximation, we set $\mathcal{K}_1 = \mathcal{K}_2 = \mathcal{K}_3 = \mathcal{K}$, corresponding to elastic isotropy. Then equation (36) becomes

$$\mathcal{F}_0(\mathbf{r}) = \frac{1}{2}\mathcal{K}\big\{\big[\nabla \,.\, \hat{\mathbf{n}}(\mathbf{r})\big]^2 + \big[\nabla \times \hat{\mathbf{n}}(\mathbf{r})\big]^2\big\}. \quad (37)$$

Symmetry considerations suggest that $\hat{\mathbf{n}}$ can be written in terms of $\Psi(\mathbf{r})$, the local azimuthal angle such that $\hat{\mathbf{n}} = (\sin\Psi, \cos\Psi, 0)$. The free energy expression thus reduces to

$$\mathcal{F}_0 = \frac{1}{2}\mathcal{K}\nabla\Psi \,.\, \nabla\Psi. \quad (38)$$

Next, we will consider the electric field of a twisted light beam at a general position vector $\mathbf{r} = (r, \phi, z)$ in cylindrical coordinates, expressed as

$$\mathbf{E}_{klp}(\mathbf{r}) = \mathbf{f}_{klp}(\mathbf{r})e^{i\Theta_{klp}(\mathbf{r})}, \quad (39)$$

where $\mathbf{f}_{klp}$ is the electric field amplitude function corresponding to the expressions given in equation (4). At frequencies far removed from a molecular resonance, the coupling of the light can be cast in terms of the dielectric properties of the liquid crystal. Ignoring any frequency dispersion, the application of the twisted light thus leads to an additional field-dependent free energy term $\mathcal{F}_{\text{int}}$ given by [46]

$$\mathcal{F}_{\text{int}} = -\frac{1}{4}\varepsilon_0\varepsilon_a(\hat{\boldsymbol{n}} \,.\, \boldsymbol{E}_{klp})(\hat{\boldsymbol{n}} \,.\, \boldsymbol{E}^*_{klp}), \quad (40)$$

where ε_a is the dielectric isotropy of the liquid crystal. Assuming that the field is plane-polarized along the x-axis, we find that total free energy can be written as

$$\mathcal{F} = \mathcal{F}_0 + \mathcal{F}_{\text{int}} = \frac{1}{2}\mathcal{K}\nabla\Psi \, . \, \nabla\Psi - \Lambda_{klp} \sin^2 \Psi(r, z) \tag{41}$$

where $\Lambda_{l,p}$ is given by

$$\Lambda_{klp} = \frac{1}{2}\varepsilon_0 \varepsilon_a |\mathbf{f}_{klp}|^2. \tag{42}$$

Using Landau's free energy formalism, it emerges that Ψ satisfies a second-order partial differential equation in two dimensions (r, z):

$$\mathcal{K}\left\{\frac{\partial^2\Psi}{\partial r^2} + \frac{1}{r}\frac{\partial\Psi}{\partial r} + \frac{\partial^2\Psi}{\partial z^2}\right\} - \Lambda_{l,p} \sin 2\Psi(r, z) = 0. \tag{43}$$

The above theory has been applied to the case of the nematic liquid crystal 5CB (pentylcyanotrphenyl) doped with an anthraquinone derivative dye (AD1). The system is modeled as an infinitely wide film of thickness d, sandwiched between two thin glass plates, those which fix the director angle along the top $z = d$ and bottom $z = 0$. The general boundary conditions are then in the form

$$\begin{aligned} \Psi(r, 0) &= \Psi_0, \\ \Psi(r, d) &= \Psi_d \end{aligned} \tag{44}$$

and

$$\left.\frac{\partial\Psi(r, z)}{\partial r}\right|_{r=\infty} = 0. \tag{45}$$

The relevant parameters in this case are $K = 0.64 \times 10^{-12}$ N, and $\varepsilon_a = 0.5832$ [47]. The thickness of the film is taken as $d = 2000\lambda$, where the wavelength is $\lambda = 600$ nm and the intensity of the light is 10^8 W m^{-2}; the beam waist is taken as $w_0 = 50\lambda$. We consider the cases of $(l, p) = (5, 0)$ and $(l, p) = (5, 1)$, but the theory can be applied to any l, p mode, and we will choose other modes for illustration.

Figures 7.10 and 7.11 display a vector field distribution and the corresponding color-coded plot of the director orientations in the r, z plane with the boundary conditions $\Psi_0 = 0$ and $\Psi_d = \pi/2$ corresponding to the situation in the absence of the twisted light. Figures 7.12 and 7.13 show the modified landscape once a twisted light beam with $l = 5$, $p = 0$ has been switched on. It is clear that there is a marked reorientation of the directors, when compared to the situation in the beam-free case. Figures 7.14 and 7.15 concern the case where $l = 5$, $p = 1$ for the applied twisted light mode. It is shown that differences in the variation of intensity of the light manifest themselves in the director reorientation.

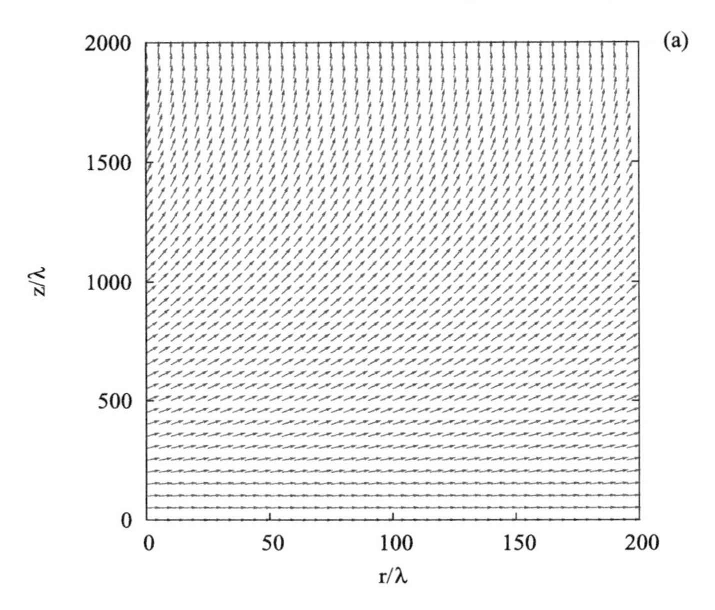

Figure 7.10 Vector field representation of the director orientation of the twisted nematic liquid crystal before application of the twisted light. The orientation at the boundaries $z = 0$ (*bottom*) and $z = d$ (*top*) are fixed as described in text.

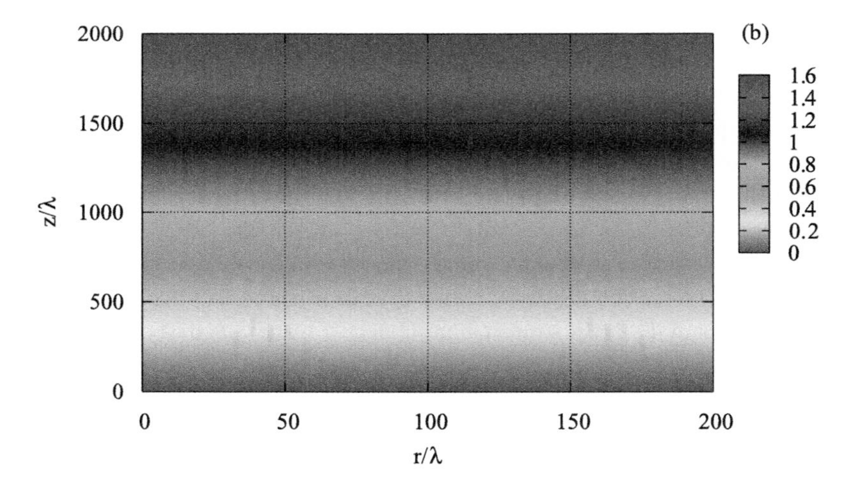

Figure 7.11 Color-coded contour representation of the same data as shown in Figure 7.10. The key represents a scale of the local director orientation angle Ψ relative to the r-axis. It spans the angular range $\Psi = 0$ at the bottom (*red*) to $\Psi = \pi/2$ at the top (*magenta*). See color insert.

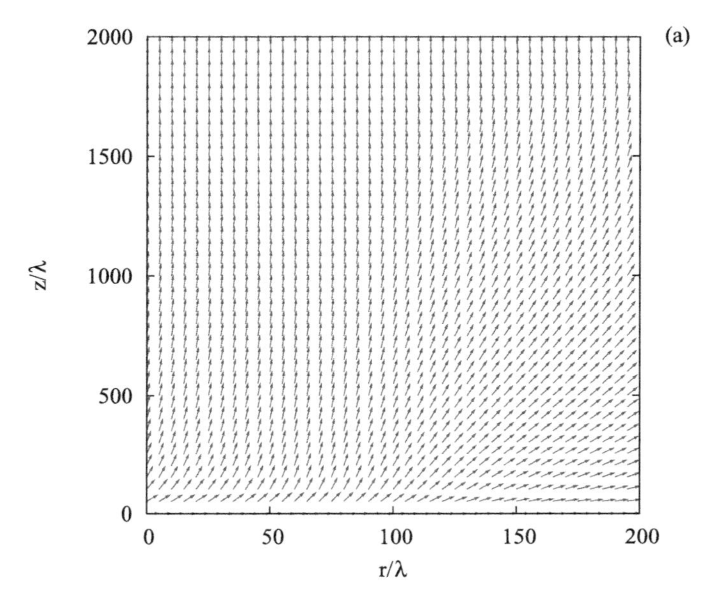

Figure 7.12 Vector field representation of the director orientation of the twisted nematic liquid crystal after the application of the twisted $LG_{l,p}$ light, where $l = 5$, $p = 0$.

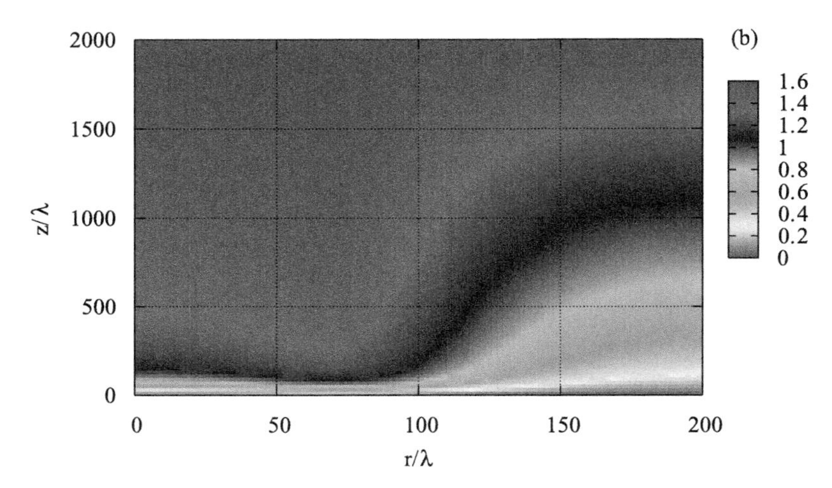

Figure 7.13 Color-coded contour representation of the same data as shown in Figure 7.12. The key represents a scale of the local director orientation angle Ψ relative to the r-axis. It spans the angular range $\Psi = 0$ at the bottom (*red*) to $\Psi = \pi/2$ at the top (*magenta*). See color insert.

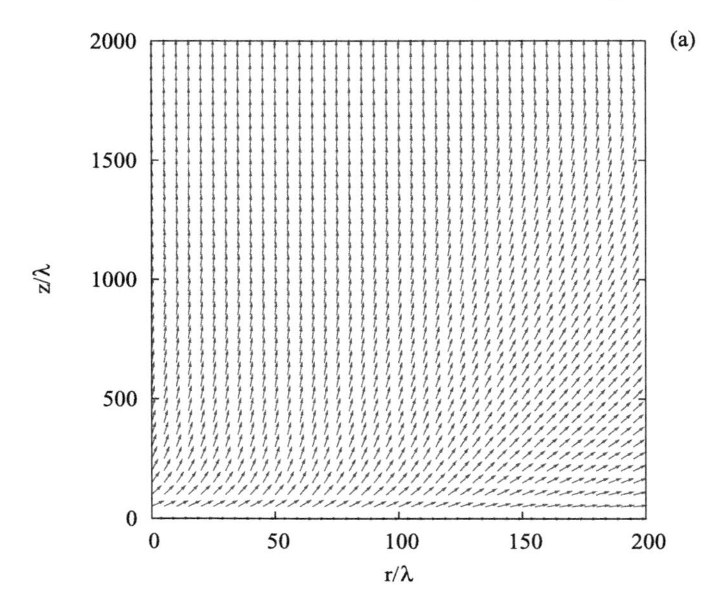

Figure 7.14 Vector field representation of the director orientation of the twisted nematic liquid crystal after the application of the twisted $LG_{l,p}$ with $l = 5$, $p = 1$. Other parameters are described in text.

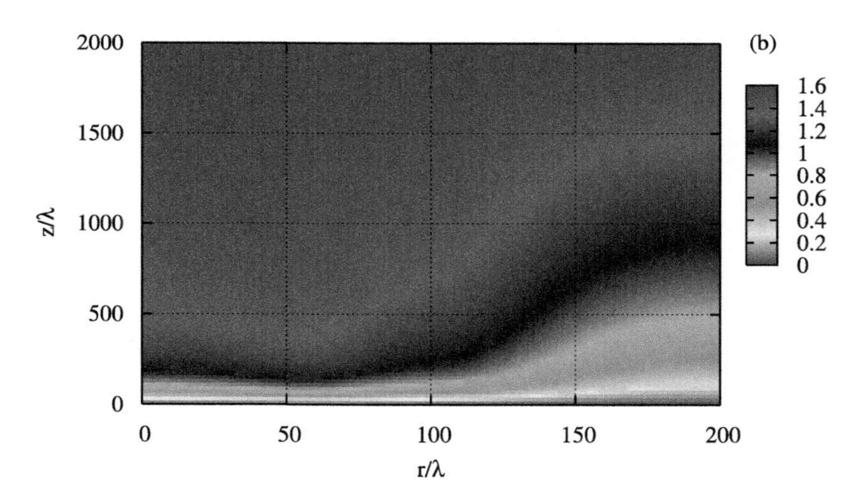

Figure 7.15 Color-coded contour representation of the same data as shown in Figure 7.14. The key represents a scale of the local director orientation angle Ψ relative to the r-axis. It spans the angular range $\Psi = 0$ at the bottom (*red*) to $\Psi = \pi/2$ at the top (*magenta*). See color insert.

7.7 COMMENTS AND CONCLUSIONS

This chapter is concerned primarily with a theory of two-level systems responding to the field of Laguerre–Gaussian light and, as a supplementary topic, the influence of such light on liquid crystals. The primary aim is to present an up-to-date account of work in this branch of optical angular momentum effects. The results for atomic, ionic, and molecular motion display novel features; not only do such particles experience modified translational forces in such fields, but the radiation forces include rotational components that are solely attributable to the OAM of the light. In the transient regime, applicable within the time interval of the order of Γ^{-1}, where Γ is the transition rate from the upper state, the particles experience time-dependent forces and torques leading to characteristic particle trajectories. In the steady-state regime, applicable for times much larger than Γ^{-1}, the particles experience time-independent forces and torques and the motion leads to cooling, trapping, and novel rotational effects in a variety of situations. The saturation light-induced torque has the simple form

$$\boldsymbol{\mathcal{T}} \approx \hbar l \Gamma \hat{z}. \tag{46}$$

The result has the simple interpretation that a single transition transfers angular momentum of magnitude $\hbar l$, and since there are Γ transitions per unit time, the product amounts to a torque of magnitude $\hbar \Gamma l$ along the beam axis $\hat{z}$. This supports the prediction that the orbital angular momentum of light is quantized in units of $\hbar$. We have also seen that the two-level system subject to such light will experience a new Doppler shift associated with the rotational component of the interaction. We have seen that this additional Doppler shift is the source of modification of the dissipative forces responsible for laser cooling. The treatments of optical molasses involving pairs of Laguerre–Gaussian beams in one, two and three dimensions involve a variety of steady-state optical forces with the propensity to produce a novel and highly distinctive behavior. The Laguerre–Gaussian light generates optical potential wells associated with the total dipole force contributions from all the beams present, while the total dissipative force provides a mechanism for cooling or heating the azimuthal motion along with the axial motion. For each pair of counterpropagating beams, the light-induced torque is doubled or annulled, depending on the relative sign of the OAM quantum number l.

It is clear that the subject of atomic and molecular manipulation using structured light, especially light carrying orbital angular momentum, is still at an early stage of development. Experimental work in which atoms, ions, and molecules are trapped in optical potentials generated by multiple beams still needs to be carried out. Although the theoretical predictions presented here involve trapping in the maximum intensity regions, similar predictions, albeit differing in detail, can be made for trapping in the dark regions of higher-order intersecting multiple beams.

ACKNOWLEDGMENTS

The research described in this chapter owes much to work with a number of colleagues and research students, including Les Allen, Vassilis Lembessis, Wei Lai, William Power, Luciana Dávila Romero, Andrew Carter, Mohammad Al-Amri, and Matt Probert. The authors wish to acknowledge their valuable contributions. Financial support from the Science and Engineering Research Council (EPSRC) is gratefully acknowledged.

REFERENCES

[1] D.G. Grier, A revolution in optical manipulation, *Nature 424* (2003) 810.
[2] See Chapters 4–6 and the special journal issue on optical tweezers *J. Mod. Opt. 50* (2003) 1501, and following pages.
[3] K. Dholakia, P. Reece, Optical manipulation takes hold, *Nano Today 1* (2006) 18.
[4] V.S. Letokhov, V.G. Minogin, Laser Light Pressure on Atoms, Gordon and Breach, New York, 1987.
[5] A.P. Kazantsev, G.I. Surdutovich, V.P. Yakovlev, Mechanical Action of Light on Atoms, World Scientific, Singapore, 1990.
[6] C.S. Adams, E. Riis, Laser cooling and trapping of neutral atoms, *Prog. Quantum Electron. 21* (1997) 1.
[7] C.N. Cohen Tannoudji, Nobel lecture: Manipulating atoms with photons, *Rev. Mod. Phys. 70* (1998) 707.
[8] C.E. Weiman, D.E. Pritchard, D.J. Weinland, Atom cooling, trapping and quantum manipulation, *Rev. Mod. Phys. 71* (1999) S253.
[9] C.J. Pethick, H. Smith, Bose–Einstein Condensation in Dilute Gases, Cambridge Univ. Press, 2002.
[10] D.L. Andrews, A.R. Carter, M. Babiker, M. Al-Amri, Transient optical angular momentum effects and atom trapping in multiple twisted beams, *Proc. SPIE Int. Soc. Opt. Eng. 6131* (2006) 613103.
[11] L.C. Dávila Romero, A.R. Carter, M. Babiker, D.L. Andrews, Interaction of Laguerre–Gaussian light with liquid crystals, *Proc. SPIE Int. Soc. Opt. Eng. 5736* (2005) 150.
[12] D.S. Bradshaw, D.L. Andrews, Interactions between spherical nanoparticles optically trapped in Laguerre–Gaussian modes, *Opt. Lett. 30* (2005) 3039.
[13] A.R. Carter, M. Babiker, M. Al-Amri, D.L. Andrews, Transient optical angular momentum effects in light-matter interactions, *Phys. Rev. A 72* (2005) 043407.
[14] A.R. Carter, M. Babiker, M. Al-Amri, D.L. Andrews, Generation of microscale current loops, atom rings and cubic clusters using twisted optical molasses, *Phys. Rev. A 73* (2006) 021401.
[15] K. Volke Sepolveda, V. Garces-Chavez, S. Chavez-Cerda, J. Arlt, K. Dholakia, Orbital angular momentum of a high-order Bessel light beam, *J. Opt. B: Quantum Semidass. Opt. 4* (2002) S82.

[16] D. McGloin, K. Dholakia, Bessel Beams: Diffraction in a new light, *Contemp. Phys. 46* (2005) 15.
[17] C. Lopez-Mariscal, J.C. Gottierrez-Vega, G. Milne, K. Dholakia, Orbital angular momentum transfer in helical Mathieu beams, *Opt. Exp. 14* (2006) 4182.
[18] A.R. Carter, L.C. Dávila Romero, M. Babiker, D.L. Andrews, M.I.J. Probert, Orientational effects of twisted light on twisted nematic liquid crystals, *Phys. B: At. Mol. Opt. Phys. 39* (2006) S523.
[19] L. Allen, M.W. Beijesbergen, R.J.C. Spreeuw, J.P. Woerdman, Orbital angular momentum of light and the transformation of Laguerre–Gaussian laser modes, *Phys. Rev. A 45* (1992) 8185.
[20] M. Babiker, W.L. Power, L. Allen, Light-induced torque on moving atoms, *Phys. Rev. Lett. 73* (1994) 1239.
[21] L. Allen, M. Babiker, W.L. Power, Azimuthal Doppler shift in light beams with orbital angular momentum, *Opt. Commun. 112* (1994) 141.
[22] L. Allen, M. Babiker, W.K. Lai, V.E. Lembessis, Atom dynamics in multiple Laguerre–Gaussian beams, *Phys. Rev. A 54* (1996) 4259.
[23] L. Allen, V.E. Lembessis, M. Babiker, Spin–orbit coupling in free-space Laguerre–Gaussian light beams, *Phys. Rev. A 53* (1996) R2937.
[24] V.E. Lembessis, A mobile atom in a Laguerre–Gaussian laser beam, *Opt. Commun. 159* (1999) 243.
[25] E. Courtade, O. Houde, J.F. Clement, P. Verkerk, D. Hennequin, Dark optical lattice of ring traps for cold atoms, *Phys. Rev. A 74* (2006) 031403.
[26] S.J. van Enk, Selection rules and centre of mass motion of ultracold atoms, *Quantum Opt. 6* (1994) 445.
[27] M. Babiker, C.R. Bennett, D.L. Andrews, L.C. Dávila Romero, Orbital angular momentum exchange in the interaction of twisted light with molecules, *Phys. Rev. Lett. 89* (2002) 143601.
[28] G. Juzeliūnas, P. Ohberg, Slow light in degenerate Fermi gases, *Phys. Rev. Lett. 93* (2004) 033602.
[29] G. Juzeliūnas, P. Ohberg, J. Ruseckas, A. Klein, Effective magnetic fields in degenerate atomic gases induced by light beams with orbital angular momenta, *Phys. Rev. A 71* (2005) 053614.
[30] D.M. Villeneuve, S.A. Asyev, P. Dietrich, M. Spanner, M.Yu. Ivanov, P.B. Corkum, Forced molecular rotation in an optical centrifuge, *Phys. Rev. Lett. 85* (2000) 542.
[31] A. Muthukrishnan, C.R. Stroud Jr., Entanglement of internal and external angular momenta of a single atom, *J. Opt. Quantum Semiclass. Opt. 4* (2002) S73.
[32] M.S. Bigelow, P. Zernom, R.W. Boyd, Breakup of ring beams carrying orbital angular momentum in sodium vapor, *Phys. Rev. Lett. 92* (2004) 083902.
[33] A.V. Bezverbny, V.G. Niz'ev, A.M. Tomaikin, Dipole traps for neutral atoms formed by nonuniformly polarised Laguerre modes, *Quantum Electron. 34* (2004) 685.
[34] A. Alexanderscu, E. De Fabrizio, D. Kojoc, Electronic and centre of mass transitions driven by Laguerre–Gaussian beams, *J. Opt. B: Quantum Semiclass. Opt. 7* (2005) 87.
[35] W. Zheng-Ling, Y. Jiang-Ping, Atomic (or molecular) guiding using a blue-detuned doughnut mode, *Chin. Phys. Lett. 22* (2005) 1386.
[36] J.W.R. Tabosa, D.V. Petrov, Optical pumping of orbital angular momentum of light in cold cesium atoms, *Phys. Rev. Lett. 83* (1999) 4967.
[37] S. Al-Awfi, M. Babiker, Field-dipole orientation mechanism and higher order modes in atom guides, *Mod. Opt. 48* (2001) 847.

[38] M.A. Clifford, J. Arlt, J. Courtial, K. Dholakia, High-order Laguerre–Gaussian laser modes for studies of cold atoms, *Opt. Commun. 156* (1998) 300.

[39] Y. Song, D. Millam, W.T. Hill III, Generation of nondiffracting Bessel beams by use of a spatial light modulator, *Opt. Lett. 24* (1999) 1805.

[40] T.A. Wood, H.F. Gleeson, M.R. Dickinson, A.J. Wright, Mechanisms of optical angular momentum transfer to nematic liquid crystalline droplets, *Appl. Phys. Lett. 84* (2004) 4292.

[41] D.P. Rhodes, D.M. Gherardi, J. Livesey, D. McGloin, H. Melville, T. Freegarde, K. Dholakia, Atom guiding along high order Laguerre–Gaussian light beams formed by spatial light modulation, *J. Mod. Opt. 53* (2006) 547.

[42] P. Domokos, H. Ritsch, Mechanical effects of light in optical resonators, *Opt. Soc. Am. B 20* (2007) 1098.

[43] B. Piccirillo, C. Toscano, F. Vetrano, E. Santamento, Orbital and spin photon angular momentum transfer in liquid crystals, *Phys. Rev. Lett. 86* (2001) 2285.

[44] B. Piccirillo, A. Vefla, E. Santamento, Competition between spin and orbital photon angular momentum transfer in liquid crystals, *J. Opt. B: Semiclass. Opt. 4* (2004) S20.

[45] M.V. Berry, Paraxial beams of spinning light in singular optics, *Proc. SPIE 3487* (1988) 6.

[46] W. Stewart, The Static and Dynamic Continuum Theory of Liquid Crystals, Taylor and Francis, London, 2004.

[47] D.O. Krimer, G. Demeter, L. Kramer, Pattern-forming instability induced by light in pure and dye-doped nematic liquid crystal, *Phys. Rev. E 66* (2002) 031707.

Chapter 8

Optical Vortex Trapping and the Dynamics of Particle Rotation

Timo A. Nieminen, Simon Parkin, Theodor Asavei, Vincent L.Y. Loke, Norman R. Heckenberg, and Halina Rubinsztein-Dunlop

University of Queensland, Australia

8.1 INTRODUCTION

Since the pioneering work by He and colleagues [34,35], the transport of optical angular momentum by optical vortex beams has been applied to the rotation of microscopic particles in optical traps. Even earlier, optical vortex traps had been proposed as a method to reduce the scattering forces that oppose optical trapping; in the ray optics picture, only high-angle rays contribute to the gradient force, and the use of a hollow beam eliminates the low-angle rays that would still contribute to the scattering force [1].

There are many interesting phenomena in trapping and micromanipulation of microscopic particles using optical vortex beams; some depend on the transport of orbital angular momentum by the beam, and others do not. It is not easy to predict the behavior of conventional optical traps other than in a very general way. To go beyond that to accurate quantitative results (and sometimes the revelation of behavior that is qualitatively surprising as well) requires computational modeling of the interaction between trapping beam and trapped particle.

While methods such as geometric optics and Rayleigh scattering are only accurate in large and small particle regimes outside the usual range of particles trapped and

STRUCTURED LIGHT AND ITS APPLICATIONS

manipulated [1,33], it is possible to use electromagnetic theory to model optical trapping in the gap between the domains of these methods [60,64–66,71]. In practice, it is possible to obtain agreement with experiment to better than one percent [43].

After an initial review of theoretical basics, and the properties of nonparaxial optical vortices, we will computationally investigate a variety of different phenomena in optical vortex trapping and micromanipulation, review related experimental results, and discuss possible further experimental work.

8.2 COMPUTATIONAL ELECTROMAGNETIC MODELING OF OPTICAL TRAPPING

The calculation of optical forces and torques is essentially an electromagnetic scattering problem—the incident field carries energy, momentum, and angular momentum toward the particle in the trap, and the superposition of the scattered and incident fields carries these away. The difference between the inward and outward fluxes gives the absorbed power, and optical force and torque. As noted above, particles typically trapped and manipulated using optical tweezers are inconveniently both too small for short wavelength approximations such as geometric optics and too large for long wavelength approximations such as Rayleigh scattering.

There are many different methods available for the computational modeling of electromagnetic problems in this intermediate size range. Perhaps the most widely used are the finite-difference time-domain method (FDTD) and the finite-element method (FEM). Due to the popularity and versatility of such methods, it is remarkable that so little use has been made of them for modeling optical trapping. On a closer examination of the problem, the reasons for this become clear. First, repeated calculations are needed for modeling trapping—perhaps a few dozen calculations to find the equilibrium position of a particle within a trap, and the spring constant, through to thousands when, for example, calculating a "map" of the force exerted on the particle as a function of axial and radial displacement in the trap. Second, the typical optical tweezers arrangement provides us with a relatively simple electromagnetic problem—there is one particle, usually spherical and on the order of a few wavelengths in size, scattering a monochromatic beam, far enough away from surfaces so that they can be safely ignored. This is not far removed from the Lorenz–Mie problem, the scattering of a plane wave by a single sphere, for which an analytical solution exists.

The basic formulation of the Lorenz–Mie solution is very simply extended to arbitrary monochromatic illumination, or even to nonspherical particles (while the theoretical extension is simple, the practical extension is another matter; some of the issues involved are considered later). Fundamentally, the Lorenz–Mie solution makes use of a discrete basis set of functions $\psi_n^{(\mathrm{inc})}$, where n is mode index labeling the functions, each of which is a divergence-free solution of the Helmholtz equation, to represent the incident field,

$$U_{\mathrm{inc}} = \sum_{n}^{\infty} a_n \psi_n^{(\mathrm{inc})}, \tag{1}$$

and $\psi_k^{(\mathrm{scat})}$ to represent the scattered wave, so that the scattered field can be written as

$$U_{\mathrm{scat}} = \sum_{k}^{\infty} p_k \psi_k^{(\mathrm{scat})}. \tag{2}$$

The expansion coefficients a_n and p_k together specify the total field external to the particle.

When the electromagnetic response of the scatterer is linear, the relationship between the incident and scattered fields must be linear, and can be written as the matrix equation

$$p_k = \sum_{n}^{\infty} T_{kn} a_n \tag{3}$$

or

$$\mathbf{P} = \mathbf{TA}. \tag{4}$$

The T_{kn}, which are the elements of the transition matrix, or system transfer matrix, often simply called the T-matrix, are a complete description of the scattering properties of the particle at the wavelength of interest. This is the foundation of the classic Lorenz–Mie solution [51,54], the extension of the Lorenz–Mie solution to arbitrary illumination, usually called generalized Lorenz–Mie theory (GLMT) [28], the extension of the Lorenz–Mie solution to nonspherical but still separable geometries such as spheroids, also usually called GLMT [31,32], and general problem considered for an arbitrary particle and arbitrary illumination, usually called the T-matrix method [55, 69,89].

When the scatterer is finite and compact, the most useful set of basis functions is vector spherical wave functions (VSWFs) [55,69,70,89]. In particular, the convergence of the VSWFs is well-behaved and known [9], and this allows the sums given above to be truncated at some finite $n_{\max}$ without significant loss of accuracy.

At this point, we can outline the basic procedure for calculating the optical force and torque acting on a particle in a trap:

1. Calculate the T-matrix

2. Calculate the incident field expansion coefficients a_n
3. Find the scattered field expansion coefficients p_k using $\mathbf{P} = \mathbf{TA}$
4. Calculate the inflow of energy, momentum, and angular momentum from the total field

For a spherical particle, the T-matrix is diagonal, and there is an analytical formula for the diagonal elements. The formula does require the calculation of spherical Hankel and Bessel functions, and practical use necessitates numerical calculation on a computer. However, even an unoptimized algorithm gives the T-matrix for a sphere 10 wavelengths in radius in under 1 second, and considering that step 1 does not need to be repeated unless calculations are performed for a different particle, it is not a computationally onerous task.

Step 2 is the most demanding step in the procedure. For a Gaussian beam, the localized approximation [29,47,56] gives reasonable results. For an arbitrary beam, our approach is to find the field that gives the best least-squares fit to the beam entering the objective lens, treating the objective lens as an ideal plane-to-spherical wavefront converter [70]. For a rotationally symmetric beam, this reduces to a one-dimensional problem, and can take on the order of a second, depending on the truncation parameter $n_{\max}$. The transformation properties of the VSWFs under rotation of the coordinate system or translation of the origin of the coordinate system [9,11,30,87] can be used to find the a_n for an arbitrary position of the focal point of the beam from the a_n for when the focal point is at the origin [16,70]. The transformation matrices for both rotations and axial translations can be efficiently computed using recursive methods [11,30,87].

Step 3 is a simple matrix-vector multiplication.

Step 4, as it requires integration of the Maxwell stress tensor and its moment, appears at first glance to be an especially time-consuming step. However, the orthogonality properties of the VSWFs can be exploited to reduce the integrals involved to sums over products of the field expansion coefficients [7,14,66].

As a result, repeated calculations can be made for the optical forces and torques within an optical trap, taking on the order of a second or a few seconds for each calculation. Thus, while general methods such as FDTD and FEM have been used for modeling optical trapping [26,90], the comparative efficiency of the GLMT/T-matrix method makes it a very attractive method.

The interested reader is invited to download our optical tweezers computational toolbox, available at http://www.physics.uq.edu.au/people/nieminen/software.html [64–66].

Apart from allowing us to calculate optical forces and torques, this method also allows us to investigate the behavior of nonparaxial beams, since we obtain a beam that is an exact solution of the vector Helmholtz equation in the process of finding the force and torque. Since the VSWFs are also angular momentum eigenfunctions, we also obtain further detail on the transport of electromagnetic angular momentum. While most readers are likely to be familiar with the basic principles of transport of spin and orbital angular momentum by paraxial laser beams, the situation is less clear for nonparaxial beams. Therefore, we will briefly review the fundamental theory of electromagnetic angular momentum, and then review the angular momentum properties of the VSWFs and the transport of angular momentum by nonparaxial optical vortices.

8.3 ELECTROMAGNETIC ANGULAR MOMENTUM

There is a certain degree of confusion and disagreement in the literature on electromagnetic angular momentum. To at least some extent, this results from the very limited coverage, if any, that angular momentum receives in typical electromagnetics texts. This situation is not helped by the various conflicting statements (including ours!) appearing in the research literature that are of unstated restricted validity, or even wrong. In addition, there are a number of points of genuine difficulty (fortunately we will find that they do not pose any problem for the case at hand). Therefore, a brief coverage of the fundamentals of angular momentum in field theory is worthwhile.

Since angular momentum is a quantity of interest because it is a conserved quantity, the proper starting place is conservation laws in classical field theory. The derivation of the conservation laws closely parallels that in classical mechanics [27], where Noether's theorem [72] is used to obtain *conserved currents* when the action integral is invariant under some transformation. Noether's theorem is often stated informally along the lines of "for every symmetry, there is a conserved quantity." Invariance with respect to translations in space leads to conservation of momentum, translations in time to conservation of energy, and invariance under rotations to conservation of angular momentum.

The following is a brief summary of the presentation given by Soper [83]. If the Lagrangian density transforms as a scalar density under homogeneous Lorentz transformations, which include rotations in 3D space as a subset, we find that the angular momentum current is

$$J_{\alpha\beta}{}^{\mu} = x_\alpha T_\beta{}^{\mu} - x_\beta T_\alpha{}^{\mu} + S_{\alpha\beta}{}^{\mu}, \tag{5}$$

where $S_{\alpha\beta}{}^{\mu}$ is the spin tensor, and the corresponding conserved tensor

$$J_{\alpha\beta} = \int \mathrm{d}^3\mathbf{x}\, J_{\alpha\beta}{}^{0} \tag{6}$$

is the angular momentum. Since T is the canonical energy–momentum tensor, the first two terms are the moment of the linear momentum density, and therefore are the orbital angular momentum. If all of the fields appearing in the Lagrangian density are scalar fields, then the last term is always zero—for scalar fields, the only type of angular momentum that can be present is orbital angular momentum.

For field theories with vector fields or higher-rank tensor fields, the last term can be nonzero, and is generally described as the spin angular momentum, or intrinsic angular momentum [39,80,83].

Since the Lagrangian density of the free field is

$$\mathcal{L} = -\frac{1}{4} F_{\alpha\beta} F^{\alpha\beta}, \tag{7}$$

where F is the electromagnetic 4-tensor, which, in turn is given in terms of the 4-potential by

$$F_{\alpha\beta} = \partial_\alpha A_\beta - \partial_\beta A_\alpha, \tag{8}$$

we can explicitly write the spin tensor in terms of fields and potentials as

$$S_{\alpha\beta}{}^{\mu} = F^{\mu}{}_{\alpha} A_\beta - F^{\mu}{}_{\beta} A_\alpha. \tag{9}$$

At this point, the spin density $s_j = \frac{1}{2}\epsilon_{jkl} S_{kl}{}^{0}$ can be shown to be

$$\mathbf{s} = \mathbf{E} \times \mathbf{A}. \tag{10}$$

This makes it clear that this is in fact spin angular momentum—there is no dependence on the choice of origin about which moments are taken; this is an *intrinsic* angular momentum density, which we can take as the very definition of "spin" density in field theories. Therefore, as unambiguously as possible, we see that the total electromagnetic angular momentum is equal to the sum of an orbital angular momentum term and a spin term.

Notably, the expression for the total angular momentum density,

$$\mathbf{j} = \mathbf{l} + \mathbf{s}, \tag{11}$$

incorporating this spin density disagrees with the commonly seen statement that the angular momentum density of an electromagnetic wave is

$$\mathbf{j} = \mathbf{r} \times (\mathbf{E} \times \mathbf{H})/c^2. \tag{12}$$

This disagreement constitutes one of the four or so long-standing controversies in classical electrodynamics, apparently first noted by Ehrenfest in the early twentieth century [37]. That the controversy has persisted for so long is due in large part to the observable consequences—such as the torque exerted on a material object—being the same whether we choose either expression (11) or (12) as the angular momentum density, for any physically realizable electromagnetic fields [37,39,66,83,91]. Therefore, the reader who is primarily interested in the concrete topic of particle rotation is readily forgiven for steering clear of discussion of this controversy.

However, expression (12) is essentially a statement that spin angular momentum does not exist; if the angular momentum density is indeed equal to $\mathbf{r} \times (\mathbf{E} \times \mathbf{H})/c^2$, it is clearly dependent on $\mathbf{r}$, and thus the choice of origin. Choosing an origin such that $\mathbf{r} = 0$ would yield an angular momentum density of zero, showing that the spin density is zero. Since there is a useful physically observable distinction between spin and orbital angular momentum (spin flux can be found by measuring the Stokes parameters of the light [14]), at least for physically realizable monochromatic fields, we will discuss this issue later in the chapter.

Since the derivation of the conservation laws of the classical electromagnetic field is one of the keystones of both classical and quantum electrodynamics, and mathematical physics more generally, there are sound theoretical reasons to prefer expression (11). However, the separation of the angular momentum density into spin and orbital parts, as embodied in this expression, is not, in general, gauge-invariant (although it is for the special case of a monochromatic field [3,14]). Therefore, it is common to transform the angular momentum density into a gauge-invariant form, by transforming the canonical energy–momentum tensor into the symmetric energy–momentum tensor [38,39]. While some authors, such as Jauch and Rohrlich [39], are careful to state that this transformation requires the vanishing of surface terms in an integral as $r \to \infty$, others do not.

The reverse of this procedure, obtaining the division of the angular momentum into spin and orbital parts from the more intuitive expression (12) was demonstrated by Humblet [37], with the same surface terms appearing.

For a physically realizable field, which is bounded in both space and time, these surface terms automatically become zero at infinity, leading to the equivalence of expressions (11) and (12) in practice, as far as the *total* angular momentum is concerned. Only for fields that do not vanish sufficiently rapidly as $r \to \infty$, such as infinite plane waves, is there a discrepancy. Therefore, expression (12) provides a simple way of calculating the total angular momentum of a realizable electromagnetic field directly from $\mathbf{E}$ and $\mathbf{H}$, even though it is not the correct angular momentum density.

We realize that at least some readers will be skeptical of our argument for the correctness of expression (11); there would hardly be a controversy if it was found convincing by all. For example, one could argue that the Noetherian conservation law only tells us the total angular momentum, and identification of the integrand with the angular momentum density is unjustified.

Since the discrepancies with physically observable effects can only arise from unphysical electromagnetic fields, such as an infinite plane wave, we invite skeptical readers to consider such an unphysical ideal case—a circularly polarized plane wave normally incident on an infinite planar medium. The torque per unit area can be calculated for two revealing cases: a semi-infinite absorbing medium, and a quarter-wave plate thickness of birefringent material. The simplest way to obtain the torque is to consider the torque due to the interaction of the incident field and the induced dipole moment per unit volume; the calculation for the second case is originally due to Sadowsky, and is given by Beth [6] in his classic paper. If the circularly polarized plane wave carries no angular momentum, where does the torque come from?

8.4 ELECTROMAGNETIC ANGULAR MOMENTUM OF PARAXIAL AND NONPARAXIAL OPTICAL VORTICES

The VSWFs have already been mentioned, and some of the details are important when it comes to considering the transport of angular momentum. When considering the transport properties of the field, it is simplest to use purely incoming (designated as type 2 VSWFs) and purely outgoing (type 1) VSWFs, which are

$$\begin{aligned} \mathbf{M}_{nm}^{(1,2)}(k\mathbf{r}) &= N_n h_n^{(1,2)}(kr)\mathbf{C}_{nm}(\theta,\phi), \\ \mathbf{N}_{nm}^{(1,2)}(k\mathbf{r}) &= \frac{h_n^{(1,2)}(kr)}{krN_n}\mathbf{P}_{nm}(\theta,\phi) \\ &\quad + N_n\left(h_{n-1}^{(1,2)}(kr) - \frac{nh_n^{(1,2)}(kr)}{kr}\right)\mathbf{B}_{nm}(\theta,\phi), \end{aligned} \tag{13}$$

where $h_n^{(1,2)}(kr)$ are spherical Hankel functions of the first and second kind, $N_n = [n(n+1)]^{-1/2}$ are normalization constants, and $\mathbf{B}_{nm}(\theta,\phi) = \mathbf{r}\nabla Y_n^m(\theta,\phi)$, $\mathbf{C}_{nm}(\theta,\phi) = \nabla \times (\mathbf{r}Y_n^m(\theta,\phi))$, and $\mathbf{P}_{nm}(\theta,\phi) = \hat{\mathbf{r}}Y_n^m(\theta,\phi)$ are the vector spherical harmonics [55,69,70,89], and $Y_n^m(\theta,\phi)$ are normalized scalar spherical harmonics. The usual polar spherical coordinates are used, where θ is the colatitude measured from the $+z$-axis, and ϕ is the azimuth, measured from the $+x$-axis toward the $+y$-axis.

Significantly, the spherical harmonics $Y_n^m(\theta,\phi)$ are separable into polar and azimuthal parts and can be written as

$$Y_n^m(\theta,\phi) = \Theta_{nm}(\theta)\Phi_m(\phi) = \Theta_{nm}(\theta)\exp(im\phi). \tag{14}$$

Since the process of obtaining the vector spherical harmonics and VSWFs involves taking partial derivatives with respect to the angular variables, this azimuthal term,

$\exp(im\phi)$, is retained throughout. As a result, there is a close relationship, discussed below, between the transport of angular momentum by VSWFs and paraxial optical vortices.

It should be noted that each mode is described by two mode indices—a radial mode index, or *degree*, n, and an azimuthal mode index, or *order*, m—and polarization—$\mathbf{M}_{nm}$ and $\mathbf{N}_{nm}$ are transverse electric (TE) and transverse magnetic (TM) modes, respectively (by "transverse" in this context, we mean that the radial vector component of the field is zero).

Since the wave functions given above are purely incoming and purely outgoing, each has a singularity at the origin. Since fields that are free of singularities are often of interest, the singularity-free *regular* vector spherical wave functions are often used:

$$\mathbf{RgM}_{nm}(k\mathbf{r}) = \frac{1}{2}\left[\mathbf{M}_{nm}^{(1)}(k\mathbf{r}) + \mathbf{M}_{nm}^{(2)}(k\mathbf{r})\right], \tag{15}$$

$$\mathbf{RgN}_{nm}(k\mathbf{r}) = \frac{1}{2}\left[\mathbf{N}_{nm}^{(1)}(k\mathbf{r}) + \mathbf{N}_{nm}^{(2)}(k\mathbf{r})\right]. \tag{16}$$

If we note the relationship between Bessel and Hankel functions, we see that we can simply write these by replacing the spherical Hankel functions in the expressions for the incoming/outgoing VSWFs with spherical Bessel functions. As these carry angular momentum both toward and away from the origin, we will not consider them further as far as transport properties are concerned. However, it is useful to note that these are the VSWFs that we will use to calculate the propagation of nonparaxial optical vortices.

At this point, is useful to consider the relationship between paraxial optical vortex beams, and the nonparaxial vortices we can represent using the VSWFs above. The most important difference is that the paraxial beams are solutions of the *scalar* paraxial wave equation; as noted above, scalar fields cannot carry spin angular momentum, and the usual clear and distinct separation between spin and orbital angular momenta in the paraxial limit follows as a necessary consequence. In turn, the spin angular momentum is added on as an extra term, usually conceptually based on the quantum mechanical description of photon spin of $\pm\hbar$, in much the same way that polarization is joined onto the scalar paraxial field, in the form of a transverse polarization vector.

As far as orbital angular momentum is concerned, the most important feature of a paraxial optical vortex beam is the azimuthal mode index ℓ, which we will refer to as the *order*, although this usage is not universal. The azimuthal mode index ℓ most notably appears in the complex exponential azimuthal phase term, $\exp(i\ell\phi)$, which gives orbital angular momentum of $\ell P/\omega$ where P is the power and ω is the optical angular frequency, or in quantum mechanical terms, $\ell\hbar$ per photon. For completeness, we can note that this orbital angular momentum of $\ell\hbar$ is *not* the total orbital angular momentum—it is the component of the total orbital angular momentum in the direction of the beam axis, with the moment taken about the beam axis. The dominant component of the total orbital angular momentum, with moments taken about an arbitrary point, would be the factor $\mathbf{r} \times \mathbf{S}_{tot}/c$, where $\mathbf{S}_{tot}/c$ is the total linear momentum flux across the beam. However, this is not usually of any great interest—the important vector component of the orbital angular momentum is that parallel to the beam axis, i.e., the orbital angular momentum about the beam axis, and it is this part of the orbital angular momentum that makes optical vortex beams special and interesting. Accordingly, we will also focus on this same aspect of the angular momentum for nonparaxial vortices as well.

As an aside, there is an interesting feature of this axial orbital angular momentum that appears to be unique to optical vortices; while the orbital angular momentum density depends on where we choose the origin to be, the orbital angular momentum flux integrated across the beam profile is independent of this choice of origin. This has been termed *intrinsic orbital angular momentum*, and it is the optical analog of the spin of a classical rigid body about its center of mass (in which case, the "spin" *density* also depends on the choice of origin, but the total "spin" angular momentum does not).

If we return now to the VSWFs, we can note that the presence of spherical harmonics means that there is an azimuthal $\exp(im\phi)$ phase term, which we can expect to be related to the angular momentum in much the same way as the $\exp(i\ell\phi)$ phase term for paraxial vortices. Indeed, we find that *both* mode indices relate to the angular momentum—the radial index n gives the magnitude of the total angular momentum, while the azimuthal index m gives the component along the z-axis. (This can be interpreted as the physical reason why the allowed values of the azimuthal index m are restricted by $-n \leqslant m \leqslant n$, since the magnitude of a single vector component of the angular momentum cannot exceed the magnitude of the angular momentum.)

While one might at first expect that the boundary conditions when we match the paraxial and nonparaxial beams might require that the azimuthal mode indices be equal, that is, that $m = \ell$, this is not the case. The paraxial beam is described by the product of a Cartesian polarization vector and the scalar paraxial field, and thus the fields are given in terms of Cartesian basis vectors. The VSWFs, on the other hand, make use of spherical basis vectors, and if we proceed from $\phi = 0$ in the xy-plane, around the z-axis to $\phi = \pi$, the radial and azimuthal basis vectors $\hat{\mathbf{r}}$ and $\hat{\phi}$ also rotate by a half-turn. Therefore, for the VSWF fields to match the paraxial fields requires an extra half-wave phase shift in the course of this half-rotation, and we find that $m = \ell \pm 1$. Physically, this corresponds to the combination of spin and orbital angular momentum. If the paraxial beam is left circularly polarized (i.e., carrying $+\hbar$ spin per photon), then we have $m = \ell + 1$. If right circular, then we have $m = \ell - 1$. If the paraxial beam is plane polarized, then treating this as a superposition of the two circular components, we see that both values of m are present.

This reminds us immediately that in the nonparaxial case, we are dealing with the total angular momentum, rather than the spin and orbital angular momenta separately. However, the physical distinction between spin and orbital angular momentum is important if we are planning to make an optical measurement of the angular momentum. Therefore, we can follow a similar procedure to that used to obtain an expression for the optical force and torque, integrate the spin flux through a spherical surface surrounding the trapped particle, and obtain an expression for the spin flux in terms of a sum of products of field expansion coefficients [14].

8.5 NONPARAXIAL OPTICAL VORTICES

In order to find the least-squares best-fitting nonparaxial beam that matches the focused "paraxial" beam (i.e., the original paraxial beam with the wavefront converted from plane to spherical by the objective lens), we begin with the far-field limit for a Laguerre–Gaussian beam in spherical coordinates [70,81],

$$U = U_0 (2\psi)^{|\ell|/2} L_p^{|\ell|}(2\psi) \exp(-\psi + \mathrm{i}\ell\phi), \tag{17}$$

where $\psi = (k^2 w_0^2 \tan^2\theta)/4$, k is the wavenumber in the medium, and $L_p^{|\ell|}$ is the generalized Laguerre polynomial. For $p = 0$, which is the only case we will consider in detail, we have $L_0^{|\ell|} = 1$.

Since we want to express the focused field in terms of VSWF expansion coefficients, we need to find a_{nm} and b_{nm} such that

$$\mathbf{E}(\mathbf{r}) = \sum_{n=1}^{n_{\max}} a_{nm} \mathbf{RgM}_{nm}(k\mathbf{r}) + b_{nm} \mathbf{RgN}_{nm}(k\mathbf{r}), \tag{18}$$

with m equal to the total angular momentum per photon of the incident paraxial beam, gives the closest possible match to that given by equation (17), converted to a vector field by an appropriate polarization vector [70]. Since the fields of Laguerre–Gaussian beams are rotationally symmetric apart from the $\exp(im\phi)$ variation, it suffices to choose a set of points along a single "line of longitude," most simply $\phi = 0$, at which to determine the least-squares fit. This makes the choice of polarization vector simple, since before the objective we have

$$\mathbf{E} = (E_x \hat{\mathbf{x}} + E_y \hat{\mathbf{y}}) U_{\text{paraxial}}, \tag{19}$$

and after the objective we have

$$\mathbf{E} = (E_\theta \hat{\theta} + E_\phi \hat{\phi})U, \tag{20}$$

since the wavefronts are now spherical, and we have $E_\theta = E_x$ and $E_\phi = E_y$ when $\phi = 0$.

Typically, we would wish to consider the most tightly focused beam achievable with a given numerical aperture. Aiming to have a parameter easily relatable to the usual criterion for optimal overfilling in optical tweezers, we can describe the degree of convergence by a convergence angle θ_{conv} such that

$$U(\theta_{\text{conv}})/U_{\text{max}} = 1/\text{e}. \tag{21}$$

For practical purposes, this means that we need to find the parameter ψ corresponding to a particular θ_{conv}, or $\psi_{\text{conv}} = \psi(\theta_{\text{conv}})$. Since U_{max} occurs at $\psi = |\ell|/2$, and ψ_{conv} lies between this value and $\psi = \infty$, this requires the numerical solution of

$$\left(2\psi_{\text{conv}}/|\ell|\right)^{|\ell|/2} \exp\left(-\psi_{\text{conv}} + |\ell|/2 + 1\right) - 1 = 0 \tag{22}$$

subject to those limits on allowable values of ψ. We include a routine for this, and an accurate polynomial approximation of the solution in our optical tweezers toolbox [65].

As the beam becomes more and more tightly focused, we expect to see more deviation from the paraxial beam. In particular, once the focal spot approaches the diffraction-limited focal spot in size, it will cease to shrink significantly as the convergence angle continues to increase. For a Gaussian beam, the resulting deviation between far-field matched paraxial fields and exact electromagnetic fields exceeds 10% for waist radii smaller than 1.2λ. In Figures 8.1 and 8.2, we show the effect of increasingly strong focusing of an LG_{03} beam, circularly polarized such that the total angular momentum is $m = 4\hbar$ per photon. Cross-sections of the beam are shown for beam convergence angles of 30°, 40°, 50°, and 60°. Figure 8.1 shows the cross-sections with whiteness linearly proportional to $|\mathbf{E}|^2$, while Figure 8.2 shows them with a logarithmic whiteness scale, highlighting the structure of the lower-energy-density regions.

A striking feature, especially visible in the logarithmic-scale plots (Figure 8.2), is the cone-tube-cone appearance of the more tightly focused beams. Physically, this results from the transport of angular momentum by the beam. Since, as noted above, the angular momentum flux of a finite beam is given by the integral of the moment of the Poynting vector over the beam profile, a sufficient moment arm to obtain the required angular momentum flux is required [13]. Essentially, once the bright ring of the beam has reached a radius equal to this minimum moment arm, its minimum size has been attained. Mathematically, this behavior results from the radial part of the

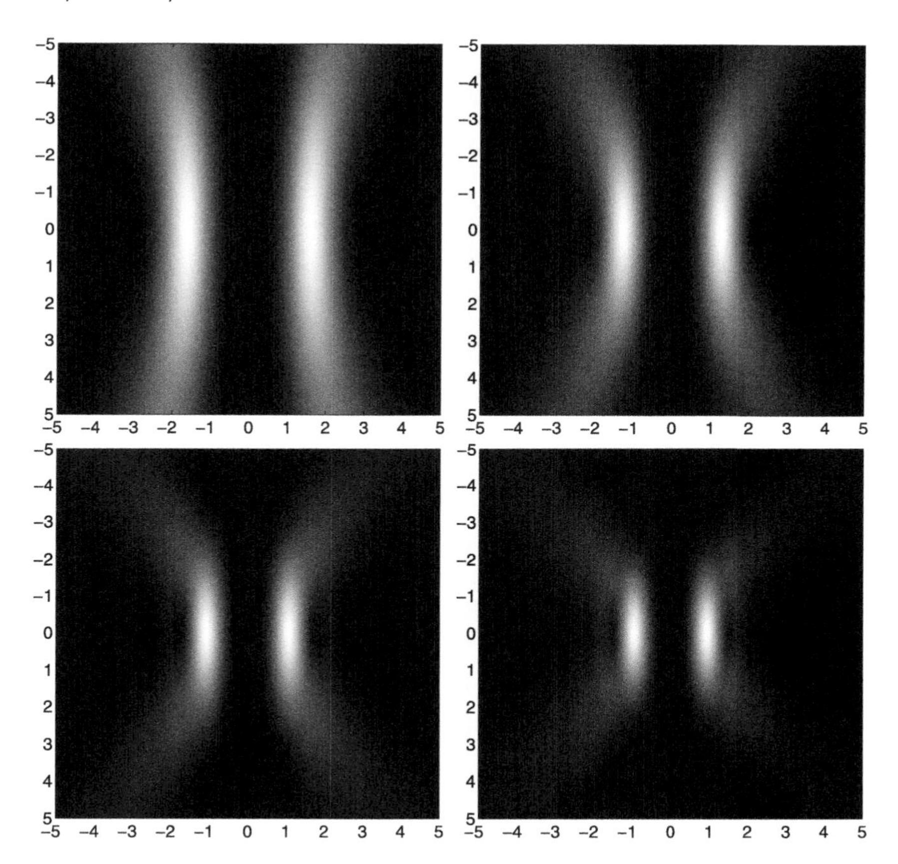

Figure 8.1 Cross-sections of increasingly tightly focused LG_{03} beams. The beams are circularly polarized such that the total angular momentum is $m = 4\hbar$ per photon. The convergence angles of the beams are 30° (*top left*), 40° (*top right*), 50° (*bottom left*), and 60° (*bottom right*). The whiteness is linearly proportional to $|\mathbf{E}|^2$. The scales show radial and axial distances in units of the wavelength of the light.

VSWFs being spherical Bessel functions, which are strongly peaked in the vicinity of $kr \approx n$, as shown in Figure 8.3.

This mathematical property of the VSWFs, coupled with the requirement that $n \geqslant m$, means that, apart from beams with very low angular momentum, the width of the most tightly focused optical vortices is linearly proportional to the total angular momentum flux of the beam. This has been noted, with some surprise, by Curtis and Grier [15] as an experimental finding. This also means that we can expect optical vortices with spin angular momentum of the same handedness as their orbital angular

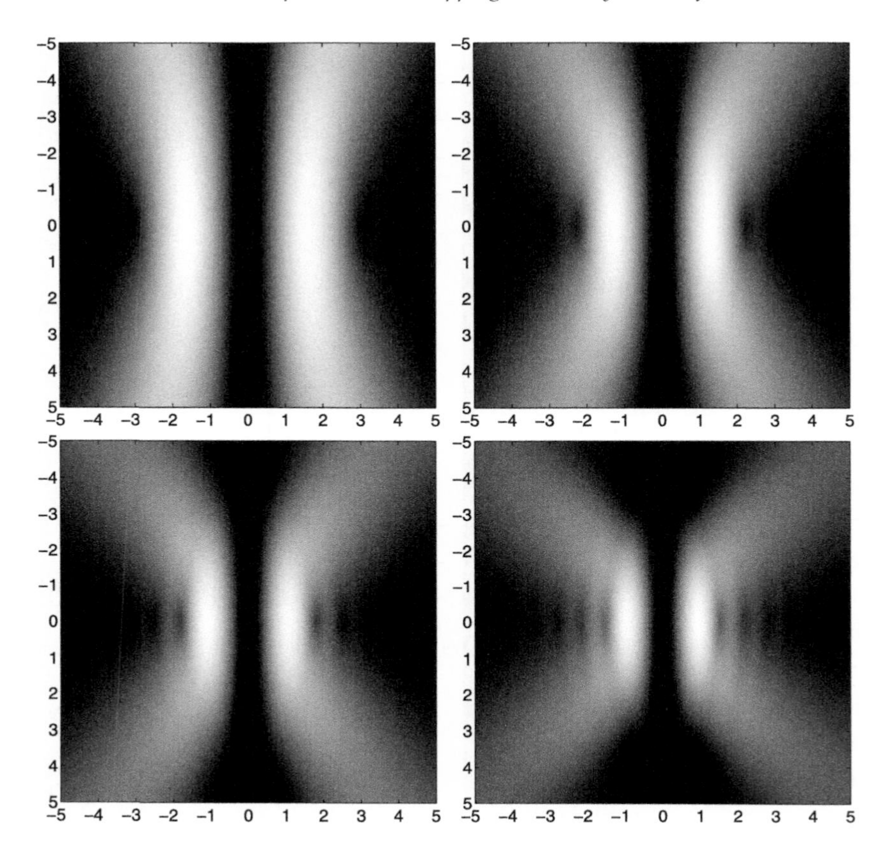

Figure 8.2 Cross-sections of increasingly tightly focused LG_{03} beams. The beams are circularly polarized such that the total angular momentum is $m = 4\hbar$ per photon. The convergence angles of the beams are 30° (*top left*), 40° (*top right*), 50° (*bottom left*), and 60° (*bottom right*). The whiteness is logarithmically proportional to $|\mathbf{E}|^2$, emphasizing the low-energy-density regions of the cross-section. The scales show radial and axial distances in units of the wavelength of the light.

momentum to have larger minimum sizes than those with spin opposing their orbital angular momentum. This effect is shown in Figure 8.4.

We can also note the close resemblance of the focal plane of the most tightly focused beam in Figure 8.2 to a Bessel beam carrying orbital angular momentum, with a bright ring surrounded by secondary rings. Mathematically, the secondary rings also result from the radial dependence of the VSWFs on spherical Bessel functions, which continue to oscillate for kr beyond the maximum peak, as shown in Figure 8.3. The behavior of Bessel beams can help us understand the nature of the cone-tube-cone struc-

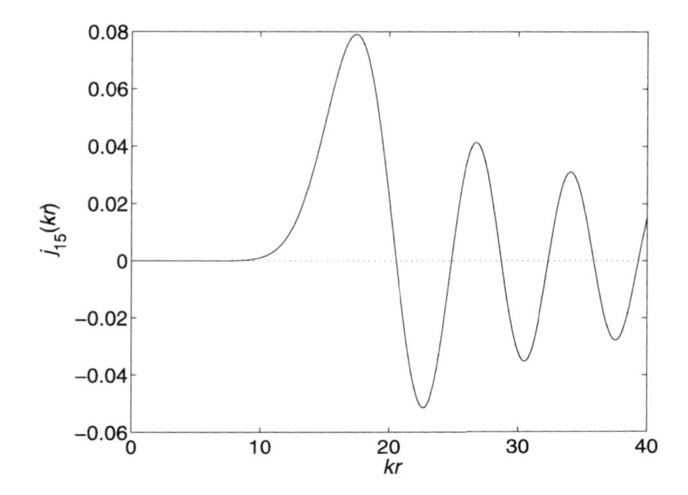

Figure 8.3 Spherical Bessel function with $n = 15$. The close-to-zero value for $kr < n$ is the mathematical explanation for the dark hollow region in optical vortex beams. Since $n \geqslant m$, this also explains the minimum size to which optical vortices can be focused. The oscillatory nature of the function outside the bright ring describes the secondary rings.

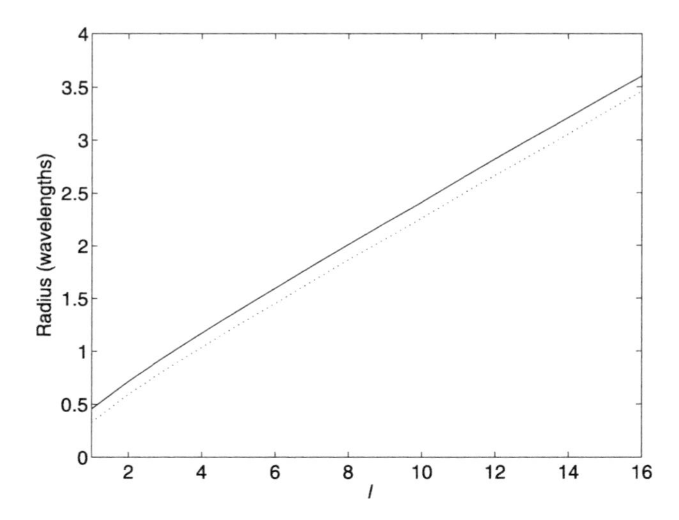

Figure 8.4 Radius of bright ring in the focal plane for tightly focused $LG_{0\ell}$ beams, for ℓ from 1 to 16. Radii are shown for beam with circular polarization adding to the orbital angular momentum (*solid line*), and partially canceling the orbital angular momentum (*dotted line*). The radii for any given ℓ differ by almost exactly $1/(2\pi)$.

ture of focused optical vortices. While Bessel beams, strictly speaking, are not physically realizable since they require an infinite energy flux, truncated approximations of Bessel beams can be generated by axicons or spatial light modulators. Unlike Bessel beams, such finite approximate Bessel beams are not propagation-invariant. However, for some distance, the central region of such beams is approximately propagation-invariant. The central part of a Bessel beam consists of either a bright spot or a ring about a phase singularity, surrounded by a succession of secondary rings; in a true Bessel beam, the power in each ring is the same, leading to the infinite power of the overall beam. As is usual in multiringed laser beams, each successive ring is a half-wave out of phase with the preceding ring (thus, each dark region between rings contains a line phase singularity), and therefore each ring prevents diffractive spreading of interior rings—only the outermost ring can effectively expand through diffraction. This diffractive expansion continues until the outermost ring has negligible power density, at which point the next inward ring is effectively the new outermost ring, and can begin to spread. This continues until only the innermost ring remains, and can itself spread. This process is shown in Figure 8.2, from the focal plane proceeding along the beam. Since an ideal spherical beam results in an energy density distribution that is symmetric about the focal plane, the convergence of the beam to the focal plane is the exact opposite of its divergence away from the focal plane.

In principle, we could also describe focused optical vortices as superpositions of vector cylindrical wave functions (that is, vector Bessel beams). However, while this would give a uniform angular momentum per photon about the beam axis, just as we get with VSWFs and paraxial Laguerre–Gaussian modes, the total angular momentum of each individual mode is infinite due to the infinite power per mode. This is in marked contrast with VSWFs and LG modes, where modes are orthogonal with respect to both power and angular momentum about the beam axis. Since vector Bessel beams have neither of these orthogonality properties, the total power and angular momentum cannot be understood in terms of sums of those of individual modes, greatly reducing their utility in such cases. However, since vector Bessel beams appear to have attracted recent interest in the optical vortex literature [88], it is worth mentioning that the sum of single TE and TM vector Bessel beams might be a useful analytical approximation of the field of a strongly focused optical vortex near the beam axis in the focal plane. Vector Bessel beams have been used for many decades in electromagnetics for the analysis of systems with cylindrical geometry, such as waveguides [84]. However, most work on optical Bessel beams appears to have used the scalar version.

We will now move on to considering the basic features of optical trapping using optical vortices.

8.6 TRAPPING IN VORTEX BEAMS

Trapping by hollow beams has attracted attention since the early years of optical tweezers due to the possibility of increased axial trapping efficiency [1]. When optical tweezers are used to trap particles other than the smallest or lowest refractive index contrast particles, the trap is distinctly asymmetric—interestingly, one of the results obtainable from all three major theoretical pictures (geometric optics, Rayleigh scattering, and exact electromagnetic theory). While in the radial direction (i.e., for displacements of the particle away from the beam axis), the restoring force is symmetric, this is not the case for axial displacements, where the gradient force acts symmetrically, but the "scattering force" acts in the direction of beam propagation. In the geometric optics picture, only the high-angle rays contribute to the gradient force, so changing from a Gaussian beam trap to a hollow beam trap eliminates the small-angle rays that contribute only to the scattering force.

However, outside the geometric optics picture, the situation is not so clear. The focal spot is necessarily larger, so one might reasonably expect axial trapping to be weaker, even if the force versus axial displacement curve is more symmetric. As improved axial efficiency is one of the major suggested reasons for optical vortex trapping, we calculate the axial force versus displacement for a number of optical vortex traps and present the results in Figures 8.5 and 8.6.

Interestingly, while improved axial trapping is found in the form of a larger maximum restoring force for displacements beyond the focal plane, this appears to result from an increased gradient force rather than any reduction in scattering force; an increase in forward force pushing the particle toward the focus is shown in Figures 8.5 and 8.6. In the case of the $n = 1.59$ particle (Figure 8.5), where this increase is greater than the improvement in the reverse restoring force, there appears to be an increase in the scattering force. Neither increase is large. For the lower contrast, and therefore less reflective, $n = 1.45$ particle, the curves for both the Gaussian and Laguerre–Gaussian beams are much more symmetric, and an increase in the scattering force will be small and have little effect.

Consideration of the beam shape suggests that an increased gradient force is likely. As the particle moves away from the focus along the axis of a tightly focused optical vortex such as those shown in Figures 8.1 and 8.2, the furthermost portion of the particle is moving into the dark region of the beam. Since, in the Gaussian beam case, the particle remains centered on the most intense portion of the cross-section, the axial gradient is smaller. Comparison of Figures 8.5 and 8.6 with the final (i.e., most tightly focused) beam cross-sections in Figures 8.1 and 8.2 shows that the maximum force pushing and pulling a particle toward the focus (when the center of the particle is

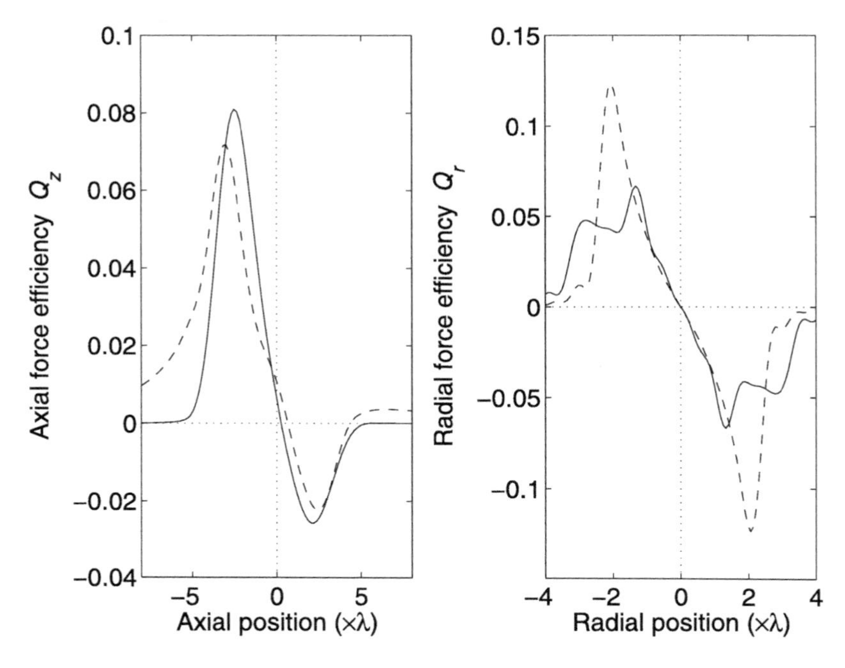

Figure 8.5 Axial and radial trapping efficiencies for LG_{03} beams, compared with a Gaussian beam trap. Both beams are circularly polarized with the same handedness as the orbital angular momentum of the LG_{03} beam. The angle of convergence of both beams is 60°. The particle is $2\lambda_{medium}$ in radius, corresponding to a diameter of 3.2 μm for trapping in water at 1064 nm. The relative refractive index corresponds to that of polystyrene ($n = 1.59$) in water. The force efficiencies for the LG_{03} beam trap (*solid lines*), and the Gaussian beam trap (*dotted lines*) are shown.

about two wavelengths away from the focus) does occur when the particle is starting to move into the dark region.

However, this improvement in axial trapping, both in terms of increased trap stiffness and increased maximum restoring force, comes at the cost of reduced radial trapping. Furthermore, the distance over which the radial force is linear with radial displacement is reduced compared to the Gaussian beam trap, reducing the utility of the trap as a radial force sensor. Considering that the generation of optical vortices from Gaussian beams generally involves loss of power, the decrease in axial force resulting from this loss of power is likely to be comparable to, or greater than, the increase resulting from the improvement in axial trapping efficiency. Therefore, an optical vortex trap as shown here does not appear to be useful simply for improved trapping. Other features of optical vortices, such as transport of orbital angular momentum, and the

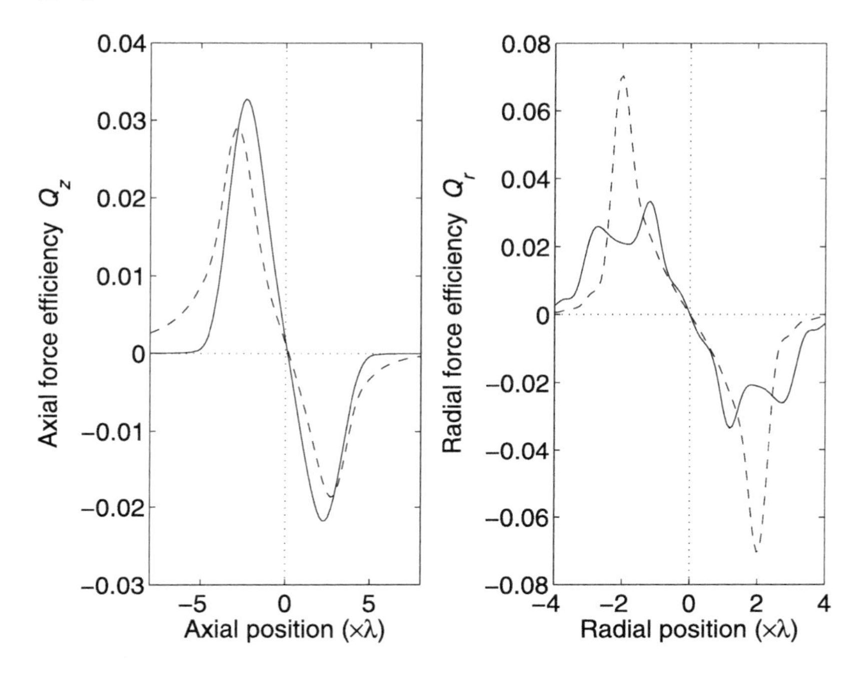

Figure 8.6 Axial and radial trapping efficiencies for LG_{03} beams (solid line), compared with a Gaussian beam trap (dashed line). The beams and particles are identical to those in Figure 8.5, except that the relative refractive index is that for PMMA ($n = 1.45$) in water.

presence of a dark hollow center of the beam may, however, make such an optical vortex trap desirable.

In Figures 8.5 and 8.6, the axial force about the equilibrium region is very linear with axial displacement. However, the diameter of the trapped particle is almost exactly the same as the length of the "tube" of the cone-tube-cone beam.

Calculation of the axial force on larger and smaller particles reveals two interesting features. For small particles, shown in Figure 8.7(a), the axial force is still linear with displacement, but the gradient force only becomes large as the particle approaches the end of the "tube." Within the tubular portion of the beam, the scattering force easily overcomes the feeble gradient force, and the particle is trapped much further beyond the focus than an identical particle trapped in a Gaussian beam.

For large particles, shown in Figure 8.7(b), the axial force ceases to be as linear as the earlier case of a particle with diameter similar to the "tube" length. As the enhancement in the force pushing the particle along the beam toward the focus is much larger than the enhancement in the reverse force, the scattering force appears to

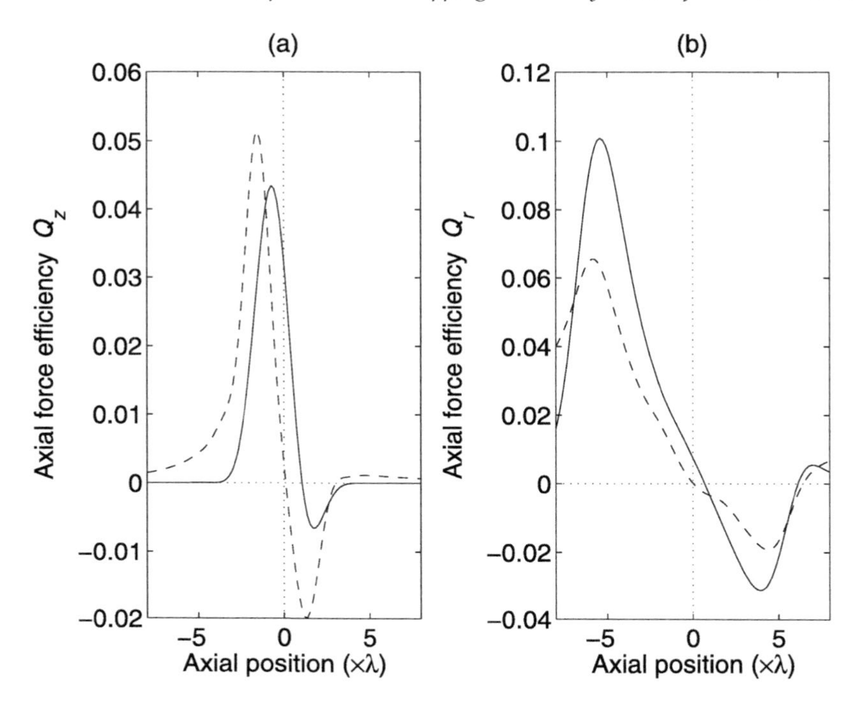

Figure 8.7 Axial trapping efficiencies for large and small particles in LG_{03} beams, compared with a Gaussian beam trap. Both beams are circularly polarized with the same handedness as the orbital angular momentum of the LG_{03} beam. The angle of convergence of both beams is 60°. The particles have radii of (a) $1\lambda_{medium}$ and (b) $4\lambda_{medium}$, corresponding to 1/2 and 2 times that in Figure 8.5. The relative refractive index corresponds to that of polystyrene ($n = 1.59$) in water. The force efficiencies for the LG_{03} beam trap (*solid lines*), and the Gaussian beam trap (*dotted lines*) are shown.

be significantly larger than in the Gaussian beam trap. This conclusion is supported by the equilibrium position being further beyond the focal plane than in the Gaussian trap. Despite this, axial trapping is still stronger than in the Gaussian trap.

However, it is known that small particles will be trapped in the bright ring rather than on the dark beam axis [21,36]. Common sense suggests that the critical particle radius separating these two behaviors should be approximately equal to the radius of the bright ring; when the particle is smaller than the bright ring, it will experience an outward radial gradient force for any displacement away from the beam axis. Calculation of the radial forces shows that the critical particle radius is very nearly equal to the beam ring radius, as shown in Table 8.1. The particles considered in Figure 8.7 are indeed trapped on the axis, although the small particles are close to the minimum size for on-axis trapping.

Table 8.1

Trapping in bright ring versus trapping on axis

	Polarized with		Polarized against	
ℓ	$r_{particle}$	r_{ring}	$r_{particle}$	r_{ring}
1	0.49	0.46	0.37	0.33
2	0.73	0.72	0.63	0.60
3	0.95	0.95	0.84	0.82
4	1.15	1.17	1.04	1.04
5	1.34	1.39	1.23	1.25

The particle radius below which the particle is trapped in the bright ring is shown. At larger radii, the particle is trapped on the beam axis. For comparison, the radius of the bright ring, which is almost identical to the particle radius separating the two trapping behaviors, is shown. The radii are given as multiples of the beam wavelength in the trapping medium. The particles have a relative refractive index equal to that of polystyrene ($n = 1.59$) in water. The beam convergence angle is 60° in all cases. The left side shows the critical radii for beams with circular polarization of the same handedness as the orbital angular momentum; the right side shows radii when polarization opposes orbital angular momentum.

At the critical radius separating the two types of radial trapping behavior, the radial force is very close to zero for displacements near the beam axis, as shown in Figure 8.8. The behavior changes very rapidly with particle size, as shown by the dot-dashed and dashed lines in Figure 8.8, which show the force on particles 1% larger than and smaller than the critical radius (*solid line*), respectively. Particles significantly smaller than the critical radius are trapped centered on the bright ring; the particle 1% smaller than the critical radius is trapped off-center, with the ring passing through the particle but not through its center.

Nonabsorbing homogeneous isotropic spherical particles on the beam axis will not experience a torque, regardless of the orbital angular momentum flux and the polarization of the beam. Smaller particles that are trapped in the bright ring *will* experience a torque about the beam axis (although not about their own centers). This results in the particles orbiting about the beam axis, in the bright ring. This has been demonstrated on a number of occasions [25,74,78].

This orbital motion results from the so-called "scattering" force (technically, both the scattering and gradient force result from scattering [62,68]), since the gradient force will only act to pull the particle into the bright ring (with no azimuthal intensity gradient around the ring, the gradient force cannot cause orbital motion). Therefore, the orbital torque should be proportional to the volume squared for Rayleigh particles (i.e., proportional to r^6). For larger particles, the bright ring will illuminate approximately a diameter, and the scattering force can be expected to be proportional to the diameter. Since the viscous drag force is approximately proportional to the radius, we

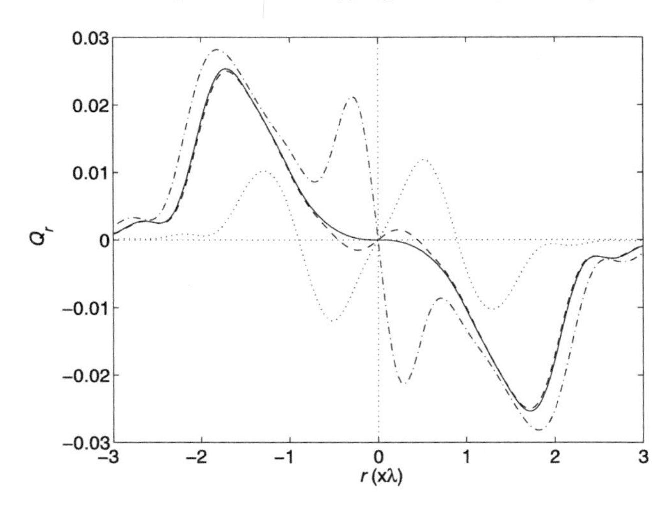

Figure 8.8 Radial force on particles of different sizes in an LG_{03} beam. The beam is circularly polarized with the same handedness as the orbital angular momentum, and has an angle of convergence of 60°. The particles have radii of $0.9483\lambda_{\text{medium}}$ (*solid line*), which is the critical radius separating trapping on-axis and trapping in the bright ring, 1% greater than the critical radius (*dot–dash line*), 1% less than the critical radius (*dashed line*), and $0.5\lambda_{\text{medium}}$ (*dotted line*). The relative refractive index corresponds to that of polystyrene ($n = 1.59$) in water. The particles are assumed to be in the focal plane.

can expect the orbital speed of the smallest trappable particles to be very small, with the speed increasing as the radius increases, until a plateau is reached. Of course, as the particle radius exceeds the critical radius, the particle will move to the beam axis, and orbital motion will cease.

These expectations are confirmed by calculation of orbital speed, shown in Figure 8.9, assuming Stokes drag (which will only approximate the drag force since the motion is not linear). The ripples on the curves arise from the reflectivity of the particle varying with radius, due to thin-film–like interference effects. Notably, the curves for the dashed and dotted–dashed lines are very similar; these correspond to LG_{03} beams, with circular polarization of opposite handedness. The dotted-line and dashed-line beams have the same total angular momentum, being LG_{01} and LG_{03} with polarization adding to and subtracting from the total angular momentum, respectively. The other two beams are LG_{03} and LG_{05} beams with identical total angular momentum obtained in the same way.

Interestingly, as the angular momentum of the beam increases, the orbital speed decreases. This effect is even more noticeable for the orbital frequency. However, the similarity of the results for the two LG_{03}, although their angular momentum is very

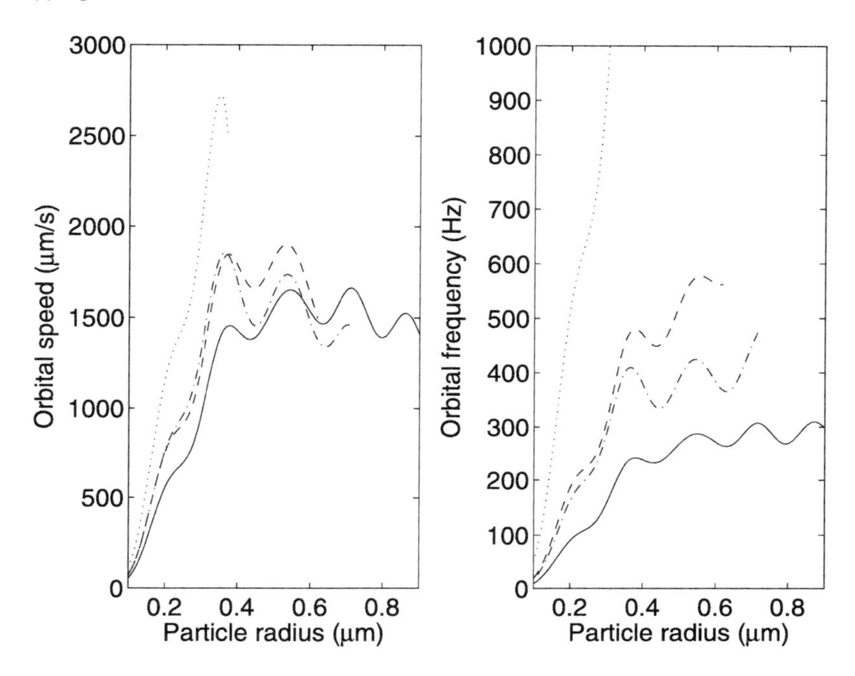

Figure 8.9 Orbital speed and frequency for particles trapped in the bright ring. The beams are selected from those shown in Table 8.1: LG_{01}, polarized with (*dotted line*); LG_{03}, polarized with (*dotted–dashed line*); LG_{03}, polarized against (*dashed line*); and LG_{05}, polarized against (*solid line*). The particles are polystyrene ($n = 1.59$) trapped in water by a 1 W beam at 1064 nm free-space wavelength. Speed and frequency scale linearly with power.

different, suggests that the angular momentum is not the major factor. Furthermore, since the results for the LG_{05} are also close to the LG_{03} curves, despite differing significantly in orbital angular momentum, the variation does not seem due to the differences in orbital angular momentum, per se. Instead, it appears to be primarily due to the radius of the bright ring of the beam (listed in Table 8.1). As the bright ring becomes larger, a particle of the same size intercepts a smaller fraction of the beam, resulting in less azimuthal force, and hence a smaller orbital speed. The frequency is affected even more, since the circumference of the orbit increases as well.

Although the speeds and frequencies shown in Figure 8.9 appear high, these are for a power of 1 W at the focus. The azimuthal force efficiencies are modest, reaching a maximum of approximately 0.005 for the largest particles. This is consistent with measurements of the transfer of orbital angular momentum to clusters of spheres arranged in a ring [76,78], taking into account that we have only considered a single sphere here.

As noted earlier, nonabsorbing homogeneous isotropic spheres trapped on the beam axis will experience no torque. This is a result of the effect of particle symmetry on angular momentum transfer. Since one interesting practical application of optical vortex trapping is the application of optical torques to both naturally occurring and artificial particles, we proceed to discuss the relationship between optical torque and symmetry in more detail.

8.7 SYMMETRY AND OPTICAL TORQUE

The shape of a trapped particle plays a central role in the transfer of angular momentum between it and the trapping beam. As mentioned above, even if the beam carries a large angular momentum flux, a spherical particle will not experience a torque about its axis in the absence of absorption. Perhaps the most important aspect of the shape of a particle with respect to the application of optical torque is its rotational symmetry. Considerations of the effect of particle symmetry on the interaction between particle and beam allows us to understand some quite general aspects of the generation of optical torque.

The coupling between incident and scattered azimuthal orders, when the fields are expressed as superpositions of VSWFs, is dominated by the rotational symmetry of the scatterer. Since the azimuthal order m is the total angular momentum per photon, this means that the rotational symmetry of the particle is the predominant influence on optical torque, in the absence of absorption. (In the presence of absorption, angular momentum can be transferred to the particle without changing the angular momentum per photon, simply by changing the photon number flux.)

The principal features of optical angular momentum transfer can be deduced from Floquet's theorem, which relates solutions of differential equations and the periodicity of boundary conditions. In the same way that when an incident plane wave with a wave vector with component k_x parallel to a grating with lattice vector $q_x = \text{period}/2\pi$ gives rise to scattered waves with parallel components of their wave vectors $k_{nx} = k_x - nq_x$, incident VSWFs with rotational phase variation of $\exp(\mathrm{i}m\phi)$, scattered by a particle with rotational symmetry $\exp(\mathrm{i}p\phi)$ result in scattered VSWFs with rotational phase variation $\exp\{\mathrm{i}(m - np)\phi\}$. For a particle lacking rotational symmetry, $p = 1$ (i.e., we need to rotate the particle by 2π radians before the original shape reappears), and an incident VSWF can, in principle, couple to all angular momenta in the scattered VSWFs. A particle with discrete rotational symmetry, on the other hand, only couples to a limited subset of scattered angular momenta, determined by its order of rotational symmetry. For example, if a particle has fourth-order discrete rotational symmetry (i.e., $p = 4$), an incident VSWF with $m = m_0$ only couples to scattered VSWFs with

$m = m_0$, $m_0 \pm 4$, $m_0 \pm 8$, $m_0 \pm 12$, and so on. The case of a rotationally symmetric particle can be seen as the limit as $p \to \infty$, which means that there is no coupling to angular momenta different from the incident angular momentum—a rotationally symmetric particle cannot alter the angular momentum per photon about its axis of symmetry [55,57,67,89]. Optical torque about the particle axis can only result from a change in photon flux—in practical terms, through absorption. Since optical tweezers require very high power densities to produce effective forces and torques, absorption easily produces high temperatures and does not appear to be a technologically useful method of optical angular momentum transfer.

Therefore, if we wish to produce structures designed to transfer angular momentum from a trapping beam, whether or not the beam carries angular momentum, the structure must deviate from continuous rotational symmetry. This asymmetry can be either microscopic in the form of electromagnetic anisotropy, such as birefringence [20], or macroscopic due to the shape of the particle. One strand of the web of optical tweezers has been directed toward the practical implementation of optical rotation of microstructures, especially specially fabricated microparticles intended for use as optically driven micromotors, micropumps, or other micromachine components. Work in this field has been recently reviewed by Nieminen and colleagues [61].

Such optically driven micromachines typically possess discrete rotational symmetry. A particle with pth-order rotational symmetry has a shape with an azimuthal component that can be described by a Fourier series of the form

$$\Phi(\phi) = \sum_{k=-\infty}^{\infty} A_k \exp(\mathrm{i}p\phi). \tag{23}$$

The lowest-order non-DC Fourier components (i.e., $k = \pm 1$) determine to which azimuthal orders of scattered VSWFs the incident wave can be coupled by the particle. As noted earlier, pth-order rotational symmetry results in an incident wave of order m_0 being coupled to scattered VSWFs with $m = m_0 - np$, for integer n.

Firstly, we can, on average, expect scattering to the lowest orders of scattering (i.e., $n = 0, \pm 1$) to be strongest. Thus, most light is likely to be scattered to scattered wave azimuthal orders $m = m_0$, $m_0 \pm p$.

Secondly, only m such that $|m| \leqslant n_{\max}$ are available, with $n_{\max}$ being determined by the radius of a sphere required to enclose the micromachine element. Therefore, if the order of symmetry p is large and the micromachine small, only a small number of azimuthal modes will be available. The ultimate case of this is the homogeneous isotropic sphere, with $p = \infty$, when only the incident $m = m_0$ is available for the scattered wave. In principle, this can be exploited to maximize torque. As a numerical example, consider a structure with $p = 8$ and size such that $n_{\max} = 6$, illuminated with a beam such that $m_0 = 4$. In this instance, the only scattered wave azimuthal orders that are available are $m = -4, 0, 4$, and we might expect such a structure to generate more torque, *cetera paribus*, than a structure that can also scatter to $m = \pm 8$ (since the scattering to $m = +8$ will probably be stronger than the scattering to $m = -8$). However, such a structure would have a maximum radius of less than a wavelength, which, apart from causing fabrication to be difficult, with subwavelength resolution being required even at the perimeter, would mean that the device would largely sit within the dark center of a typical tightly focused optical vortex. Alternative methods of illumination, such as a Gaussian or similar beam perpendicular to the device symmetry axis illuminating one side [41], can avoid this, but will not necessarily result in greater efficiency. In the long run, the maximum torque is proportional to the particle radius [13]. Thus, designing the particle itself to more efficiently scatter preferentially into either higher m (i.e., more positive or less negative) or lower m (i.e., less positive or more negative) orders is desirable.

Consequently, it is useful to note that, thirdly, the symmetry of scattering to positive and negative orders of scattering (i.e., to $m < m_0$ and $m > m_0$) is dependent of the chirality of the particle shape. If the particle is achiral—mirror symmetric about a plane containing the axis of rotational symmetry—the coupling from $m_0 = 0$ to $\pm m$ will be identical, since these modes are mirror-images of each other. For a chiral particle, however, the coupling to these mirror-image modes will not be identical; even in this case of illumination with zero angular momentum modes (i.e., $m_0 = 0$), optical torque will be generated. If the particle is chiral such that scattering from $m_0 = 0$ to $+m$ is favored, then, even for incident modes with $m_0 \neq 0$, scattering to higher (i.e., more positive) m will generally be favored. In this way, a particle can be optimized for the generation of torque with a particular handedness, leading to greater efficiency when this handedness is the same as that of the angular momentum carried by the driving beam.

Fourthly, we can note another factor that is to some extent a generalization of the second result from symmetry above. Since, for any given particle, there are always more modes available with low $|m|$ than with high $|m|$, scattering from a given m_0 to modes with $m = \pm\delta$ will usually be such that modes with $|m| < |m_0|$ will receive more power than those with $|m| > |m_0|$. As a result, an arbitrary particle placed in illumination carrying angular momentum, but otherwise arbitrary, will usually experience a torque.

Why, then, is optical torque regarded as unusual, if most randomly chosen particles should experience torques? Although a torque might exist, it might also be very small; if smaller than torques associated with the rotational Brownian motion of the particle, the optical torque will not be noticeable. Rotation should be more readily observable in environments of low viscosity, such as particles in gas. Under such circumstances, rotation is observed [18], although convection and thermophoresis may well be the dominant effects. Torques on random particles can also be maximized by a high refractive index, increasing the reflectivity [42]. However, most naturally occurring light is predominantly unpolarized, and as such can be represented as sums of VSWFs with $m = \pm 1$. The same is the case for linearly polarized light. It is especially in this case that torques can result most readily in alignment rather than continuous spinning, which can also mask the presence of an optical torque.

Noting that light with a mixture of azimuthal modes with $m = \pm 1$ has modes separated by $\Delta m = 2$, particles with 2nd-order rotational symmetry ($p = 2$) will couple between these orders. In these circumstances, interference effects between light scattered, for example from $m_0 = -1$ to $m = +1$ and light scattered from $m_0 = +1$ also to $m = +1$, becomes important. The phase of the scattered light depends upon the rotational orientation of the particle, and therefore the final amplitudes in each scattered azimuthal order depend on the particle orientation. As a result, the optical torque also depends on the orientation, often leading to the existence of a stable equilibrium in which the torque is zero—particles will be aligned in a particular direction, depending on the relative phases of the incident $m = \pm 1$ modes. The relative phase of the incident modes determines the plane of polarization, and therefore the alignment direction depends on the direction of linear polarization. Noting that essentially an elongated particle has a $p = 2$ component to its shape, we can expect elongated particles to, in general, tend to align to the plane of polarization of light incident perpendicularly to their long axis [5,7]. Similar considerations apply for illumination with shaped beams of light with a mixture of azimuthal modes with $\Delta m = 2$; in particular, this will be the case for beams with elongated focal spots [19,75].

Therefore, we can summarize some general principles for the design of optically driven micromachines:

- Size matters! The maximum angular momentum available from the beam is proportional to the radius of the particle, assuming that the entire beam can be focused onto the particle [13]. If the beam is larger, the portion that misses the particle cannot contribute to the torque. However, as well as maximizing the overlap between the beam and particle, it is also important to maximize the angular momentum content of the beam. This can be achieved in either of two ways; first, by choosing a wavelength such that the diffraction-limited Gaussian spot of a circularly polarized beam is the same size as the particle (since the angular momentum flux is P/ω, reducing the wavelength maximizes the angular momentum), and second, by exploiting orbital angular momentum. The use of orbital angular momentum maximizes the angular momentum content *at a particular wavelength*, allowing a single set of optical components to be used for micromachines of various sizes. We can also note that the viscous drag on a rotating sphere is proportional to r^3. Thus, the highest rotation speeds will typically result from smaller micromachines. The torque, on the other hand, will increase with increasing size.

- The rotational symmetry of the particle can be chosen so as to optimize the torque. The ideal choice depends upon the angular momentum of the incident beam—for an incident beam of azimuthal order m_0, pth-order rotational symmetry with $p = m_0$ appears to be a good choice, allowing coupling of the incident beam to the lowest angular momentum modes possible, with $m = 0$. More generally, $m_0 \leqslant p \leqslant 2m_0$ should give good performance. For smaller p, the difference between the magnitudes of the $n = \pm 1$ orders is less than $2m_0$, and thus the difference in coupling is likely to be smaller, and for greater p, all scattered orders have a magnitude of their angular momentum greater than that of the incident beam, which is likely to increase coupling to the nontorque-producing zeroth order scattered modes ($n = 0$).
- Chiral particles can be produced, allowing rotation even by incident beams carrying little or even no angular momentum. Greater torque will also result, compared with a similar particle, when illuminated by high angular momentum light of the appropriate handedness. The price that is paid is that rotation in one direction is preferred; if one desires equal performance in both directions, then an achiral device is necessary.
- The coupling between incident and scattered VSWFs is essentially a vector version of the coupling between an incident paraxial Laguerre–Gaussian mode and diffracted modes due to a hologram; in both cases the phase variation is of the form $\exp(im\phi)$, and the effect of the symmetry of the particle or hologram on the coupling is the same. This allows a simple conceptual model of optically driven micromachines: microholograms.

The principles outlined in this section have been applied to the design of optically driven micromachine elements and have proved successful [44]. These principles can also be recognized in successful devices that appear to be based upon geometric optics principles, or even trial and error [61]. Figure 8.10 shows two very simple $p = 4$ devices based upon these principles. The achiral cross rotor is designed to be rotated by an incident beam carrying angular momentum, while the chiral rotor will rotate in an incident linearly polarized Gaussian beam. Viewing these structures as microholographic elements, they can be seen as binary phase approximations of interference patterns between paraxial Gaussian and LG_{04} beams. If both have planar wavefronts, then the achiral rotor will result; if the Gaussian beam has curved wavefronts the chiral rotor will result. This suggests that the direction of rotation of the chiral rotor can be reversed by reversing the curvature of an incident Gaussian beam; this effect has been observed by Galajda and Ormos [23]. Simple structures such as those shown in Figure 8.10 have also been fabricated and tested by Ukita and Kanehira [85]. The

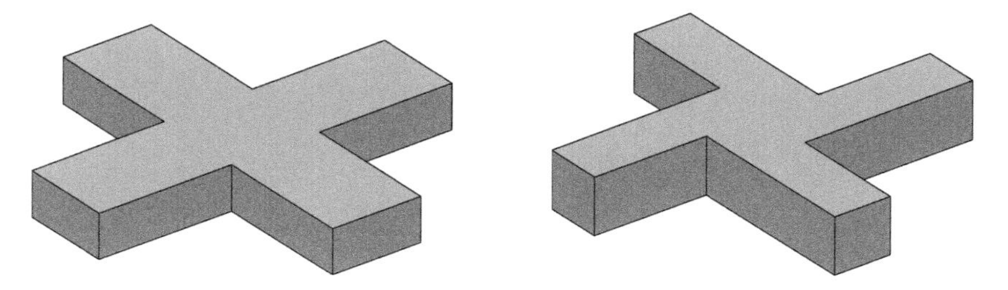

Figure 8.10 Simple designs for achiral (left) and chiral (right) optically driven micromachines, based upon the design principles outlined in text. They can be considered as binary phase holograms. The centers are solid for structural integrity.

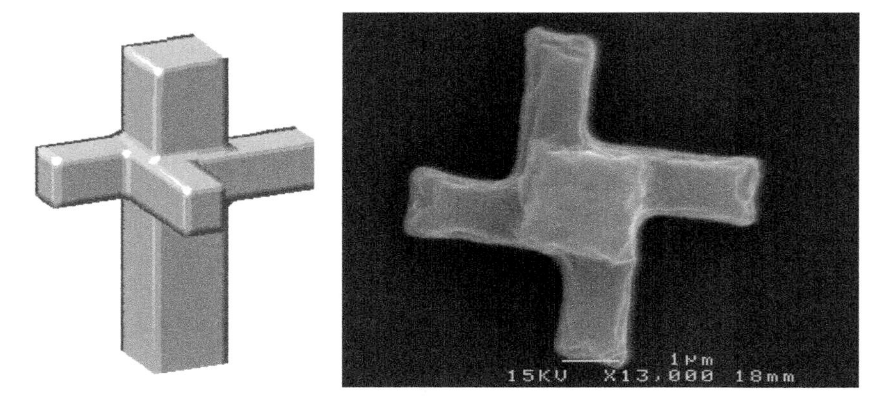

Figure 8.11 Stalked cross microrotor. This is similar to the chiral design shown in Figure 8.10, but is equipped with a central stalk so as to maintain the proper orientation when three-dimensionally trapped by optical tweezers. A computer-generated model (*left*) and a scanning electron micrograph of an actual device produced by two-photon photopolymerization (*right*) are shown.

microhologram picture of these devices suggests that the optimum thickness is that which produces a half-wave phase difference between light that passes through the arms of the structure and light that passes between them.

However, simple planar structures such as these have a major flaw if three-dimensional trapping of them is intended—flat structures will tend to align with their short axis perpendicular to the beam axis [5,10,24]. An additional structure—a central stalk—can be introduced to maintain the proper orientation. A design of this type is shown in Figure 8.11.

Figure 8.12 Scanning electron micrograph of a diffractive optical element designed to produce a beam with $8\hbar$ angular momentum per photon.

A useful method of producing such devices is two-photon photopolymerization [22,44,53]. A chiral stalked cross produced by this method is also shown in Figure 8.11.

This type of microfabrication can also be used to incorporate the production of beams carrying orbital angular momentum into an integrated device [44]. A microscopic diffractive optical element of this type is shown in Figure 8.12. Since this type of structure alters the angular momentum content of a beam that passes through it, it experiences a reaction torque—such a structure can also be used as a microrotor in its own right [86]. Again, this type of rotor can be viewed as a chiral holographic element.

While a number of optically driven rotating devices conforming to the general principles outlined in this section have been experimentally demonstrated, analysis of such structures has predominantly been carried out by the use of geometric optics. Since wave effects (such as the phase difference between different regions of the transmitted wave) are clearly important from the microhologram viewpoint, there is a great deal of useful work to be performed in the way of computational modeling of such devices. However, the geometries generally make the methods that can be used to calculate optical forces and torques on simple structures such a spheres, spheroids, cylinders, or similar, inapplicable. While direct methods such as the finite-difference time-domain method (FDTD) have been used [12], the repeated calculations required to characterize the behavior of a rotor within an optical trap places strong demands on efficient repeated calculations. Therefore, it appears to be useful to combine such general methods with the T-matrix method described earlier [48–50,63].

One disadvantage resulting from the use of orbital angular momentum to drive rotation is that while spin angular momentum is relatively easy to measure optically [7,8,14,45,58], it is rather more difficult to optically measure orbital angular momentum. In principle, it should be possible to do so by a variety of methods, such as using holograms as mode filters [52,77], interferometry using Dove prisms to introduce a phase shift [46], or measurement of the rotational frequency shift [2,4]. While such methods have been demonstrated to accurately measure optical torque [77], accurate optical measurement of orbital torque within optical tweezers has proved elusive. This appears to largely result from sensitivity to alignment, both transverse and angular, and aberrations.

However, a method of estimating the orbital torque has been demonstrated [76,78], wherein the rotation rate and spin torque are measured for the same object being rotated by left- and right-circularly polarized beams, and a linearly polarized beam, all carrying the same orbital angular momentum. If the spin torque is similar in magnitude (the handedness will be opposite) for the two circularly polarized cases, and the difference in rotation rate as compared with the linearly polarized beam is also the same for the two, then we can safely assume that, as a reasonable approximation, the orbital torque is the same in all cases. The variation in rotation rate with the spin then allows the viscous drag to be measured. The orbital torque can then be estimated from the rotation rate in the linearly polarized beam.

The orbital torque was measured using this method for planar clusters of two, three, and four spheres rotated by LG_{02}, LG_{03}, and LG_{04} beams [76,78]. The torque efficiencies are shown in Figure 8.13. As noted earlier, these measured torque efficiencies are

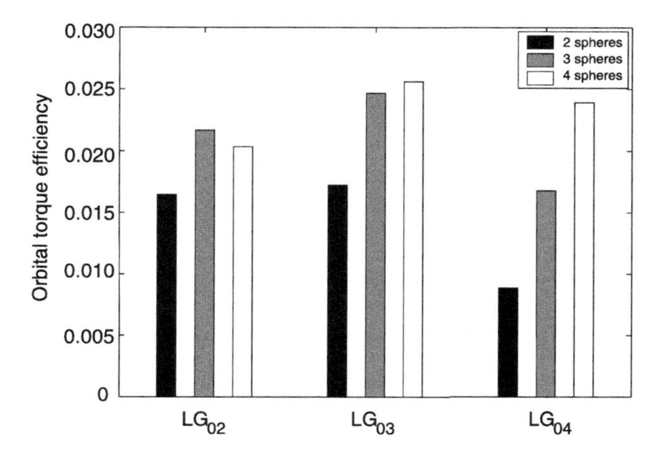

Figure 8.13 Orbital torque efficiencies for planar clusters of spheres rotated by focused Laguerre–Gaussian beams of varying orbital angular momenta.

comparable with the efficiency of 0.005 calculated above for a single sphere. These results also support the general principles of coupling between angular momentum orders outlined above. For example, with the LG_{04}, the highest torque efficiency occurs for four spheres, when the first-order scattering couples to the lowest possible angular momenta. The variation of efficiency with beam mode also supports this. Analysis of these results is complicated by the fact that the incident beam is not a pure angular momentum mode when linearly polarized; ideally, many more particle-mode combinations should be tested.

Finally, we will move on to consider two special cases of optical vortices—optical vortices carrying no angular momentum, and longitudinal optical vortices.

8.8 ZERO ANGULAR MOMENTUM OPTICAL VORTICES

Noting that Gaussian beams consisting of a single azimuthal mode—circularly polarized beams—carry angular momentum of $\pm\hbar$ per photon, we next turn to ask what kind of beam might carry zero angular momentum. In the context of optical vortices, the answer is clear—an LG_{01} beam with circular polarization opposing the orbital angular momentum [82].

Two other types of beams also meet these criteria of zero angular momentum and single-modedness: radially and azimuthally polarized beams. Radially and azimuthally polarized beams are optical vortices, but are usually considered to be "polarization vortices," rather than the more usual "phase vortex" optical vortex. As all three types of beams can be represented in terms of VSWFs with $m = 0$, there must be a close relationship between them.

An azimuthally polarized beam has no radial component of its electric field (in spherical coordinates), and is therefore a pure TE mode. By symmetry, the radially polarized beam is a pure TM mode. The LG mode, on the other hand, with radial components of both **E** and **H**, with electric and magnetic fields on an equal footing, is a superposition of TE and TM modes of amplitudes of equal magnitude. The counter polarized LG_{01} mode is therefore a combination of radially and azimuthally polarized beams.

In the same way that Laguerre–Gaussian beams attracted attention due to the possibility of improved axial trapping due to the absence of low-angle rays, radially and azimuthally polarized beams have also generated interest. In addition, the tight focal spots produced [17,79] suggest that tighter confinement is possible when trapping nanoparticles. Finally, since tightly focused radially polarized beams have a large longitudinal electric field component on the beam axis, where the Poynting vector is

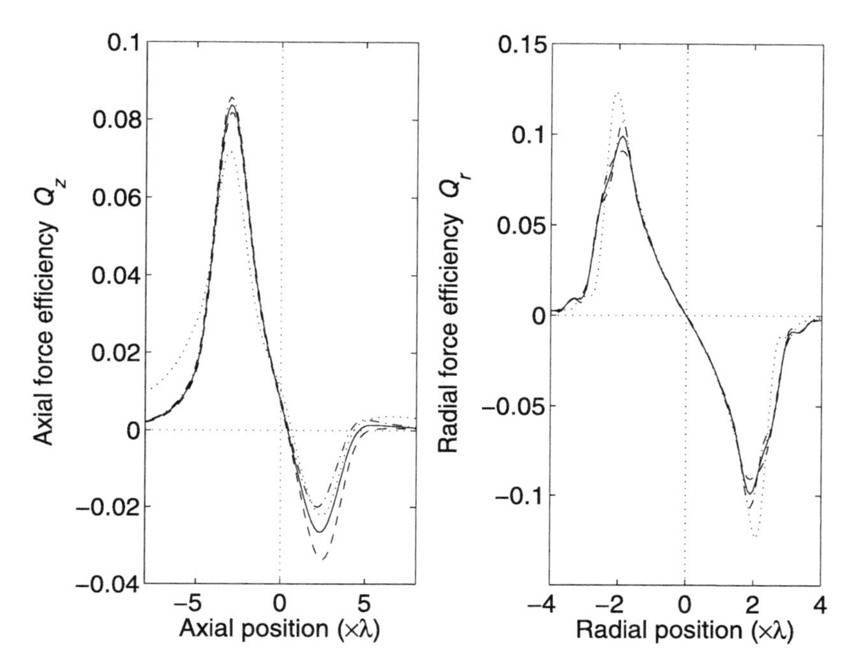

Figure 8.14 Axial and radial trapping efficiencies for radially polarized (*dashed line*), azimuthally polarized (*dot–dash line*), and counter-circularly polarized LG_{01} (*solid line*) beam traps, compared with a Gaussian beam trap (*dotted line*). The angle of convergence of all beams is 60°. The particle is $2\lambda_{\text{medium}}$ in radius, corresponding to a diameter of 3.2 μm for trapping in water at 1064 nm. The relative refractive index corresponds to that of polystyrene ($n = 1.59$) in water.

zero, they offer the prospect of improved trapping of reflecting or absorbing particles.

Figure 8.14 shows the axial and radial force efficiencies for traps based on these three related zero angular momentum optical vortices. The radially polarized beam shows a large increase in the maximum reverse axial force, the LG_{01} beam a smaller increase, and the azimuthally polarized beam a decrease compared to a Gaussian beam trap. All three beams show a similar increase in the maximum forward axial force.

Kawauchi and colleagues [40] calculated a similar pattern of behavior using geometric optics, explaining the improvement due, the dark center, and for the radially polarized beam a larger fraction of p-polarization. Here, unlike the case for angular momentum carrying Laguerre–Gaussian beams, we find agreement between the approximate and exact results.

8.9 GAUSSIAN "LONGITUDINAL" OPTICAL VORTEX

Since an ideal lens is rotationally symmetric, it cannot alter the angular momentum per photon (as measured about the axis of symmetry of the lens) of a beam that passes through it. Therefore, a paraxial Gaussian beam carrying $\hbar$ spin angular momentum per photon and zero orbital angular momentum (after being focused by a high numerical aperture lens) must still carry $\hbar$ angular momentum per photon.

However, a simple semiclassical picture of spin transport by photons suggests that not all of this angular momentum can be spin after the beam has been focused. In the far field, where the converging beam is a spherical wave, the field is locally a plane wave. However, everywhere away from the beam axis, the direction of propagation is not parallel to the beam axis. Since a photon in a plane-wave mode has spin either parallel to or antiparallel to the direction of propagation, the spin of each photon cannot be parallel to the beam axis. Since only the component of the spin of the photon that is parallel to the beam axis contributes to the total spin angular momentum flux (since the other components will cancel when integrated across the beam), the total spin must be less than $\hbar$ per photon.

Since the scalar far-field amplitude of the focused Gaussian beam is equal to

$$U = U_0 \exp\{-(k^2 w_0^2 \tan^2\theta)/4\}, \tag{24}$$

or

$$U = U_0 \exp(-\tan^2\theta/\tan^2\theta_0), \tag{25}$$

where θ_0 is the beam convergence angle, we can readily find an expression for the total spin by integrating over the beam profile. If the beam is circularly polarized everywhere in the far field, we can integrate over a hemisphere to obtain

$$S_z = A/P, \tag{26}$$

where

$$A = \int_0^{\pi/2} \exp(-2\tan^2\theta/\tan^2\theta_0) \sin\theta \cos\theta \, \mathrm{d}\theta, \tag{27}$$

$$P = \int_0^{\pi/2} \exp(-2\tan^2\theta/\tan^2\theta_0) \sin\theta \, \mathrm{d}\theta, \tag{28}$$

and S_z is the component of the total spin per photon parallel to the beam axis. These expressions can be readily integrated numerically, giving the dependence of spin on beam convergence angle shown in Figure 8.15. A rigorous electromagnetic calculation can also be carried out, which yields exactly the same numerical result [59].

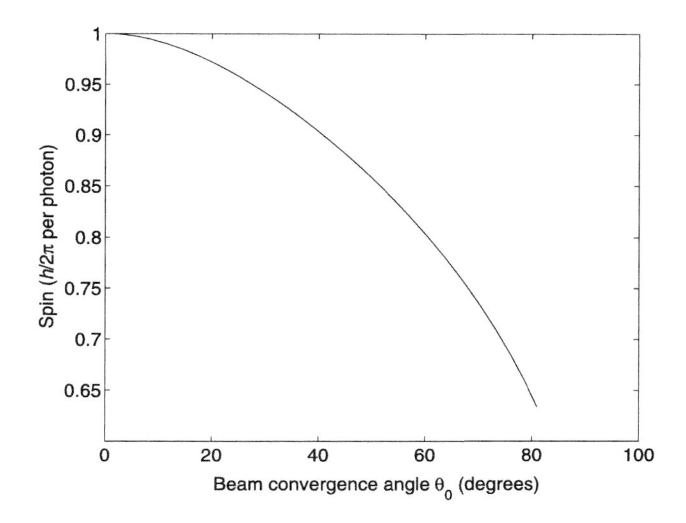

Figure 8.15 Spin angular momentum of a focused TEM_{00} Gaussian beam.

Since the spin angular momentum is less than $\hbar$ per photon, while the total angular momentum is still equal to $\hbar$ per photon, the beam necessarily carries orbital angular momentum, equal to the difference.

The conversion from spin to orbital angular momentum at the lens is at first surprising, since it is usual to use deliberate departure from rotational symmetry to generate orbital angular momentum. The generation of orbital angular momentum usually involves a torque; in this case, the torque parallel to the beam axis is zero. However, since spin density is independent of the choice of origin about which moments are taken and the orbital angular momentum density is not, there must be a choice-of-origin–dependent torque *density* on the lens. If we consider a ray parallel to the beam axis being focused by the lens, we can see that the momentum of the ray changes. Therefore, there must be a reaction force acting on the lens at the point where the ray passes through, and therefore, a torque density depending upon the choice of the origin results. We can even consider the presence of this torque density is what *requires* orbital angular momentum to be generated, in order to conserve angular momentum in all coordinate systems. Consideration of the symmetry of this reaction force shows that the components of the total force acting on the lens perpendicular to the beam axis must be zero, and the component of the total torque about any axis parallel to the beam axis must also be zero. As a result, the component of the total angular momentum flux of the focused beam parallel to the beam axis must also be independent of the choice of origin, despite the dependence of the angular momentum density of this choice. This independence is one of the characteristic features of optical vortices [73].

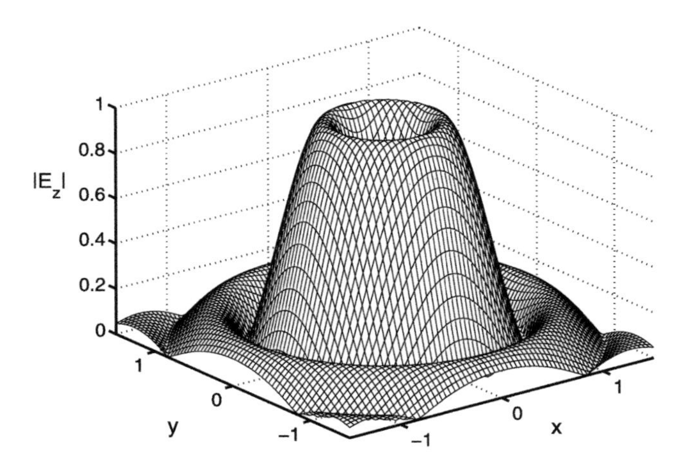

Figure 8.16 Longitudinal optical vortex. The vector component of the amplitude parallel to the beam axis has the form of an optical vortex, with phase variation $\exp(i\phi)$.

Therefore, a closer look at a focused circularly polarized Gaussian beam is warranted in order to discover a possible optical vortex hiding within such a nonvortical beam. Direct calculation of the amplitude, and the individual Cartesian vector components thereof, shows that the components of the field perpendicular to the beam axis are approximately Gaussian, as expected. However, the component parallel to the beam axis, which can be $1/3$ the magnitude of the transverse components for an easily achievable tightly focused beam, has the appearance of a typical optical vortex; this is shown in Figure 8.16.

Further consideration of the field of a tightly focused linearly polarized Gaussian beam in which there are longitudinal components of opposite phase on either side of the beam axis in the plane of polarization shows that this is exactly what we should expect. The pattern of the longitudinal field components of the focused linearly polarized beam are similar to the profile of a TEM_{01} beam. Just as we can add two orthogonally linearly polarized beams a quarter-wave out of phase to produce a circularly polarized beam, we can add TEM_{01} and TEM_{10} beams a quarter-wave out of phase to produce an LG_{01} beam. Thus, we obtain a longitudinal optical vortex.

Again, we can see this result in the mathematics of VSWFs. Both the paraxial and focused circularly polarized Gaussian beams consist of VSWFs with $m = 1$. Thus, the fields all share an azimuthal phase variation $\exp(i\phi)$. This variation does not give rise to a vortex structure of the transverse components since the radial and azimuthal basis vectors themselves rotate by 2π as we circle the beam axis; this is why a nonvortex field such as a plane wave parallel to the z-axis requires that only $m = \pm 1$ modes are present. The polar basis vector, on the other hand, does not change direction as we proceed around the beam axis in the focal plane; in this plane, $\hat{\theta} = -\hat{\mathbf{z}}$. This results in the longitudinal field having a vortex structure, while the transverse field does not.

8.10 CONCLUSION

Computational modeling of optical traps is a convenient method to illustrate and study their properties. Models are particularly useful for optical vortex traps, as the intricate details of such traps are often difficult to disentangle in experimental work. An example of such detail is the trapping efficiency of Laguerre–Gaussian beams compared to the standard Gaussian trap. We have shown that while the gradient force can be increased, the scattering force remains the same. In this case, the use of Laguerre–Gaussian beams does not result in much change in the trapping efficiency.

The trapping positions of different-sized particles in Laguerre–Gaussian beam traps were also investigated to establish when a particle is trapped centered on the beam axis and when it is trapped in the bright ring of the optical vortex, which causes the particle to orbit about the beam axis. This orbital motion was studied, and it was shown that the geometry of the trap could be more important than its orbital angular momentum content for torque generation.

Symmetry is perhaps the most important aspect of torque in optical tweezers. Treating particles as microholograms or Laguerre–Gaussian mode-converters is convenient when considering the symmetry of objects trapped in vortex beams.

We have also discussed some zero angular momentum trapping beams that have some interesting properties, such as the possibility of enhanced trapping of reflective or absorbing particles in a radially polarized beam due to the longitudinal fields at the focus of the trap.

REFERENCES

[1] A. Ashkin, Forces of a single-beam gradient laser trap on a dielectric sphere in the ray optics regime, *Biophys. J. 61* (1992) 569–582.

[2] R.d'E. Atkinson, Energy and angular momentum in certain optical problems, *Phys. Rev. 47* (1935) 623–627.

[3] S.M. Barnett, Optical angular-momentum flux, *J. Opt. B: Quantum Semiclassical Opt. 4* (2002) S7–S16.

[4] I.V. Basistiy, V.V. Slyusar, M.S. Soskin, M.V. Vasnetsov, A.Y. Bekshaev, Manifestation of the rotational Doppler effect by use of an off-axis optical vortex beam, *Opt. Lett. 28* (14) (2003) 1185–1187.

[5] S. Bayoudh, T.A. Nieminen, N.R. Heckenberg, H. Rubinsztein-Dunlop, Orientation of biological cells using plane polarised Gaussian beam optical tweezers, *J. Mod. Opt. 50* (10) (2003) 1581–1590.

[6] R.A. Beth, Mechanical detection and measurement of the angular momentum of light, *Phys. Rev. 50* (1936) 115–125.

[7] A.I. Bishop, T.A. Nieminen, N.R. Heckenberg, H. Rubinsztein-Dunlop, Optical application and measurement of torque on microparticles of isotropic nonabsorbing material, *Phys. Rev. A 68* (2003) 033802.

[8] A.I. Bishop, T.A. Nieminen, N.R. Heckenberg, H. Rubinsztein-Dunlop, Optical microrheology using rotating laser-trapped particles, *Phys. Rev. Lett. 92* (19) (2004) 198104.

[9] B.C. Brock, Using vector spherical harmonics to compute antenna mutual impedance from measured or computed fields, Sandia report SAND2000-2217-Revised, Sandia National Laboratories, Albuquerque, New Mexico, 2001.

[10] Z. Cheng, P.M. Chaikin, T.G. Mason, Light streak tracking of optically trapped thin microdisks, *Phys. Rev. Lett. 89* (10) (2002) 108303.

[11] C.H. Choi, J. Ivanic, M.S. Gordon, K. Ruedenberg, Rapid and stable determination of rotation matrices between spherical harmonics by direct recursion, *J. Chem. Phys. 111* (1999) 8825–8831.

[12] W.L. Collett, C.A. Ventrice, S.M. Mahajan, Electromagnetic wave technique to determine radiation torque on micromachines driven by light, *Appl. Phys. Lett. 82* (16) (2003) 2730–2732.

[13] J. Courtial, M.J. Padgett, Limit to the orbital angular momentum per unit energy in a light beam that can be focussed onto a small particle, *Opt. Commun. 173* (2000) 269–274.

[14] J.H. Crichton, P.L. Marston, The measurable distinction between the spin and orbital angular momenta of electromagnetic radiation, *Electronic J. Differential Equations Conf. 04* (2000) 37–50.

[15] J.E. Curtis, D.G. Grier, Structure of optical vortices, *Phys. Rev. Lett. 90* (13) (2003) 133901.

[16] A. Doicu, T. Wriedt, Computation of the beam-shape coefficients in the generalized Lorenz–Mie theory by using the translational addition theorem for spherical vector wave functions, *Appl. Opt. 36* (1997) 2971–2978.

[17] R. Dorn, S. Quabis, G. Leuchs, Sharper focus for a radially polarized light beam, *Phys. Rev. Lett. 91* (23) (2003) 233901.

[18] F. Ehrenhaft, Rotating action on matter in a beam of light, *Science 101* (1945) 676–677.

[19] A. Forrester, J. Courtial, M.J. Padgett, Performance of a rotating aperture for spinning and orienting objects in optical tweezers, *J. Mod. Opt. 50* (10) (2003) 1533–1538.

[20] M.E.J. Friese, T.A. Nieminen, N.R. Heckenberg, H. Rubinsztein-Dunlop, Optical alignment and spinning of laser trapped microscopic particles, *Nature 394* (1998) 348–350, *Nature 395* (1998) 621, Erratum.

[21] K.T. Gahagan, G.A. Swartzlander Jr., Optical vortex trapping of particles, *Opt. Lett. 21* (1996) 827–829.

[22] P. Galajda, P. Ormos, Complex micromachines produced and driven by light, *Appl. Phys. Lett. 78* (2001) 249–251.

[23] P. Galajda, P. Ormos, Rotors produced and driven in laser tweezers with reversed direction of rotation, *Appl. Phys. Lett. 80* (24) (2002) 4653–4655.

[24] P. Galajda, P. Ormos, Orientation of flat particles in optical tweezers by linearly polarized light, *Opt. Exp. 11* (5) (2003) 446–451.

[25] V. Garcés-Chávez, D. McGloin, M.J. Padgett, W. Dultz, H. Schmitzer, K. Dholakia, Observation of the transfer of the local angular momentum density of a multiringed light beam to an optically trapped particle, *Phys. Rev. Lett. 91* (9) (2003) 093602.

[26] R.C. Gauthier, Computation of the optical trapping force using an FDTD based technique, *Opt. Exp. 13* (10) (2005) 3707–3718.

[27] H. Goldstein, C. Poole, J. Safko, Classical Mechanics, 3rd ed., Addison–Wesley, San Francisco, 2002.

[28] G. Gouesbet, G. Grehan, Sur la généralisation de la théorie de Lorenz–Mie, *J. Opt. (Paris) 13* (2) (1982) 97–103.

[29] G. Gouesbet, J.A. Lock, Rigorous justification of the localized approximation to the beam-shape coefficients in generalized Lorenz–Mie theory. II. Off-axis beams, *J. Opt. Soc. Am. A 11* (9) (1994) 2516–2525.

[30] N.A. Gumerov, R. Duraiswami, Recursions for the computation of multipole translation and rotation coefficients for the 3-D Helmholtz equation, *SIAM J. Sci. Comput. 25* (4) (2003) 1344–1381.

[31] Y. Han, G. Gréhan, G. Gouesbet, Generalized Lorenz–Mie theory for a spheroidal particle with off-axis Gaussian-beam illumination, *Appl. Opt. 42* (33) (2003) 6621–6629.

[32] Y. Han, Z. Wu, Scattering of a spheroidal particle illuminated by a Gaussian beam, *Appl. Opt. 40* (2001) 2501–2509.

[33] Y. Harada, T. Asakura, Radiation forces on a dielectric sphere in the Rayleigh scattering regime, *Opt. Commun. 124* (1996) 529–541.

[34] H. He, M.E.J. Friese, N.R. Heckenberg, H. Rubinsztein-Dunlop, Direct observation of transfer of angular momentum to absorptive particles from a laser beam with a phase singularity, *Phys. Rev. Lett. 75* (1995) 826–829.

[35] H. He, N.R. Heckenberg, H. Rubinsztein-Dunlop, Optical particle trapping with higher-order doughnut beams produced using high efficiency computer generated holograms, *J. Mod. Opt. 42* (1) (1995) 217–223.

[36] N.R. Heckenberg, M.E.J. Friese, T. Nieminen, H. Rubinsztein-Dunlop, Mechanical effects of optical vortices, in: M. Vasnetsov, K. Staliunas (Eds.), *Optical Vortices*, in: *Horizons in World Physics*, vol. 228, Nova Science, Commack, New York, 1999, pp. 75–105.

[37] J. Humblet, Sur le moment d'impulsion d'une onde électromagnétique, *Physica 10* (7) (1943) 585–603.

[38] J.D. Jackson, Classical Electrodynamics, 3rd ed., John Wiley, New York, 1999.

[39] J.M. Jauch, F. Rohrlich, The Theory of Photons and Electrons, 2nd ed., Springer, New York, 1976.

[40] H. Kawauchi, K. Yonezawa, Y. Kozawa, S. Sato, Calculation of optical trapping forces on a dielectric sphere in the ray optics regime produced by a radially polarized laser beam, *Opt. Lett. 32* (2007) 1839–1841.

[41] L. Kelemen, S. Valkai, P. Ormos, Integrated optical motor, *Appl. Opt. 45* (12) (2006) 2777–2780.

[42] M. Khan, A.K. Sood, F.L. Deepak, C.N.R. Rao, Nanorotors using asymmetric inorganic nanorods in an optical trap, *Nanotechnology 17* (2006) S287–S290.

[43] G. Knöner, S. Parkin, T.A. Nieminen, N.R. Heckenberg, H. Rubinsztein-Dunlop, Measurement of the index of refraction of single microparticles, *Phys. Rev. Lett. 97* (2006) 157402.

[44] G. Knöner, S. Parkin, T.A. Nieminen, V.L.Y. Loke, N.R. Heckenberg, H. Rubinsztein-Dunlop, Integrated optomechanical microelements, *Opt. Exp. 15* (9) (2007) 5521–5530.

[45] A. La Porta, M.D. Wang, Optical torque wrench: Angular trapping, rotation, and torque detection of quartz microparticles, *Phys. Rev. Lett. 92* (19) (2004) 190801.

[46] J. Leach, M.J. Padgett, S.M. Barnett, S. Franke-Arnold, J. Courtial, Measuring the orbital angular momentum of a single photon, *Phys. Rev. Lett. 88* (25) (2002) 257901.

[47] J.A. Lock, G. Gouesbet, Rigorous justification of the localized approximation to the beam-shape coefficients in generalized Lorenz–Mie theory. I. On-axis beams, *J. Opt. Soc. Am. A 11* (9) (1994) 2503–2515.

[48] V.L.Y. Loke, T.A. Nieminen, T. Asavei, N.R. Heckenberg, H. Rubinsztein-Dunlop, Optically driven micromachines: Design and fabrication, in: M. Mishchenko, G. Videen (Eds.), *Tenth Conference on Light Scattering by Nonspherical Particles*, International Centre for Hect and Mass Transfer, Ankara, 2007, pp. 109–112.

[49] V.L.Y. Loke, T.A. Nieminen, N.R. Heckenberg, H. Rubinsztein-Dunlop, Exploiting symmetry and incorporating the T-matrix in the discrete dipole approximation (DDA) method, in: T. Wriedt, A. Hoekstra (Eds.), *Proceedings of the DDA-Workshop*, Institut für Werkstofftechnik, Bremen, 2007, pp. 10–12.

[50] V.L.Y. Loke, T.A. Nieminen, S.J. Parkin, N.R. Heckenberg, H. Rubinsztein-Dunlop, FDFD/T-matrix hybrid method, *J. Quant. Spectrosc. Radiat. Transfer 106* (2007) 274–284.

[51] L. Lorenz, Lysbevægelsen i og uden for en af plane Lysbølger belyst Kugle, *Vidensk. Selsk. Skrifter 6* (1890) 2–62.

[52] A. Mair, A. Vaziri, G. Weihs, A. Zeilinger, Entanglement of the orbital angular momentum states of photons, *Nature 412* (2001) 313–316.

[53] S. Maruo, K. Ikuta, H. Korogi, Force-controllable, optically driven micromachines fabricated by single-step two-photon microstereolithography, *J. Microelectromech. Syst. 12* (5) (2003) 533–539.

[54] G. Mie, Beiträge zur Optik trüber Medien, speziell kolloidaler Metallösungen, *Ann. Phys. 25* (3) (1908) 377–445.

[55] M.I. Mishchenko, Light scattering by randomly oriented axially symmetric particles, *J. Opt. Soc. Am. A 8* (1991) 871–882.

[56] A.A.R. Neves, A. Fontes, L.A. Padilha, E. Rodriguez, C.H. de Brito Cruz, L.C. Barbosa, C.L. Cesar, Exact partial wave expansion of optical beams with respect to an arbitrary origin, *Opt. Lett. 31* (16) (2006) 2477–2479.

[57] T.A. Nieminen, Comment on: "Geometric absorption of electromagnetic angular momentum," C. Konz, G. Benford, *Opt. Commun. 235* (2004) 227–229.

[58] T.A. Nieminen, N.R. Heckenberg, H. Rubinsztein-Dunlop, Optical measurement of microscopic torques, *J. Mod. Opt. 48* (2001) 405–413.

[59] T.A. Nieminen, N.R. Heckenberg, H. Rubinsztein-Dunlop, Angular momentum of a strongly focussed Gaussian beam, arXiv:physics/0408080.

[60] T.A. Nieminen, N.R. Heckenberg, H. Rubinsztein-Dunlop, Computational modelling of optical tweezers, *Proc. Soc. Photo. Instrum. Eng. 5514* (2004) 514–523.

[61] T.A. Nieminen, J. Higuet, G. Knöner, V.L.Y. Loke, S. Parkin, W. Singer, N.R. Heckenberg, H. Rubinsztein-Dunlop, Optically driven micromachines: progress and prospects, *Proc. Soc. Photo. Instrum. Eng. 6038* (2006) 237–245.

[62] T.A. Nieminen, G. Knöner, N.R. Heckenberg, H. Rubinsztein-Dunlop, Physics of optical tweezers, in: M.W. Berns, K.O. Greulich (Eds.), *Laser Manipulation of Cells and Tissues*, in: *Methods in Cell Biology*, vol. 82, Elsevier, 2007, pp. 207–236.

[63] T.A. Nieminen, V.L.Y. Loke, A.M. Brańczyk, N.R. Heckenberg, H. Rubinsztein-Dunlop, Towards efficient modelling of optical micromanipulation of complex structures, *PIERS Online 2* (5) (2006) 442–446.

[64] T.A. Nieminen, V.L.Y. Loke, G. Knöner, A.M. Brańczyk, Toolbox for calculation of optical forces and torques, *PIERS Online 3* (3) (2007) 338–342.

[65] T.A. Nieminen, V.L.Y. Loke, A.B. Stilgoe, G. Knöner, A.M. Brańczyk, Optical tweezers computational toolbox 1.0, http://www.physics.uq.edu.au/people/nieminen/software.html, 2007.

[66] T.A. Nieminen, V.L.Y. Loke, A.B. Stilgoe, G. Knöner, A.M. Brańczyk, N.R. Heckenberg, H. Rubinsztein-Dunlop, Optical tweezers computational toolbox, *J. Opt. A: Pure Appl. Opt. 9* (2007) S196–S203.

[67] T.A. Nieminen, S.J. Parkin, N.R. Heckenberg, H. Rubinsztein-Dunlop, Optical torque and symmetry, *Proc. Soc. Photo. Instrum. Eng. 5514* (2004) 254–263.

[68] T.A. Nieminen, H. Rubinsztein-Dunlop, N.R. Heckenberg, Calculation and optical measurement of laser trapping forces on nonspherical particles, *J. Quant. Spectrosc. Radiat. Transfer 70* (2001) 627–637.

[69] T.A. Nieminen, H. Rubinsztein-Dunlop, N.R. Heckenberg, Calculation of the T-matrix: general considerations and application of the point-matching method, *J. Quant. Spectrosc. Radiat. Transfer 79–80* (2003) 1019–1029.

[70] T.A. Nieminen, H. Rubinsztein-Dunlop, N.R. Heckenberg, Multipole expansion of strongly focussed laser beams, *J. Quant. Spectrosc. Radiat. Transfer 79–80* (2003) 1005–1017.

[71] T.A. Nieminen, H. Rubinsztein-Dunlop, N.R. Heckenberg, A.I. Bishop, Numerical modelling of optical trapping, *Comput. Phys. Commun. 142* (2001) 468–471.

[72] E. Noether, Invariante Variationsprobleme, *Nachrichten von der Gesselschaft der Wissenschaften zu Göttingen, Mathematisch-Physikalische Klasse 1918* (1918) 235–257, English translation by M.A. Tavel *Transport Theory and Statistical Mechanics 1* (3) (1971) 183–207.

[73] A.T. O'Neil, I. MacVicar, L. Allen, M.J. Padgett, Intrinsic and extrinsic nature of the orbital angular momentum of a light beam, *Phys. Rev. Lett. 88* (2002) 053601.

[74] A.T. O'Neil, M.J. Padgett, Three-dimensional optical confinement of micron-sized metal particles and the decoupling of the spin and orbital angular momentum within an optical spanner, *Opt. Commun. 185* (2000) 139–143.

[75] A.T. O'Neil, M.J. Padgett, Rotational control within optical tweezers by use of a rotating aperture, *Opt. Lett. 27* (9) (2002) 743–745.

[76] S.J. Parkin, G. Knöner, T.A. Nieminen, N.R. Heckenberg, H. Rubinsztein-Dunlop, Torque transfer in optical tweezers due to orbital angular momentum, *Proc. Soc. Photo. Instrum. Eng. 6326* (2006) 63261B.

[77] S.J. Parkin, T.A. Nieminen, N.R. Heckenberg, H. Rubinsztein-Dunlop, Optical measurement of torque exerted on an elongated object by a noncircular laser beam, *Phys. Rev. A 70* (2004) 023816.

[78] S. Parkin, G. Knöner, T.A. Nieminen, N.R. Heckenberg, H. Rubinsztein-Dunlop, Measurement of the total optical angular momentum transfer in optical tweezers, *Opt. Exp. 14* (15) (2006) 6963–6970.

[79] S. Quabis, R. Dorn, M. Eberler, O. Glöckl, G. Leuchs, Focusing light to a tighter spot, *Opt. Commun. 179* (2000) 1–7.

[80] F. Rohrlich, Classical Charged Particles, Addison–Wesley, Reading, MA, 1965.

[81] A.E. Siegman, Lasers, Oxford Univ. Press, Oxford, 1986.

[82] N.B. Simpson, K. Dholakia, L. Allen, M.J. Padgett, Mechanical equivalence of spin and orbital angular momentum of light: an optical spanner, *Opt. Lett. 22* (1997) 52–54.

[83] D.E. Soper, Classical Field Theory, Wiley, New York, 1976.

[84] J.A. Stratton, Electromagnetic Theory, McGraw–Hill, New York, 1941.

[85] H. Ukita, M. Kanehira, A shuttlecock optical rotator—its design, fabrication and evaluation for a microfluidic mixer, *IEEE J. Selected Topics Quantum Electron. 8* (1) (2002) 111–117.

[86] H. Ukita, K. Nagatomi, Theoretical demonstration of a newly designed micro-rotator driven by optical pressure on a light incident surface, *Opt. Rev. 4* (4) (1997) 447–449.

[87] G. Videen, Light scattering from a sphere near a plane interface, in: F. Moreno, F. González (Eds.), *Light Scattering from Microstructures*, in: *Lecture Notes in Physics*, vol. 534, Springer-Verlag, Berlin, 2000, pp. 81–96.

[88] K. Volke-Sepulveda, E. Ley-Koo, General construction and connections of vector propagation invariant optical fields: TE and TM modes and polarization states, *J. Opt. A: Pure Appl. Opt. 8* (2006) 867–877.

[89] P.C. Waterman, Symmetry, unitarity, and geometry in electromagnetic scattering, *Phys. Rev. D 3* (1971) 825–839.

[90] D.A. White, Vector finite element modeling of optical tweezers, *Comput. Phys. Commun. 128* (2000) 558–564.

[91] R. Zambrini, S.M. Barnett, Local transfer of optical angular momentum to matter, *J. Mod. Opt. 52* (8) (2005) 1045–1052.

Chapter 9

Rotation of Particles in Optical Tweezers

Miles Padgett and Jonathan Leach

University of Glasgow, UK

9.1 INTRODUCTION

Optical tweezers were pioneered by Ashkin and colleagues in the 1980s [4], working in Bell Labs. They utilized the gradient force associated with a very tightly focused laser beam to create a potential well, such that any nearby dielectric particle experiences an attractive force toward the beam focus. In competition with the gradient force are the scattering force, which acts in the direction of propagation of the light, and the force due to gravity. For micron-sized particles, a single focused laser beam can produce a gradient force that is greater than the scattering force and force due to gravity and hence is sufficient to create a 3D trap. The required laser power is minimized by the buoyancy generated by immersing the particles in a surrounding fluid, which through Stoke's drag forces also acts to make the trap stable.

Within a ray-optic picture, the trapping force arises from a refraction of the light by the microscopic particle. Implicit in this refraction is a redirection of the light's linear momentum and hence a reaction force acting on the particle. The effect of the reaction force is to draw the particle to the focus of a beam where the redirection of the light is minimized. In addition to its linear momentum, light can carry angular momentum. This can be spin angular momentum, which is dependant upon the polarization of the light, or orbital angular momentum, which is dependent upon the light beam's phase structure. Just as the transfer of linear momentum from light to a particle results in optical trapping, it is also possible to transfer angular momentum from light to a particle, resulting in rotation. The controlled trapping and rotation of micron-sized particles has generated a wealth of research dedicated to the underlying principles of the structure of light and the details of momentum transfer. Rotation in optical tweezers has also found applications ranging from biology to microfluidics [18].

The details of light's momentum can be understood by considering the linear and angular momentum together, rather than as independent properties. A generic description can be made by recognizing that at any point, the light field has a Poynting vector associated with it [3]. In isotropic media, the Poynting vector indicates the direction of both the energy and the momentum flow. Summing the axial components of this vector gives the linear momentum flow, whereas summing the transverse components, acting about radius vectors with respect to a defined axis, gives the angular momentum of the beam. Depending upon the exact nature of the interaction leading to particle rotation, it can be convenient to calculate the torque either in terms of a transfer of angular momentum or linear momentum acting as a radius vector. However, fundamentally the origin of the light's various momentum components is the same, no matter what the detailed method of momentum transfer; it can always be linked to the magnitude and direction of the Poynting vector.

Although not uniquely a property of individual photons, it is convenient to express the light's momentum as a per photon quantity. In free space, the linear momentum per photon is $\hbar k$, and the torque applied to an object can be maximized by applying this linear momentum in an azimuthal direction at the perimeter of the object. Assuming the object to be of radius r, the maximum torque and hence angular momentum transfer is therefore $\hbar kr$. In practice this will be further reduced by the numerical aperture of the coupling optics, restricting the skew angle of the Poynting vector [10]. For a wavelength-sized particle, the maximum angular momentum transfer is of order $\hbar$, the spin angular momentum of the photon. For larger particles, the maximum torque scales with the radius, but note also that larger particles have a larger moment of inertia and are subject to larger drag forces, scaling with the fifth and third power of the radius, respectively. Hence, optically induced rotation is only really effective for objects in the micron size range.

9.2 USING INTENSITY SHAPED BEAMS TO ORIENT AND ROTATE TRAPPED OBJECTS

In early work, Ashkin had observed that the inherent alignment of objects with the beam symmetry could cause rod-shaped bacteria to rotate upright, aligning vertically along the trapping axis of the beam [5]. Here it is energetically favorable for the object, which is elongated, to align its long axis with the direction of propagation of the light. If it is the case that the light is deliberately brought to a focus in an asymmetric pattern,

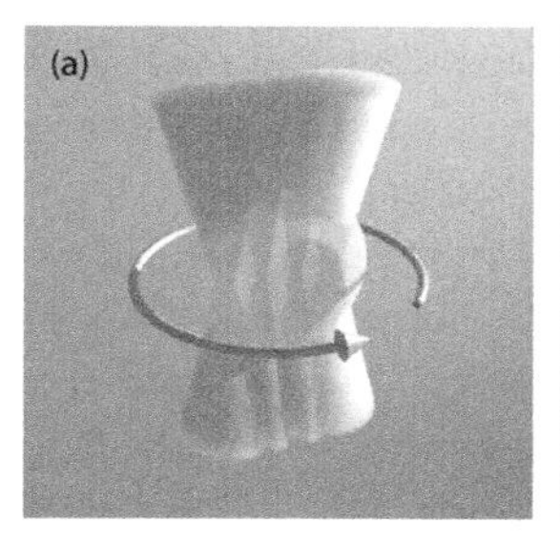

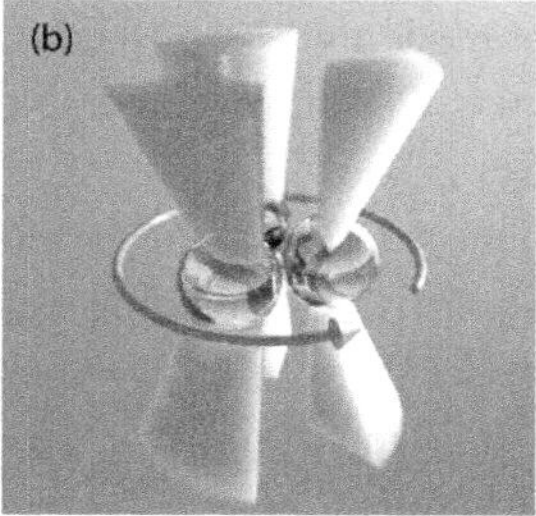

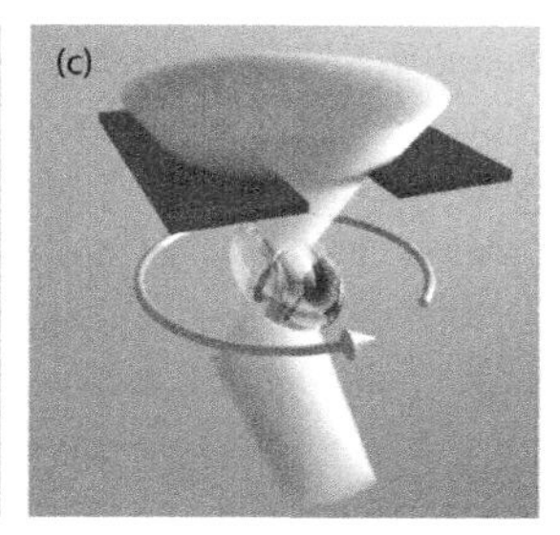

Figure 9.1 Rotation in optical tweezers using the rotation of intensity shaped beams. Sato and colleagues (a) used a Hermite–Gaussian laser mode to trap and rotate a red blood cell, Paterson and colleagues (b) used a spiral interference pattern, and O'Neil and colleagues (c) used a rotating mechanical aperture to shape a Gaussian beam. See color insert.

rather than a circularly symmetry focus, nonspherical objects will align themselves in the beam. It follows that rotation of an asymmetric trapping pattern results in rotation of the trapped object.

Deliberate rotation within optical tweezers has a history dating back to 1991 when Sato and colleagues used a high-order Hermite–Gaussian mode in an optical tweezers to trap a red blood cell [38], as shown in Figure 9.1(a). They showed that the gradient force acted to draw the cell into the high-intensity region of the beam. The rectangular symmetry of both the mode and the cell meant that the long axis of the cells naturally aligned with the long axis of the beam cross-section. Varying the orientation of a rectangular aperture within the laser cavity set the angle of the mode. Rotating the aperture caused the mode to rotate, giving a corresponding rotation of the cell. Although simple in concept, the rotation of a laser beam exactly about its own axis is more complicated than might be expected. Any asymmetric beam can be reoriented by passage through a Dove prism, but such systems are extremely difficult to align if any lateral translation is to be eliminated. However, a number of groups have successfully used various techniques to create noncircularly symmetric trapping beams that can be rotated in a controlled fashion to similarly set the orientation, or rotation, of a trapped asymmetrical object.

In 1994, Harris created angular patterns by interfering a Laguerre–Gaussian laser beam with its own mirror image [19]. Interestingly, although neither recognized nor relevant to that work, it is also the Laguerre–Gaussian beams that are the best known examples of those carrying orbital angular momentum. In 2001, Dholakia and colleagues used the same interference approach to create interference patterns within an optical tweezers [35], as shown in Figure 9.1(b). They showed that the resulting spiral interference pattern could be rotated by changing the path difference between the two beams and then used to rotate a Chinese hamster chromosome. This approach was adaptable in that changing the mode indices of the Laguerre–Gaussian mode could set the rotational symmetry of the interference pattern, thereby optimizing its shape to that of the object.

Another method of producing an asymmetric beam is to introduce a rectangular aperture into the optical path prior to the trapping beam's entry into the tweezers, thereby shaping the focused spot into an ellipse, Figure 9.1(c). Mounting the aperture within a rotatable stage that is mounted on an x/y translation stage means that its rotation axis can be aligned to the beam axis. As demonstrated in 2002, a rotating rectangular aperture mounted in the tweezing beam has been used to rotate assemblies of silica spheres that could be both trapped in three-dimensions and rotated synchronously with the aperture [34]. This simple method for rotational control does not require high-order modes, interferometric precision, or computer-controlled optical modulators.

9.3 ANGULAR MOMENTUM TRANSFER TO PARTICLES HELD IN OPTICAL TWEEZERS

In the previous section, the applied forces and resulting torques arise directly from the gradient force generated from an asymmetry of the trapping beam, or asymmetric configuration of trapping beams. However, rotation is also possible using beams with a circularly symmetric intensity distribution where it is the angular momentum of the trapping beam that is transferred. It is now well recognized that light beams carry both spin and orbital angular momentum [1,2]. The spin angular momentum arises from the beam's polarization state and has long been known to have a value of $\pm\hbar$ depending upon whether it is right- or left-hand circularly polarized. The orbital angular momentum is completely independent of this and arises from the beam's phase structure. Beams with helical phasefronts, described by $\phi = \exp(i\ell\theta)$, have a significant azimuthal component to their Poynting vectors and an associated orbital angular momentum corresponding to $\ell\hbar$ per photon.

In the 1930s, Beth used the spin angular momentum associated with circularly polarized light to drive the torsional motion of a wave-plate suspended by a quartz fiber. However, the detection of this motion required measurements of great precision [6]. But, as mentioned previously, the moment of inertia of a uniform body scales with the fifth power of its linear dimension, hence smaller particles require much less torque to accelerate.

In 1995, Rubinsztein-Dunlop and colleagues demonstrated that the orbital angular momentum carried by a Laguerre–Gaussian beam could be transferred to an absorbing particle trapped in an optical tweezers, making it spin at several Hertz [20].

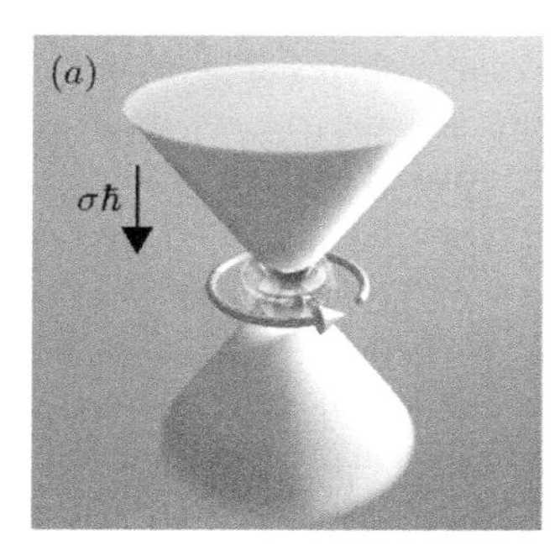

Figure 9.2 Angular momentum transfer in optical tweezers. Rotation of a trapped object can be induced by (a), the transfer of spin angular momentum using a circularly polarized beam or (b), the transfer of orbital angular momentum using a beam such as a high-order Laguerre–Gaussian or Bessel beam. See color insert.

This ground-breaking work was the first demonstration that light's angular momentum could induce a visible rotation, albeit on a microscopic object. With respect to absorption of the light, both the orbital and spin angular momentum should be transferable to the trapped object. Circularly polarizing the Laguerre–Gaussian trapping beam means that it carries spin angular momentum, which, depending upon the relative handedness of the two terms, can add to or subtract from the orbital component. Changing the relative handedness of the components speeds up or slows down the rotation [14] or, in the case of a Laguerre–Gaussian beam with $\ell = 1$, and hence an orbital angular momentum the same magnitude as the spin angular momentum, causes the particle to stop [39].

The absorption of light by the trapped particle will transfer both the spin and orbital angular momentum components with equal efficiency. However, rather than absorption, it is sufficient simply to change the angular momentum state of the light upon transmission through the particle. In Beth's experiment, the torque arose because the suspended wave, plate changed angular momentum of the transmitted light, without any absorption. Materials such as calcite have a high birefringence such that a thickness of a few microns is sufficient to act as a quarter-wave-plate, changing the polarization state of the transmitted light from circular to linear. Consequently, a micron-sized birefringent particle trapped in a circularly polarized light beam experiences a torque comparable to the total angular momentum flux in the beam, resulting in rotation rates up to 100 s Hz [15]. One might imagine a similar experiment for orbital angular momentum, where a microscopic object acts to transform the orbital angular momentum state of the beam. We will discuss one potential example later in the chapter.

So far we have restricted these angular momentum-based interactions to particles that are trapped on the beam axis and are larger than the beam diameter. In these cases, both the transfer of spin and orbital angular momentum result in the particle spinning about the axis of the beam. A more interesting comparison occurs when the particle is small with respect to the size of the beam and displaced from the beam axis. All beams with an $\exp(i\ell\phi)$ phase structure (e.g., a Laguerre–Gaussian beam) and an associated orbital angular momentum of $\ell\hbar$ per photon possess a phase singularity on the beam axis and hence no on-axis intensity. In 2001, metal particles were observed to circulate around the external perimeter of a Laguerre–Gaussian mode in a direction set by the orbital angular momentum of the beam and independent of its spin angular momentum state [33]. In 2002, a circularly polarized $\ell = 8$ Laguerre–Gaussian beam was used to confine small dielectric particles by the gradient force within the bright ring [32]. When the particle was birefringent, then the local transformation of the polarization state resulted in a torque acting on the particle causing it to spin about its own axis. By contrast, the azimuthal component of the light's Poynting vector, arising from its orbital angular momentum, meant that light scattering from the particle resulted in it orbiting around the beam axis. In these experiments, where the small particle is displaced from the axis of the laser beam, the transfer of spin and orbital angular momentum transfer are distinguishable, producing a spinning and orbiting motion of the particle, respectively. Similar results have also been obtained for high-order Bessel beams, the multiple bright rings of which allowed quantitative confirmation that the rotation rate as a function of distance from the beam center had a dependence consistent with the orbital angular momentum of the beam [17].

9.4 OUT OF PLANE ROTATION IN OPTICAL TWEEZERS

Following some early studies in 1999 [36], in 2002 Grier and colleagues revolutionized optical tweezers by showing that commercially available spatial light modulators (SLMs) could be used to generate many optical traps simultaneously, and that these traps could be repositioned at video frame rates [12]. One obvious application of these SLMs was to create two or more traps, positioned close together to trap and orientate larger asymmetric objects. Moving the traps in a circular trajectory would then impart a corresponding rotation to the trapped object [7]. Because the SLM can introduce both axial and lateral displacement to the trap position, the circular trajectory and corresponding rotation can be about any axis, including an axis that is perpendicular to the optical axis of the tweezers [7]. An interesting observation inherent in this process is that in an optical tweezers, two objects can be trapped and positioned immediately above one another. The very high numerical aperture of the objective lens means that the shadow created by one particle does not extend far, even in the axial direction.

SLMs can also be used to create multiple traps, arbitrarily distributed about the 3D volume, the dimensions of which are set by the spatial resolution of the SLM.

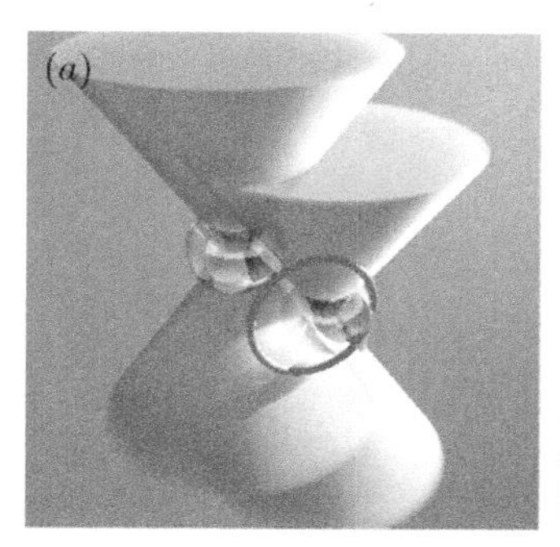

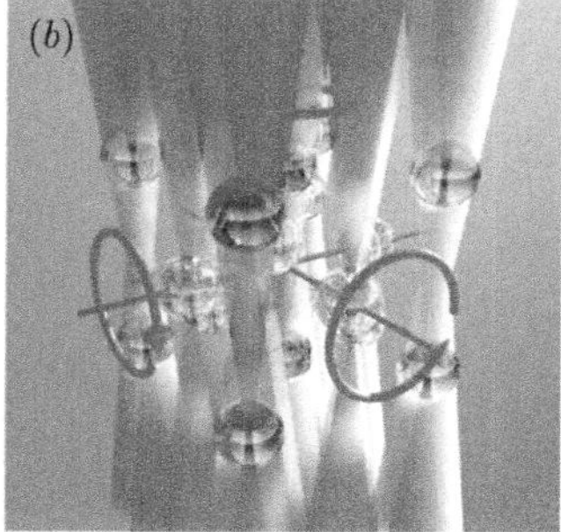

Figure 9.3 Out of plane rotation in optical tweezers. (a) Bingelyte and colleagues used a spatial light modulator to generated two independent optical traps that were controlled independently. (b) Sinclair and colleagues used a similar setup to generate a diamond unit cell using silica particles. In these examples, the rotation axis can be perpendicular to the optical axis of the trapping beams. See color insert.

As mentioned earlier, for small numbers of particles these trap positions can be set independently of each other without undue concern of shadowing effects. In this way, several layers of particles can be assembled into crystal [40] and quasi-crystal [37] configurations, and subsequently rotated about any axis of choice, as shown in Figure 9.3.

9.5 ROTATION OF HELICALLY SHAPED PARTICLES IN OPTICAL TWEEZERS

Another approach to inducing rotation within optical tweezers is to use a conventional, circularly symmetric trapping beam containing no angular momentum and then rely upon the shape of the trapped object to scatter light in an azimuthal direction such that the object suffers a recoil torque. Alternatively, one can regard the object as being a simple form of mode converter, transforming the incident beam containing no angular momentum into a scattered beam that does. This method is analogous to that of a windmill, where it is the shape of the blades that determines the sense and scale of the torque. In 1994, four armed rotors 10s microns in diameter were fabricated from silicon oxide and then trapped and set into rotation within an optical tweezers [21]. The sense of the rotation was determined by the handedness of the rotor construction. An important point is that unlike most micromachines that suffer from frictional wear, the three-dimensional trapping eliminated the need for a mechanical axle or other constraint.

Various other techniques for forming asymmetric structures have been demonstrated, including using partly silvered beads [27] and, more recently, making the

Figure 9.4 A helically shaped object placed in an optical tweezers will introduce orbital angular momentum to the transmitted or scattered light. The corresponding reaction torque will result in rotation of the object. See color insert.

objects within the tweezers itself using two-photon polymerization [16]. In the last example, multiple rotors were fabricated and positioned so that their vanes engaged to function as gear teeth. Driving either of them caused both to rotate; it was a multicomponent micromachine driven by light. Most recently, two-photon polymerization has been employed to create explicit mode converters based upon miniature spiral phase plates where the helicity of the microcomponent has been designed to give close-to-perfect transformation of the trapping beam into a Laguerre–Gaussian mode [24]. The change in the angular momentum state of the light led to a corresponding rotation of the micromode converter. Two-photon polymerization has also been used to fabricate a micrometer-sized mechanical motor driven by light from an integrated waveguide within the sample [22].

9.6 APPLICATIONS OF ROTATIONAL CONTROL IN OPTICAL TWEEZERS

Many of the original studies on rotation within optical tweezers were motivated by the desire to understand light's optical angular momentum. However, the ability to orientate and rotate micron-sized objects at rates of 100s of Hertz suggests various applications that have been and are currently being investigated. Applications of rotating particles in optical tweerers range from microscopic rheology, where precise measurements of fluid viscosity can be made, to lab-on-a-chip technologies, where optical tweezers are used as pumps and actuators.

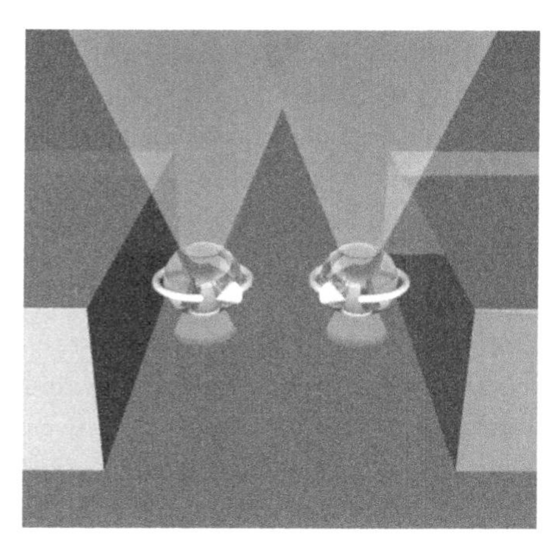

Figure 9.5 Applications of rotational control in optical tweezers. Leach and colleagues used counter-rotating vaterite spheres placed in a microfluidic channel to generate an optically driven micropump. See color insert.

Following their initial pioneering of the transfer of spin angular momentum to birefringent particles, Rubinsztein-Dunlop and colleagues developed a technique for producing near-spherical particles of birefringent vaterite and measuring precisely their rotational speed by monitoring the polarization modulation frequency of the transmitted light [23,29]. Being perfectly spherical means that the Stokes' drag force can be precisely calculated, and when combined with the measured rotation rate, used to derive a highly accurate measure of the viscosity of the surrounding fluid [9]. Most recently, this has been applied to measuring the viscosity of tear fluid, a measurement difficult by any other means, but extremely important in the diagnosis of various eye conditions (see Chapter 10).

The miniaturization of analytical processing is part of the lab-on-a-chip revolution of biological and chemical measurement. Fluid manipulation on a scale of 10s micron provides faster mixing (providing faster throughput) and creates the opportunity for performing multiple processes in parallel. In 2002 Marr and colleagues demonstrated a new concept for fluid pumps based on optical tweezers. Translation and rotation of series of spheres within a microfluidic chamber or channel established a fluid flow along the channel [41]. Holographic optical tweezers can also be used to measure the speed of the resulting fluid flow with high accuracy, repeatedly releasing and then re-trapping a probe particle and using a high-speed camera to record the initial movement of the particles within the flow [13].

As an alternative to direct control of the particle position, angular momentum transfer is an obvious route to creating flow in the surrounding fluid. Beyond the trapping of a single particle in the bright ring of a Laguerre–Gaussian mode, which that then orbits the beam axis due to an orbital angular momentum exchange, many particles can be added to the ring such that they all circulate around the beam axis [11]. Combing two rows of such beams circulating in opposite directions creates a net fluid flow down the middle, leading to transport of both the fluid itself and any particles it contains [25]. Similarly, the spin angular momentum of the light can be used to rotate multiple birefringent particles. Counterrotating a pair of such particles using right and left circularly polarized beams creates a net fluid flow between them. This is dramatically enhanced by placing the rotating pair of particles within a microfluidic channel, which cuts out the backflow around the outside [26]. Even when optimized, the fluid flow is only of order of 100s cubic microns per second. This is tiny compared to a conventional external pump, but may be of use for delivering small controlled amounts of reactant. In another approach, the two-photon polymerization discussed previously has been used to created a pump chamber and rotators that are then driven by a scanning laser beam and can pump fluid at a controlled rate of picoliters per minute [28].

Another possibility for fabricating microcomponents from birefringent material is to employ the fabrication process itself to create the birefringence with which the polarization state of the light can interact. An early example of this was the trapping and orientation of long glass rods by a polarized beam [8]. In one view, the orientation occurred through the gradient force that drew the rods to align with the high intensity axis of the beam. In an equivalent view, the elongation of the object meant that the polarization state of the trapping light was modified by the scattering process. The rods acted as a primitive birefringent waveplate, resulting in an angular momentum transfer. Micron-sized gear wheels have also been fabricated with grooves in the flat surfaces. The microscaled grooved surface is form-birefringent, such that the polarization state of the reflected light is again modified and an angular momentum is exchanged [31]. These cogs can be both trapped and driven with light, and positioned to engage with other cogs to create the basis of a pump or other devices.

Like optical tweezers themselves, rotational control within tweezers has progressed a pace, driven by developments within many different research groups. Since the initial curiosity-based demonstration of a rotating red blood cell by Sato and colleagues in 1992, subsequent investigations have targeted understanding the basic angular momentum properties of light [32]. In parallel with this work has been the application of rotational control to the driving of micromachines [30], cogs, and others to form optically actuated pumps, or in the direct measurement of physical quantities like viscosity.

REFERENCES

[1] L. Allen, M.W. Beijersbergen, R.J.C. Spreeuw, J.P. Woerdman, Orbital angular-momentum of light and the transformation of Laguerre–Gaussian laser modes, *Phys. Rev. A 45* (1992) 8185–8189.

[2] L. Allen, M.J. Padgett, M. Babiker, The orbital angular momentum of light, *Prog. Opt. 39* (1999) 291–372.

[3] L. Allen, M.J. Padgett, The Poynting vector in Laguerre–Gaussian beams and the interpretation of their angular momentum density, *Opt. Commun. 184* (2000) 67–71.

[4] A. Ashkin, J.M. Dziedzic, J.E. Bjorkholm, S. Chu, Observation of a single-beam gradient force optical trap for dielectric particles, *Opt. Lett. 11* (1986) 288–290.

[5] A. Ashkin, J.M. Dziedzic, T. Yamane, Optical trapping and manipulation of single cells using infrared-laser beams, *Nature 330* (1987) 769–771.

[6] R.A. Beth, Mechanical detection and measurement of the angular momentum of light, *Phys. Rev. 50* (1936) 115–125.

[7] V. Bingelyte, J. Leach, J. Courtial, M.J. Padgett, Optically controlled three-dimensional rotation of microscopic objects, *Appl. Phys. Lett. 82* (2003) 829–831.

[8] A.I. Bishop, T.A. Nieminen, N.R. Heckenberg, H. Rubinsztein-Dunlop, Optical application and measurement of torque on microparticles of isotropic nonabsorbing material, *Phys. Rev. A 68* (2003) 033802.

[9] A.I. Bishop, T.A. Nieminen, N.R. Heckenberg, H. Rubinsztein-Dunlop, Optical microrheology using rotating laser-trapped particles, *Phys. Rev. Lett. 92* (2004) 198104.

[10] J. Courtial, M.J. Padgett, Limit to the orbital angular momentum per unit energy in a light beam that can be focussed onto a small particle, *Opt. Commun. 173* (2000) 269–274.

[11] J.E. Curtis, D.G. Grier, Structure of optical vortices, *Phys. Rev. Lett. 90* (2003) 133901.

[12] J.E. Curtis, B.A. Koss, D.G. Grier, Dynamic holographic optical tweezers, *Opt. Commun. 207* (2002) 169–175.

[13] R. Di Leonardo, J. Leach, H. Mushfique, J.M. Cooper, G. Ruocco, M.J. Padgett, Multipoint holographic optical velocimetry in microfluidic systems, *Phys. Rev. Lett. 96* (2006) 134502.

[14] M.E.J. Friese, J. Enger, H. Rubinsztein-Dunlop, N.R. Heckenberg, Optical angular-momentum transfer to trapped absorbing particles, *Phys. Rev. A 54* (1996) 1593–1596.

[15] M.E.J. Friese, T.A. Nieminen, N.R. Heckenberg, H. Rubinsztein-Dunlop, Optical alignment and spinning of laser-trapped microscopic particles, *Nature 394* (1998) 348–350.

[16] P. Galajda, P. Ormos, Complex micromachines produced and driven by light, *Appl. Phys. Lett. 78* (2001) 249–251.

[17] V. Garces-Chavez, D. McGloin, M.J. Padgett, W. Dultz, H. Schmitzer, K. Dholakia, Observation of the transfer of the local angular momentum density of a multiringed light beam to an optically trapped particle, *Phys. Rev. Lett. 91* (2003) 093602.

[18] D.G. Grier, A revolution in optical manipulation, *Nature 424* (2003) 810–816.

[19] M. Harris, C.A. Hill, J.M. Vaughan, Optical helices and spiral interference fringes, *Opt. Commun. 106* (1999) 161–166.

[20] H. He, M.E.J. Friese, N.R. Heckenberg, H. Rubinsztein-Dunlop, Direct observation of transfer of angular-momentum to absorptive particles from a laser-beam with a phase singularity, *Phys. Rev. Lett. 75* (1995) 826–829.

[21] E. Higurashi, H. Ukita, H. Tanaka, O. Ohguchi, Optically induced rotation of anisotropic micro-objects fabricated by surface micromachining, *Appl. Phys. Lett. 64* (1994) 2209–2210.

[22] L. Kelemen, S. Valkai, P. Ormos, Integrated optical motor, *Appl. Opt. 45* (2006) 2777–2780.

[23] G. Knöner, S. Parkin, N.R. Heckenberg, H. Rubinsztein-Dunlop, Characterization of optically driven fluid stress fields with optical tweezers, *Phys. Rev. E 72* (2005) 031507.

[24] G. Knöner, S. Parkin, T.A. Nieminen, V.L.Y. Loke, N.R. Heckenberg, H. Rubinsztein-Dunlop, Integrated optomechanical microelements, *Opt. Exp. 15* (2007) 5521–5530.

[25] K. Ladavac, D.G. Grier, Microoptomechanical pumps assembled and driven by holographic optical vortex arrays, *Opt. Exp. 12* (2004) 1144–1149.

[26] J. Leach, H. Mushfique, R. di Leonardo, M. Padgett, J. Cooper, An optically driven pump for microfluidics, *Lab on a Chip 6* (2006) 735–739.

[27] Z.P. Luo, Y.L. Sun, K.N. An, An optical spin micromotor, *Appl. Phys. Lett. 76* (2000) 1779–1781.

[28] S. Maruo, H. Inoue, Optically driven micropump produced by three-dimensional two-photon microfabrication, *Appl. Phys. Lett. 89* (2006) 3.

[29] T.A. Nieminen, N.R. Heckenberg, H. Rubinsztein-Dunlop, Optical measurement of microscopic torques, *J. Mod. Opt. 48* (2001) 405–413.

[30] T.A. Nieminen, J. Higuet, G. Knoner, V.L.Y. Loke, S. Parkin, W. Singer, N.R. Heckenberg, H. Rubinsztein-Dunlop, Optically driven micromachines: Progress and prospects, *Proc. Soc. Photo. Opt. Instrum. Eng. 6038* (2006) 237–245.

[31] S.L. Neale, M.P. MacDonald, K. Dholakia, T.F. Krauss, All-optical control of microfluidic components using form birefringence, *Nature Mater. 4* (2005) 530–533.

[32] A.T. O'Neil, I. MacVicar, L. Allen, M.J. Padgett, Intrinsic and extrinsic nature of the orbital angular momentum of a light beam, *Phys. Rev. Lett. 88* (2002) 053601.

[33] A.T. O'Neil, M.J. Padgett, Three-dimensional optical confinement of micron-sized metal particles and the decoupling of the spin and orbital angular momentum within an optical spanner, *Opt. Commun. 185* (2000) 139–143.

[34] A.T. O'Neil, M.J. Padgett, Rotational control within optical tweezers by use of a rotating aperture, *Opt. Lett. 27* (2002) 743–745.

[35] L. Paterson, M.P. MacDonald, J. Arlt, W. Sibbett, P.E. Bryant, K. Dholakia, Controlled rotation of optically trapped microscopic particles, *Science 292* (2001) 912–914.

[36] M. Reicherter, T. Haist, E.U. Wagemann, H.J. Tiziani, Optical particle trapping with computer-generated holograms written on a liquid-crystal display, *Opt. Lett. 24* (1999) 608–610.

[37] Y. Roichman, D.G. Grier, Holographic assembly of quasicrystalline photonic heterostructures, *Opt. Exp. 13* (2005) 5434–5439.

[38] S. Sato, M. Ishigure, H. Inaba, Optical trapping and rotational manipulation of microscopic particles and biological cells using higher-order mode Nd-yag laser-beams, *Electron. Lett. 27* (1991) 1831–1832.

[39] N.B. Simpson, K. Dholakia, L. Allen, M.J. Padgett, Mechanical equivalence of spin and orbital angular momentum of light: an optical spanner, *Opt. Lett. 22* (1997) 52–54.

[40] G. Sinclair, P. Jordan, J. Courtial, M. Padgett, J. Cooper, Z.J. Laczik, Assembly of 3-dimensional structures using programmable holographic optical tweezers, *Opt. Exp. 12* (2004) 5475–5480.

[41] A. Terray, J. Oakey, D.W.M. Marr, Microfluidic control using colloidal devices, *Science 296* (2002) 1841–1844.

Chapter 10

Rheological and Viscometric Methods

Simon J.W. Parkin, Gregor Knöner, Timo A. Nieminen, Norman R. Heckenberg, and Halina Rubinsztein-Dunlop

University of Queensland, Australia

10.1 INTRODUCTION

Microrheology concerns the mechanical properties of media on the microscopic scale. Some areas of interest within this field are about viscoelasticity within biological cells [43,10] and the properties of polymer solutions where measurements of small volumes are advantageous for high throughput screening [4]. Microrheology techniques range from magnetic tweezers [8] to video particle tracking [28]. For a comprehensive review of these techniques and others, refer to Waigh [44]. In this chapter, we will consider microrheological techniques that involve the manipulation of microscopic particles within a fluid using optical tweezers and in particular optical traps formed by laser beams carrying angular momentum. The fluid surrounding the particle plays a significant role in the properties of the optical trap; refractive index, temperature, and viscosity all affect trapping. It is logical that a well-characterized trap can, therefore, be used to study these fluid properties.

Optical tweezers apply a trapping force to a particle via transfer of linear momentum, angular momentum can also be imparted from the light to the particle. In special cases, this transfer of angular momentum can be independent of the fluid properties that contribute to the trap strength, as the applied torque depends upon the intrinsic properties of the particle itself and its interaction with the trapping laser beam. This allows for fluid properties to be studied without the need for complicated optical trap characterization.

STRUCTURED LIGHT AND ITS APPLICATIONS

Force measurements on biological systems have probably been the most impressive application of optical tweezers. Precise calibration of the optical forces within an optical trap has allowed for the quantitative study of molecular motors [39,22,11], the transcription of DNA [45] and protein folding [18,42]. These are just a few examples of the microbiological systems that have been studied. Polystyrene or silica microspheres are usually used as handles for such experiments, and the force that the optical trap exerts on these probe particles is typically calibrated prior to the experiment. Either moving the particle with known velocity through the viscous medium surrounding it or measuring the displacement of the particle caused by thermal fluctuations within the optical trap allows this calibration, provided certain parameters are known.

Viscosity is one such parameter. By combining different calibration techniques, viscosity can be removed as a required parameter and can be measured instead [12,5]. Such methods have been used to map out viscosity gradients [26] and to measure viscoelasticity of polymer solutions over a range of frequencies [37]. In the work by Starrs and Bartlett [37] two-point microrheology was conducted using a dual-beam optical trap. This allows the probe particle–fluid coupling to be studied, which is a factor that needs to be considered in polymer solutions [25,9]. A method that combines driving a sphere with a known frequency as well as measuring its displacement due to Brownian motion is described by Tolić-Nørrelykke and colleagues [41]. This allows the optical trap to be calibrated in a single step without knowledge of the surrounding fluid's viscosity. The calibrated trap could then be used to probe viscosity provided that the dimensions of the probe particle are known precisely. Exact modeling of optical tweezers can also allow for experimental parameters used for calibration to be measured instead [20].

Another approach that does not require a complicated calibration procedure is to analyze rotational rather than translational motion in the optical trap. Chapter 9 in this book is dedicated to a discussion of the rotation of particles in optical tweezers [31], and for a further review refer to Parkin and colleagues [33]. One convenient method is to trap a birefringent crystal, which changes the polarization state of the trapping laser, causing angular momentum to be transferred from the laser beam to the particle [14]. Liquid crystals form birefringent spheres that can be rotated in the same manner [17,16], and have been used for viscosity measurements [29]. Spherical particles are desirable as the viscous drag is well known provided the particle diameter and rotation rate can be measured [24]. To eliminate complications in the flow, ideally a rigid sphere, rather than a liquid crystal droplet, can be used as the probe of the surrounding fluid's viscosity [2]. This kind of microviscometer is discussed in detail in this chapter. Passive microrheology is also possible with rotational motion by tracking the rotational Brownian motion of disk-shaped objects [6] or birefringent crystals [23]. This method has also been extended to measure viscoelasticity [7].

However, we have limited this chapter to viscosity measurements with particles actively driven by exchange of angular momentum with the laser beam. This is done in a straightforward manner with circularly polarized light and birefringent spheres. We will also consider the possibility of using transfer of orbital angular momentum from the trapping laser to drive the probe particle for viscosity measurements.

10.2 OPTICAL TORQUE MEASUREMENT

The angular momentum of light is discussed elsewhere in this book. In this chapter, we will consider in detail how angular momentum of light is transferred to an optically trapped particle and how the resulting torque can be measured in order to measure the rheological properties of the surrounding medium. By way of introduction, we will briefly summarize the salient facts about optical angular momentum.

Light can carry angular momentum in two forms: spin angular momentum and/or orbital angular momentum. Spin angular momentum is associated with the light's polarization. A circularly polarized beam carries $\pm\hbar$ of angular momentum per photon (left or right circularly polarized) and corresponds to an electric field that rotates at the optical frequency about the direction of propagation. A linearly polarized beam can be thought of as containing equal numbers of left and right circularly polarized photons and so has an average spin angular momentum of $0\,\hbar$ per photon. Orbital angular momentum is due to the spatial structure of the field. The angular momentum resulting from this structure is apparent when considering a Laguerre–Gaussian mode of light, an example of an orbital angular momentum carrying field provided its azimuthal index is nonzero. The helical phase fronts demonstrate the rotation of the energy about the center of the mode as the beam propagates. The Laguerre–Gaussian modes have $l\hbar$ per photon of orbital angular momentum associated with them (where l is the azimuthal index of the mode).

A distinction between spin angular momentum and orbital angular momentum is worth pointing out at this stage. The spin angular momentum density, by definition, is independent of the chosen origin of the coordinate system. The orbital angular momentum density, however, does depend upon the origin as the angular momentum is about that point [36].

10.2.1 Measuring Spin Angular Momentum

We are interested in applying torques to objects trapped in optical tweezers. However, in order for this technique to be used to quantitatively study a physical system, the applied torque needs to be measured. The principle of such a measurement is simply that the light imparts angular momentum to the particle, therefore the difference in angular momentum content of the incident beam and the transmitted beam will yield the value of the torque applied to the particle. In order to find the spin angular momentum carried by a beam, we consider its composition in terms of two components: left and right circular polarizations. An arbitrary beam can be represented as the sum of these components. The left circularly polarized component consists of photons with $+\hbar$ of angular momentum, while the right component has photons with $-\hbar$. The relative powers of these two components define the spin angular momentum of the beam, which can be formalized by defining a degree of circular polarization,

$$\sigma_s = \frac{P_L - P_R}{P}, \tag{1}$$

where P_L and P_R are the powers associated with the left and right circularly polarized components, respectively, and P is the total laser power. As the beam passes through a birefringent particle, a change in either of these two components, P_L and P_R, results in a change in the total angular momentum flux of the beam and the transfer of angular momentum from the beam to the particle. The resulting reaction torque on the particle is given by

$$\tau_s = \frac{\Delta\sigma_s P}{\omega}, \tag{2}$$

where $\Delta\sigma_s$ is the change in the degree of circular polarization as the beam passes through the particle and ω is the angular frequency of the light [30].

Knowledge of the torque applied to the probe particle allows the local environment to be studied. In particular, if the particle is trapped and rotating steadily, the viscous drag torque acting must be equal to the applied torque. Therefore, provided there exists an expression for the drag torque in terms of measurable quantities, the viscosity of the surrounding fluid can be determined. In experiments that will be outlined in this chapter, a spherical probe particle is rotated by transfer of spin angular momentum, which is possible due to the birefringence of its crystalline structure. The viscous drag torque on a rotating sphere is well known and is given by

$$\tau_d = 8\pi\eta a^3 \Omega, \tag{3}$$

where η is the viscosity of the surrounding liquid, a is the radius of the sphere, and Ω is its angular velocity [24]. This expression assumes low Reynolds number flows and that the surrounding fluid is Newtonian, which is valid for fluids such as water and methanol on the size scale relevant to optical tweezers. The viscous drag torque is equal to the optically applied torque ($\tau_d = \tau_o$), which gives an expression for the

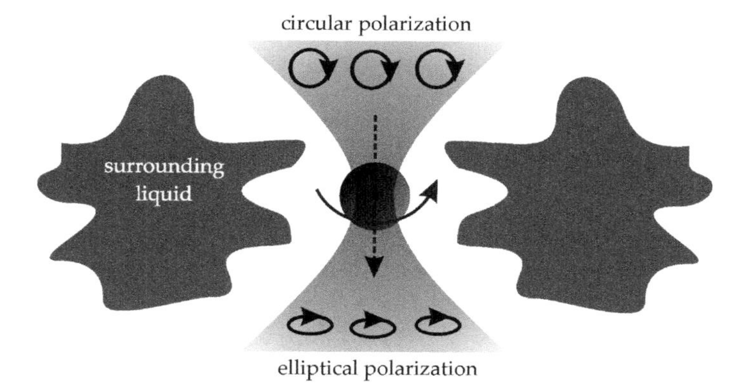

Figure 10.1 An optical torque is applied to the trapped sphere by transfer of spin angular momentum from the laser light to the sphere. Measuring how the sphere changes the polarization of the light determines the applied torque to the particle, which allows the response of the surrounding fluid to be quantified.

fluid's viscosity,

$$\eta = \frac{\Delta\sigma_s P}{8\pi a^3 \Omega\omega}, \tag{4}$$

where all these quantities can be experimentally measured. Figure 10.1 shows how torque is applied to a probe particle and how the response of the fluid can be investigated in this way. In the terms usually used in viscometry, the following is true for a Newtonian fluid: $\eta = S/R$, where S is the shear stress and R is shear rate. In order to determine viscosity, a stress is applied and the shear rate is measured. Here an applied torque causes a shear stress in the fluid, and the response of the fluid is investigated by measuring the rotation rate of the probe particle, which corresponds to shear rate.

10.2.2 Measuring Orbital Angular Momentum

Measuring the orbital angular momentum is similar in concept to measuring the spin angular momentum as, again, we are interested in the change of the angular momentum content of the beam after it passes through the trapped object. We can do this once again by decomposing the beam into orthogonal components with well-defined angular momentum. However, one important difference is that instead of having to determine the amplitudes of just two polarization components, we have now to determine the amplitudes of a potentially very large set of Laguerre–Gaussian (LG) modes. The LG modes are analogous to the orthogonal circular polarizations in that they form a basis set with well-defined angular momenta. An arbitrary beam can be decomposed into LG modes, and the amplitudes of the modes will determine the overall orbital angular momentum content of the beam. We can define a degree of orbital angular momentum,

$$\sigma_o = \left(\sum_{l=1}^{\infty} l P_l - \sum_{l=-1}^{-\infty} |l| P_l \right) / P, \tag{5}$$

where P_l is the power in an LG mode with an azimuthal index of l. P is the total power, which is equivalent to the sum of all the power in all the LG modes. The torque applied to a trapped object can now be found by using equation (2), except $\Delta\sigma$ would instead be the change in the degree of orbital angular momentum content of the beam.

While this LG mode decomposition has been demonstrated using paraxial beams [34], the method turns out to be problematic for the case of an arbitrary beam passed through optical tweezers, as we will discuss later in the chapter. Therefore, another method has been developed to measure orbital angular momentum in optical tweezers. This method uses the relatively simple spin angular momentum measurement to infer the amount of orbital angular momentum transferred. For a Newtonian fluid and in the laminar flow regime, the following is true

$$\tau_{\text{total}} = \tau_s + \tau_o = K\Omega, \tag{6}$$

where Ω is the rotation rate of the trapped particle and K is a constant. Therefore, by varying the torque due to the spin, one can determine the constant K and the orbital component of the torque [32]. In principle, objects that allow efficient transfer of orbital angular momentum can then be used to study the viscosity of the surrounding fluid. There is a complication, as equation (3) will no longer apply because the particle is unlikely to be spherical, although computational fluid dynamics can be used to determine the drag torque on arbitrarily shaped objects.

10.3 A ROTATING OPTICAL TWEEZERS-BASED MICROVISCOMETER

In this section we describe a microviscometer based upon the transfer and measurement of spin angular momentum to spherical birefringent particles of a material called *vaterite*. Vaterite is a calcium carbonate crystal that can form spheres approximately 2–10 μm in diameter under the right growth conditions [2]. We will also describe how to experimentally measure orbital angular momentum transfer in optical tweezers and discuss its potential to function as a microviscometer.

10.3.1 Experimental Setup for a Spin-based Microviscometer

The experimental setup we use is shown in Figure 10.2. The trapping laser is a Nd:YAG laser, with a wavelength of 1064 nm and a power output of 800 mW. The power is controlled using a half wave plate and a polarizing beam splitter cube. In order to achieve the best convergence of the laser beam in the focus, two lenses expand the beam so that it fills the back aperture of a high numerical aperture objective (NA = 1.3, Olympus P100). A quarter wave plate immediately before the objective makes the beam circularly polarized. A high numerical aperture (NA = 1.4) oil immersion condenser collects the transmitted trapping laser light. The sample is held on a piezo driven xyz translation stage. After the condenser, the light is directed to a circular polarization analyzing system that consists of a quarter waveplate followed by a polarizing beam splitter cube with two photodetectors at the outputs of the cube. A third photodetector is positioned to collect a tiny amount of light transmitted through a mirror, which has been linearly polarized by the mirror.

Data is collected by digitizing the analog outputs from the three photodetectors using a National Instruments data acquisition card. Typical signals for these detectors are shown in Figure 10.2. The signal from detector 3 (Figure 10.2(b)) gives the rotation rate of the particle. Laser light transmitted by the particle is, in general, ellip-

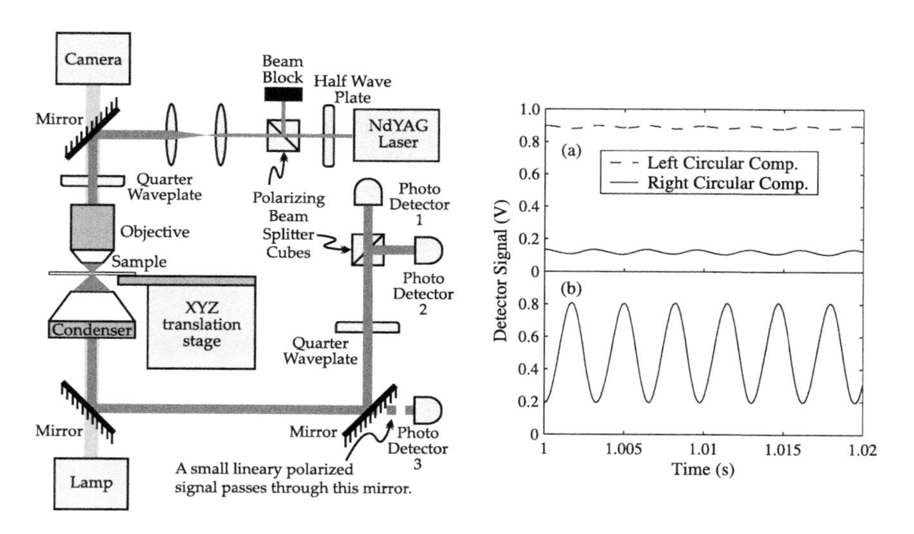

Figure 10.2 An optical tweezers setup used to make viscosity measurements (left). Typical signals from the three photodetectors used to measure the rotation rate of the particle and the optically applied torque are shown (right).

tically polarized. The ellipse rotates with the optical axis of the vaterite particle, which causes the intensity variation detected by the photodiode. This variation is sinusoidal due to the even rotation of the particle. The rotation rate of the particle is one-half the measured frequency due to the twofold symmetry that exists between the rotating linearly polarized component of the beam exiting the particle and the linear polarization detector. The signals from detectors 1 and 2 (Figure 10.2(a)) represent the magnitude of two orthogonal circularly polarized components of the transmitted laser beam. To find the angular momentum content of the beam, the components are most easily measured in the circular basis. A quarter waveplate aligned with its fast axis at 45° to the axes of a polarizing beam splitter cube ensures that the two photodetectors measure the respective left and right circularly polarized components of the beam. Thus, the degree of circular polarization can be easily determined.

In order to measure viscosity, the parameters that need to be experimentally determined, according to equation (4), are the laser power, the change in the degree of circular polarization, as well as the particle's radius and rotation rate. Rotation rate and polarization are measured by the three detectors as previously described. Laser power is measured by a photo detector that is positioned outside the microscope and that is calibrated such that it corresponds to the laser power in the focus. For the degree of accuracy required, the simplest calibration was to measure the viscosity of liquids of known viscosity and then solve equation (4) for power. We find about 50% of the power entering the objective reaches the focus, as expected [38]. In fact, this is a convenient method to measure the power in the focus, which is otherwise difficult to determine directly. The particle diameter was measured by touching the sphere against a second sphere of similar radius and measuring the center-to-center distance of the two spheres. The simple geometry allows the radius of the sphere of interest to be calculated precisely.

10.3.2 Results and Analysis

The viscosity of a sample of methanol, as a function of laser power, is shown in Figure 10.3. It is clear from this plot that the measured viscosity is not constant and, instead, decreases with increasing laser power. The form of the fit is

$$\eta = \eta_0\left(\frac{1}{1+\frac{\beta}{\alpha}\tau}\right), \tag{7}$$

where η_0 is the room temperature viscosity, τ is the optically applied torque (laser power $\times$ change in polarization), and α and β are constants. The methods to determine η_0, α, and β are described later in this chapter. The possible causes of such

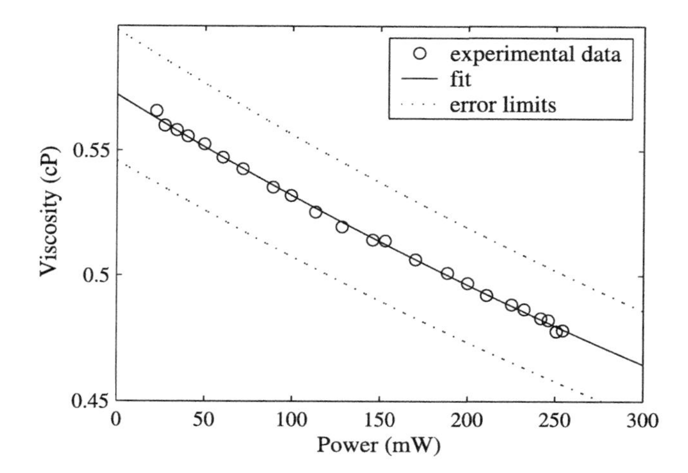

Figure 10.3 Viscosity of methanol, measured as a function of laser trapping power. The diameter of the vaterite particle used here is 3.2 μm.

behavior are shear thinning or a heating effect. By increasing the laser power, the optically applied torque is increased and so the shear rates within the fluid also increase. If the fluid had a shear rate dependent viscosity, as has been reported for fluids such as polymer solutions [1], this could explain the trend observed in the plot. However, methanol, the fluid being measured, is Newtonian and does not exhibit such phenomena as shear thinning. The other possibility is a heating effect, which would occur if either the fluid or the probe particle absorbed slightly at the wavelength of the trapping laser. Increasing the laser power would increase the energy absorbed and therefore would cause a temperature increase in the fluid surrounding the probe particle. The viscosity of most fluids decreases with increasing temperature, which would fit with the observed trend. The interpretation of Figure 10.3 becomes clear if heating due to the trapping laser occurs. This is because localized heating due to the highly focused laser beam means that a nonuniform temperature distribution will be set up in the fluid, which in turn leads to a viscosity distribution, such that a single value for viscosity no longer accurately describes the fluid. In order to analyze the results, we will consider the experimentally measured parameters: the optically applied torque and the rotation rate of the probe particle (Figure 10.4).

According to equation (4), the rotation rate should depend linearly on the optical torque. However, it is evident from Figure 10.4 that there is a nonlinear component of the response to increasing the optical torque. The form of the fit is

$$\Omega(\tau) = \alpha\tau + \beta\tau^2, \tag{8}$$

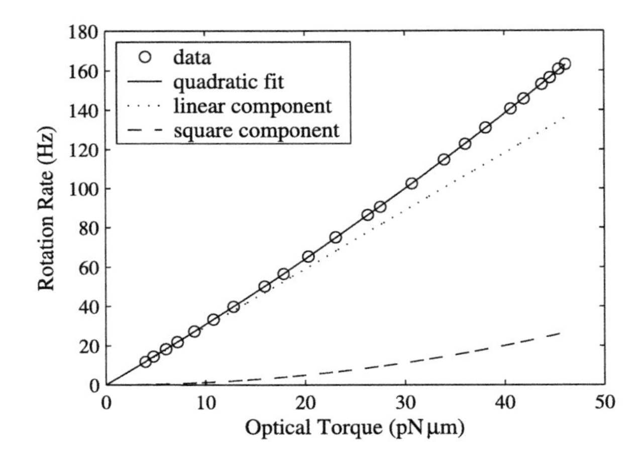

Figure 10.4 Rotation rate of the trapped vaterite particle as a function of the optically applied torque.

where α and β are constants. (This quadratic fit yields the parameters, α, β and the room temperature viscosity, used for the fit described by equation (7).) This fit represents a first-order approximation to the nonlinearity; however, it fits the data well, suggesting the approximation is valid for this range of optical powers. The linear component of this fit yields the room temperature viscosity of the surrounding fluid, as can be seen in equation (4). The square component is not accounted for, but it could be used to determine the heating caused by the trapping laser. However, we find the heating directly from the optical torque and a model that accounts for the nonuniform viscosity distribution (which arises from the temperature variation) in the surrounding fluid.

10.3.2.1 A shell model of viscosity distribution

In order to model the inhomogeneous viscosity of the fluid surrounding the probe particle, we will use a shell model. The model works for a spherically symmetric system, which we argue is the case in our experiments. This is because absorption must occur mainly within the probe particle as absorption in the surrounding fluid is insufficient to explain our results [35]. Therefore, in this model the surface temperature of the particle is uniform, and the steady state temperature of the fluid around the sphere as a function of the distance to the center of the sphere (r) is

$$T(r) = \frac{\gamma}{r} + T_0, \tag{9}$$

where

$$\gamma = \frac{H}{4\pi\kappa}, \tag{10}$$

and the constant γ can be determined by the heat flux, H, and thermal conductivity κ. The room temperature, as $r \to \infty$, is T_0. We can approximate this temperature variation, and hence viscosity variation, by setting up a series of concentric shells. Within the layer of fluid between the shells, the viscosity is uniform. The velocity profile between two shells is given by

$$\mathbf{v}_{in}(\mathbf{r}) = \frac{R_1^3 R_2^3}{R_2^3 - R_1^3}\left[\left(\frac{1}{|r|^3} - \frac{1}{R_2^3}\right)\boldsymbol{\Omega}_1 \times \mathbf{r} - \left(\frac{1}{|r|^3} - \frac{1}{R_1^3}\right)\boldsymbol{\Omega}_2 \times \mathbf{r}\right], \tag{11}$$

where R_1 is the radius of the inner shell and $\boldsymbol{\Omega}_1$ is its rotation rate, and R_2 is the radius of the outer shell and $\boldsymbol{\Omega}_2$ is its rotation rate [24]. Using the velocity profile, the viscous stress tensor on the surface of the inner shell, where $r = R_1$, can be found using

$$\sigma'_{ik} = \eta_{\text{loc}}\left(\frac{\delta v}{\delta r} - \frac{v}{r}\right), \tag{12}$$

where η_{loc} is the viscosity of the fluid in between the shells. This tensor corresponds to the frictional force per unit area on the surface of the inner shell due to the viscous drag of the liquid. The drag torque on this shell can be found by integrating over the surface of the shell [19]

$$\tau = 8\pi\eta_{\text{loc}}(\Omega_2 - \Omega_1)\frac{R_1^3 R_2^3}{R_2^3 - R_1^3}. \tag{13}$$

This expression holds for all the shells in this model. For an infinite number of very closely spaced shells, the torque becomes:

$$\tau = 8\pi\eta(r)r^6\frac{d\Omega}{d(r^3)}. \tag{14}$$

We do not have an analytical form for $\eta(r)$. However, for well-characterized fluids we have a numerical form for $\eta(T)$, which yields $\eta(r)$ from equation (10). The rotation rate of the particle is equal to the rotation rate of the fluid at the radius of the particle (a), which is given by

$$\Omega = \frac{\tau}{8\pi}\int_{r=\infty}^{r=a} \frac{1}{\eta(r)} d\left(1/r^3\right). \tag{15}$$

We do not have an analytical expression for $\eta(T)$, so we will calculate this integral numerically. Both the rotation rate Ω, and the optically applied torque τ are measured experimentally, so a fitting algorithm is used to determine a value for γ in equation (10), which gives the correct viscosity distribution, $\eta(r)$, in equation (15). Figure 10.5 shows the effect that laser light absorption has on the temperature distribution around the particle and the effect of this on the rotation rate of the particle.

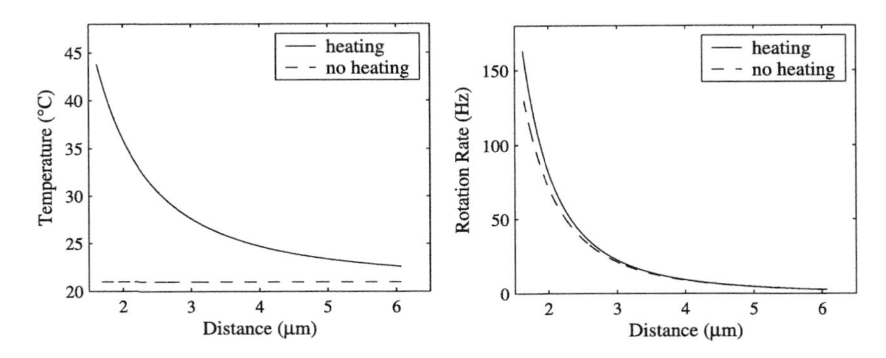

Figure 10.5 The calculated temperature distribution within the fluid is shown as a function of distance from the center of the probe particle (left). The corresponding rotation rate of the fluid in the equatorial plane of the rotating particle is also shown (right).

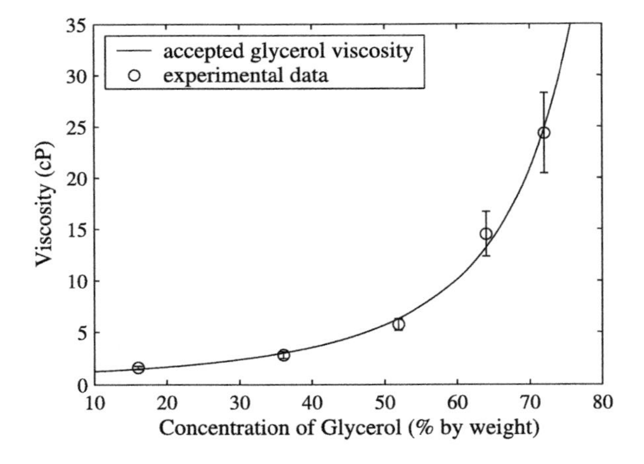

Figure 10.6 The viscosity of glycerol solutions, as measured by our microviscometer, as a function of the concentration of glycerol in the solution.

10.3.2.2 Viscosity of glycerol solutions

To demonstrate the ability of our microviscometer to measure the viscosity of solutions with viscosities varying by almost two orders of magnitude, we prepared a series of solutions with different concentrations of glycerol. Glycerol was chosen as its viscosity as a function of concentration is well characterized and has a large dynamic range. The results of this are shown in Figure 10.6, where excellent agreement is demonstrated. The uncertainty largely stems from the uncertainty in the sphere diameter.

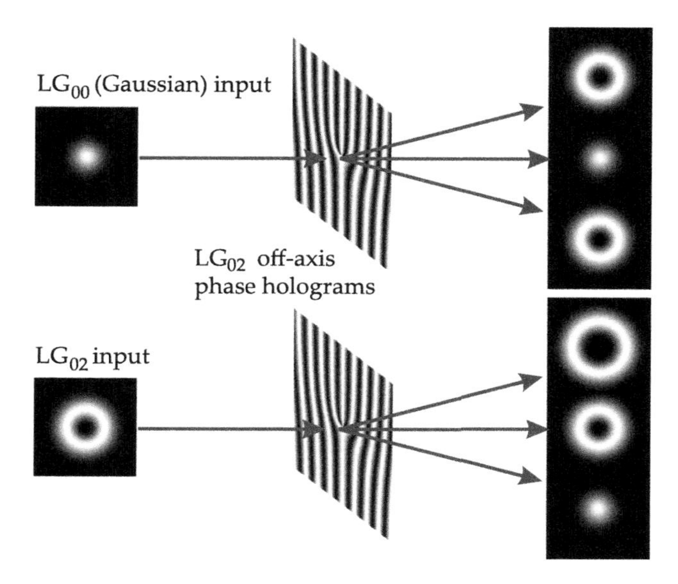

Figure 10.7 The output from an LG_{02} phase hologram for two different input beams.

10.3.3 Orbital Angular Momentum Used for Microviscometry

In order to produce orbital angular momentum and to measure the amount of orbital angular momentum transferred to an object in the optical trap, a few changes need to be made to the setup in Figure 10.2. First, a phase hologram that generates an LG mode needs to be inserted in the beam path before the objective. As an off-axis hologram is typically used, the first diffracted order needs to be selected for the input to the optical trap. Second, the circular polarizer analyzer after the microscope is replaced by another phase hologram followed by a camera to image the output of this analyzing hologram. An example of the phase hologram is shown in Figure 10.7, with different input modes and the corresponding output modes. This figure also presents the principle of the LG modal decomposition of an arbitrary beam. If a Gaussian mode is detected in the first diffracted order, then the input beam, or at least a component of the beam, matches the mode of the hologram. More precisely, they both have the same azimuthal index. A nonzero central intensity in the first diffracted order is representative of a Gaussian mode present there. By exchanging the phase hologram for LG mode holograms with different azimuthal index, the presence of the mode can be detected by again measuring the central intensity. In this way, the modal composition of the beam can be determined and thus the orbital angular momentum measured.

In the experimental setup, the phase holograms can be generated by a spatial light modulator [15] or can be contact printed onto a holographic plate [34]. The spatial light modulator could be advantageous for this application as it allows for easy switching between different LG holograms. To detect the Gaussian component within the first diffracted order from the analyzing hologram, two methods have been used. By coupling the diffracted order into a single mode fiber, and measuring the output power from the fiber using a photodetector, one can detect the Gaussian component. This method has been used in the field of quantum information to measure photons with entangled LG modes [27]. A second technique is to use a CCD camera to image the output of the hologram and determine the intensity measured by the single pixels corresponding to the center of the first diffracted orders [34]. This method has been used to measure the torque applied to an elongated phase object by a paraxial elliptical laser beam, with the axis of the ellipse at different angles to the elongated dimension of the phase object. However, the real test of this technique is in optical tweezers, where orbital angular momentum of the trapping beam can be used to actually rotate an asymmetric particle. A simple asymmetric object is two polystyrene beads trapped two dimensionally against a microscope slide. The microscope slide prevents the two beads from aligning vertically in the trap, which world otherwise eliminate the asymmetry with the beam axis. In an experiment carried out by the authors, an LG_{02} mode was used to trap and rotate two beads, each 2 micrometers in diameter. The modal composition of the beam was measured without any particles trapped (Figure 10.8(a)), and with the two beads trapped (Figure 10.8(b)). In the case without beads trapped, one can clearly see that only an LG_{02} mode is present in the beam, as this is the only mode where a Gaussian beam (nonzero central intensity) is present. However, the mode pictures become confusing when particles are present, and just from inspection of the mode pictures, it is not obvious what modes are present. However, by subtracting the center pixels of the modes without a particle present from the center pixels of the modes when the particles are trapped, we can see the change in the modal composition of the beam, which is shown in the plot in Figure 10.8.

An analysis of the results shown in Figure 10.8 reveals the problems with this method. The mode decomposition suggests the unlikely result that the beam has gained orbital angular momentum from the particles, as more power has coupled to higher-order modes than to lower-order modes. If one looks carefully at the mode pictures, it becomes apparent that if the center of the mode is shifted slightly or determined inaccurately, then the reading from the chosen pixel can vary a great deal due to the rapidly varying intensities within the center regions of the mode pictures. Alignment of the holograms and the optical trap itself can affect the measurement, and it is certainly a source of error in this experiment. In fact, any method that attempts to make a direct measurement of orbital angular momentum in optical tweezers can be

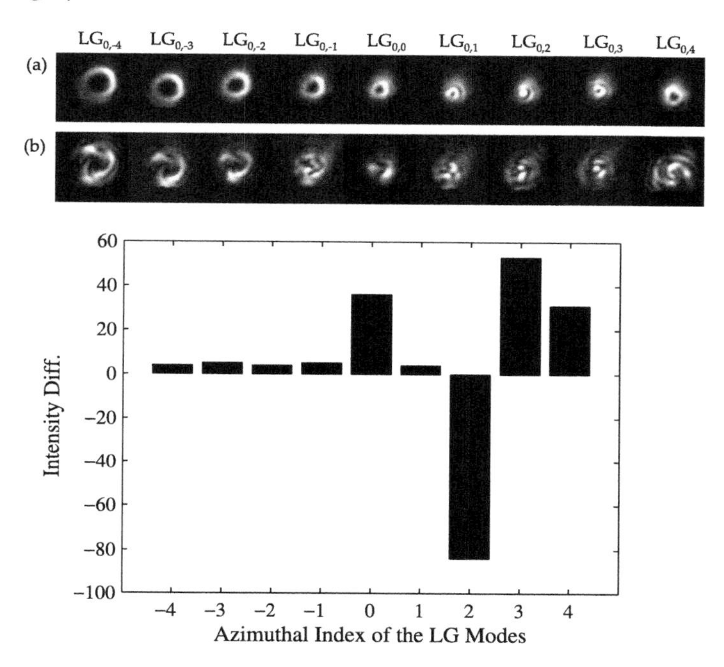

Figure 10.8 The output from the analyzing hologram placed after the optical tweezers setup that allows the modal composition of the beam to be measured. A Gaussian present in the mode picture means that the corresponding mode is present. For example, in the first row of mode pictures, (a), when no trapped particles are present, the only component of the beam is an LG_{02} mode. In the second row, (b), when two particles are trapped and rotating, it is not clear which modes are present. However, the difference in the intensities of the center pixels (second row minus the first row ((b)–(a))) plotted below the mode pictures clearly shows other modes present in the beam.

susceptible to alignment due to the orbital angular momentum's dependence upon the origin of the coordinate system, unlike the case for spin angular momentum.

The spin angular momentum measurement is relatively simple compared to the mode decomposition, and can be used to determine orbital angular momentum transfer in optical tweezers, as described early in this chapter. In order to find the torque due to orbital angular momentum from equation (6), the polarization state of the trapping light needs to be varied, and then the effect on the rotation rate needs to be measured. The result of such an experiment is shown in Figure 10.9. In this case, an LG_{02} mode trapping laser was used, and the polarizations used were left circular, linear, and right circular. In this case, the particles were trapped against a microscope slide, which would have a significant effect on the viscous drag torque on the particle. This makes viscosity measurements difficult. However, the situation can be improved by using

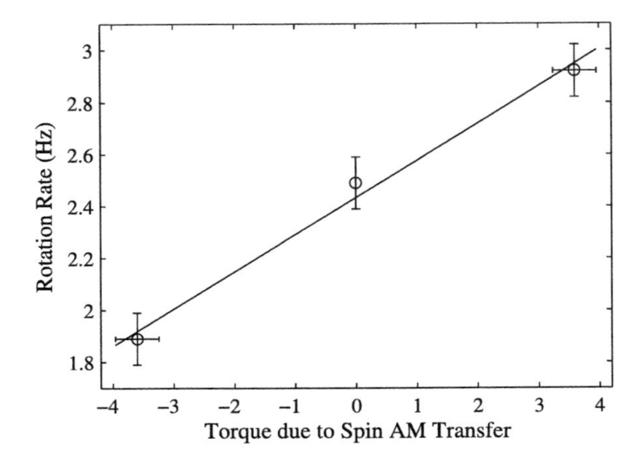

Figure 10.9 The rotation rate of two beads trapped in an LG_{02} mode versus the contribution from the spin angular momentum transferred to the particle. The units of the x-axis are $10^{-3}\hbar$ per photon.

photopolymerized structures, like the cross described by Knöner and colleagues [21], which can be three dimensionally trapped.

10.4 APPLICATIONS

10.4.1 Picolitre Viscometry

The measurement of viscosity within microscopic structures is an interesting application for microrheological techniques. Our microviscometer is well suited to such applications in view of the very localized flow that is generated due to the rotational motion of the probe particle [19]. An example of such a structure, whose mechanical properties are of wide interest, is a biological cell. A first approximation to a cell is a lipid membrane that has formed a spherical vesicle, often called a *liposome*. In a series of experiments, vaterite particles were added to a solution of cholesterol, hexane, and water being used to generate liposomes, which sometimes led to a vaterite particle being engulfed by a vesicle a few microns in diameter. The viscosity of the fluid within the very localized volume was then measured by optically applying a torque to the vaterite and measuring its rotation rate [2], which is the method described in this chapter. In this way the fluid within the micelle was found to be hexane, showing that the vesicle was only a sort of micelle with a lipid monolayer rather than the desired bilayer liposome. This experiment represents a measurement of the viscosity of a picoliter volume of fluid.

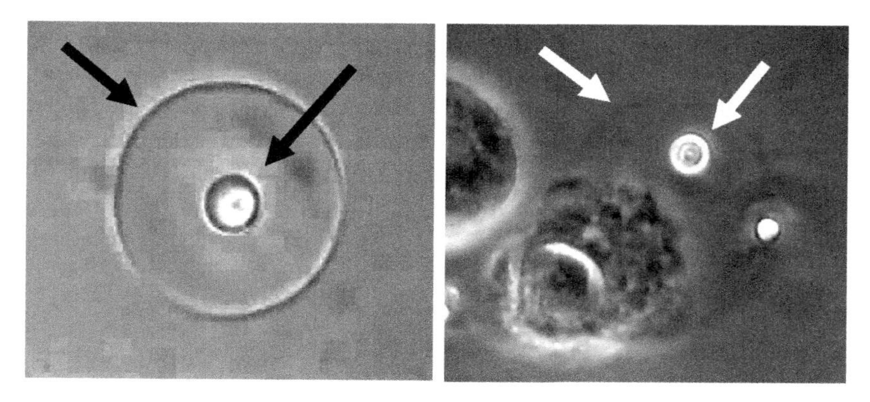

Figure 10.10 Vaterite in a micelle (left), 16.7 μm in diameter, and in a macrophage cell (right), approximately 20 μm in diameter. In both cases, the left arrow points to the membrane and the right arrow points to the vaterite particle.

We have also demonstrated a further step toward intracellular viscosity measurements by inserting a vaterite particle through the cellular membrane of a macrophage cell, as shown in Figure 10.10. In this experiment, a femtosecond laser was focused on the membrane of the cell. This caused the cell to blister or form a bleb that is just visible in Figure 10.10, which is a phase contrast image. The vaterite particle was then trapped with optical tweezers and, by positioning the focus of the femtosecond laser at the leading edge of the vaterite particle, a hole was cut in the membrane and the vaterite was pushed inside the membrane. The hole sealed behind the vaterite particle so that it was enclosed in the membrane. The vaterite was then spun by changing the polarization of the trapping beam to circular polarization, and thus the viscosity could be measured.

10.4.2 Medical Samples

Another aspect of microrheology is the study of samples where only microliter volumes are available. This is in contrast to studies of cells, as although the microvisometer probes a very small volume, the sample is often prepared from milliliter volumes of cells in buffer solution. However, some medical samples such as eye fluid can only be collected in microliter volumes. Without stimulating a tear response, about 1–5 μL of eye fluid can be collected [40]. In a proof of principle experiment, vaterite particles were trapped and rotated within the eye fluid remaining on a recently removed contact lens that had been worn for a few hours beforehand. The lens was placed on a microscope slide, and dried vaterite particles were then placed on the lens using the tip of a brass wire. A coverslip was put on top of the lens, and the sample was then put in the optical tweezers microscope. The amount of eye fluid was estimated to be less than a microliter, so the depth of the sample was not much larger than the vaterite particles themselves, and viscosity measurements would be affected by the surfaces in this case (the contact lens and the coverslip). However the fact that the vaterite could be trapped and rotated demonstrates the potential of our microviscometer to make measurements within microliter or less sample volumes. Microcapillaries can be used to extract eye fluid, which, when transferred to a microscope slide, would allow for the preparation of a deeper sample and thus a more reliable measurement of viscosity.

10.4.3 Flow Field Measurements

A rotating sphere is a convenient way to generate a steady and simple flow in a fluid. A vaterite particle, trapped away from surfaces such as the microscope slide, can create this ideal flow. The measurement of this flow field was the focus of the work by Knöner and colleagues [19]. In these experiments, a dual beam optical trap was used. The first trap was used to trap the vaterite particle and generate the flow, and the second trap was used to map out the flow field using a polystyrene bead, 1 μm in diameter, as a probe particle. (Two methods were used to measure the flow in the equatorial plane of the rotating vaterite.) The first was simply to release the probe particle at different distances from the vaterite particle and track the motion of the particle, using the video recording of the microscope image to determine its velocity and hence the velocity of the liquid. The second method was to measure the displacement of the probe particle in the second trap due to the viscous drag caused by the rotating vaterite particle. The quadrant photodetector used to measure the displacement was calibrated using the microscope stage to move the particle through the liquid at known velocities. The two techniques were found to agree with one another.

The flow of water around a vaterite, 3.6 μm in diameter, was measured, and the flow as a function of distance from the particle is shown in Figure 10.11. The velocity profile is exactly as expected in this case and really demonstrates how the spinning vaterite can be used to generate model flows. However, we see a deviation from this model behavior in some fluids, as shown in Figure 10.12. The fluid is hyaluronic acid, which is an example of a non-Newtonian fluid [13]. The non-Newtonian behavior of the fluid is due to the polymers in solution that give the fluid elasticity at high driving frequencies and also cause a decrease in viscosity at higher shear rates. However, neither of these regimes are accessed in this experiment [19]. Instead, the authors suggest that a depletion layer exists around the spinning particle, which is consistent with other work [25]. By combining the optical measurement of the torque on the spinning particle with the measurement of the flow field it generates in the surrounding

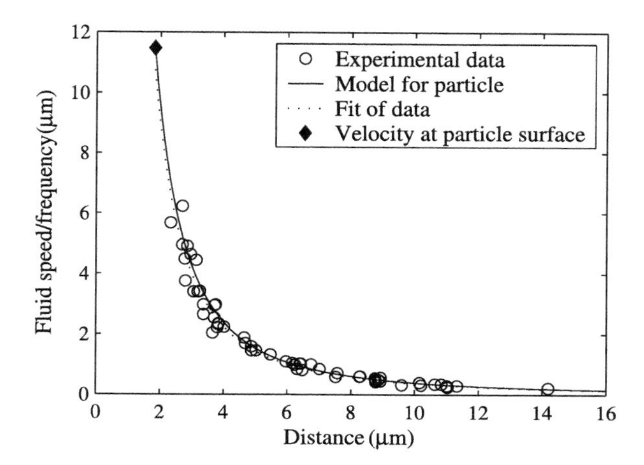

Figure 10.11 The flow created in the equatorial plane of a vaterite particle spinning in water. The experimental data agrees very well with the theoretically predicated flow.

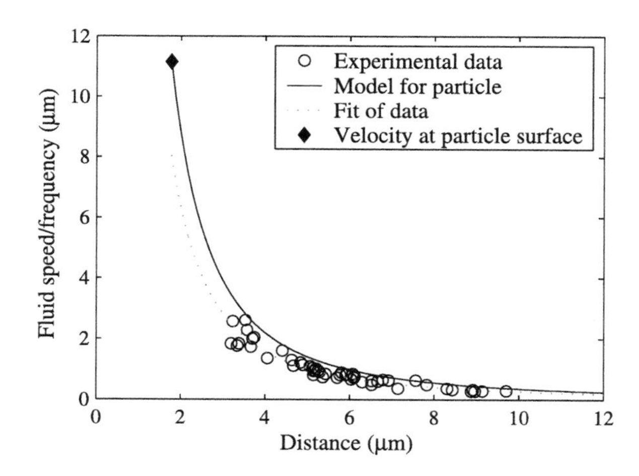

Figure 10.12 The flow created in the equatorial plane of a vaterite particle spinning in hyaluronic acid. In this case, the experimental data deviates from the theoretically predicted curve. This result is explained by a layer that forms around the vaterite particle that is depleted of the polymers within the hyaluronic acid solution, which has a lower viscosity than the bulk of the fluid.

fluid, one can correct the viscosity measurement and characterize the properties of the depletion layer itself.

The probe particle–fluid coupling is a factor that always needs to be considered in microrheology experiments. One approach to this problem is to use more than one probe particle, which has the effect of assessing the fluid between the probe particles rather that just the fluid immediately around the particle [25,37]. The flow field measurements achieve a similar result by measuring the flow at numerous distances around the spinning particle. Another approach to this problem is to consider the surface chemistry of the probe particle. The hydrophobicity and surface charge can be varied to see their effect on the particle–fluid coupling.

The localized flow created by the spinning vaterite particle can also be used to study micro-biological systems, as described by Botvinic and colleagues [3]. In these experiments, the response of a cell to a shear flow created by a trapped vaterite particle was observed. The optically applied torque allowed the shear forces on the cell be characterized.

10.5 CONCLUSION

In this chapter, we have described methods to measure optical angular momentum transfer in optical traps. Both spin angular momentum and orbital angular momentum can be transferred, but we have concentrated on spin transfer, as it is more easily implemented for viscosity measurements. Vaterite particles are suitable for this application as they are strongly birefringent and have a simple spherical geometry, which means significant torques can be applied to the particle and the viscous drag on the particle follows a simple relation.

The advantage of the microviscometry technique described in this chapter over other techniques is that a very localized region can be actively probed. The technique does not require a complicated trap stiffness calibration as the optical torque is measured directly. The localization means that picoliter volumes can be probed, which is on the same size scale as the interior of biological cells. The shear rate can be varied in experiments by varying the trapping laser power, which could potentially allow shear induced effects in non-Newtonian fluids to be studied. The technique could also be extended to study the frequency response of these fluids by applying time varying torques to the probe particle. The phase of the fluid's response to such a driving torque would then allow viscoelasticity to be measured.

REFERENCES

[1] H.A. Barnes, J.F. Hutton, K. Walters, An Introduction to Rheology, Elsevier, Amsterdam, 1989.

[2] A.I. Bishop, T.A. Nieminen, N.R. Heckenberg, H. Rubinsztein-Dunlop, Optical microrheology using rotating laser-trapped particles, *Phys. Rev. Lett. 92* (2004) 198104.

[3] E.L. Botvinick, G. Knöner, H. Rubinsztein-Dunlop, M.W. Berns, Ultra-localized flow fields applied to the cell surface, *Proc. Soc. Photo. Instrum. Eng. 6326* (2006) 26–30.

[4] V. Breedvald, D.J. Pine, Microrheology as a tool for high-throughput screening, *J. Mater. Sci. 38* (2003) 4461–4470.

[5] A. Buosciolo, G. Pesce, A. Sasso, New calibration method for position detector for simultaneous measurements of force constants and local viscosity in optical tweezers, *Opt. Commun. 230* (2004) 357–368.

[6] Z. Cheng, P.M. Chaikin, T.G. Mason, Light streak tracking of optically trapped thin microdisks, *Phys. Rev. Lett. 89* (2002) 108303.

[7] Z. Cheng, T.G. Mason, Rotational diffusion microrheology, *Phys. Rev. Lett. 90* (2003) 018304.

[8] F.H.C. Crick, A.F.W. Hughes, The physical properties of cytoplasm, *Exp. Cell Res. 1* (1950) 37–80.

[9] J.C. Crocker, M.T. Valentine, E.R. Weeks, T. Gisler, P.D. Kaplan, A.G. Yodh, D.A. Weitz, Two-point microrheology of inhomogeneous soft materials, *Phys. Rev. Lett. 85* (2000) 888–891.

[10] B.R. Daniels, B.C. Masi, D. Wirtz, Probing single-cell micromechanics in vivo: The microrheology of *C. elegans* developing embryos, *Biophys. J. 90* (2006) 4712–4719.

[11] J.T. Finer, R.M. Simmons, J.A. Spudich, Single myosin molecule mechanics: Piconewton forces and nanometre steps, *Nature 368* (1994) 113–119.

[12] E.L. Florin, A. Pralle, E.H.K. Stelzer, J.K.H. Hörber, Photonic force microscope calibration by thermal noise analysis, *Appl. Phys. A 66* (1998) S75–S78.

[13] E. Fouissac, M. Milas, M. Rinaudo, Shear-rate, concentration, molecular weight, and temperature viscosity dependences of hyaluronate, a wormlike polyelectrolyte, *Macromolecules 26* (1993) 6945–6951.

[14] M.E.J. Friese, T.A. Nieminen, N.R. Heckenberg, H. Rubinsztein-Dunlop, Optical alignment and spinning of laser-trapped microscopic particles, *Nature 394* (1998) 348–350.

[15] G. Gibson, J. Courtial, M.J. Padgett, Free-space information transfer using light beams carrying orbital angular momentum, *Opt. Exp. 12* (2004) 5448–5456.

[16] H.F. Gleeson, T.A. Wood, M. Dickinson, Laser manipulation in liquid crystals: an approach to microfluidics and micromachines, *Philos. Trans. R. Soc. A 364* (2006) 2789–2805.

[17] S. Juodkazis, M. Shikata, T. Takahashi, S. Matsuo, H. Misawa, Size dependence of rotation frequency of individual laser trapped liquid crystal droplets, *Jpn. J. Appl. Phys. 38* (1999) L518–L520.

[18] M.S.Z. Kellermayer, S.B. Smith, H.L. Granzier, C. Bustamante, Folding-unfolding transitions in single titin molecules characterized with laser tweezers, *Science 276* (1997) 1112–1116.

[19] G. Knöner, S. Parkin, N.R. Heckenberg, H. Rubinsztein-Dunlop, Characterization of optically driven fluid stress fields with optical tweezers, *Phys. Rev. E 72* (2005) 031507.

[20] G. Knöner, S. Parkin, T.A. Nieminen, N.R. Heckenberg, H. Rubinsztein-Dunlop, Measurement of the index of refraction of single microparticles, *Phys. Rev. Lett. 97* (2006) 157402.

[21] G. Knöner, S. Parkin, T.A. Nieminen, V.L.Y. Loke, H. Heckenberg, N.R. Rubinsztein-Dunlop, Integrated optomechanical microelements, *Opt. Exp. 15* (2007) 5521–5530.

[22] S.C. Kuo, M.P. Sheetz, Force of single kinesin molecules measured with optical tweezers, *Science 260* (1993) 232–234.

[23] A. La Porta, M.D. Wang, Optical torque wrench: Angular trapping, rotation, and torque detection of quartz microparticles, *Phys. Rev. Lett. 92* (2004) 190801.

[24] L.D. Landau, E.M. Lifshitz, Fluid Mechanics, 2nd ed., Butterworth–Heinemann, Oxford, 1987.

[25] A.J. Levine, T.C. Lubensky, One- and two-particle microrheology, *Phys. Rev. Lett. 85* (2000) 1774.

[26] R. Ługowski, B. Kołodziejczyk, Y. Kawata, Application of laser-trapping technique for measuring the three-dimensional distribution of viscosity, *Opt. Commun. 202* (2002) 1–8.
[27] A. Mair, A. Vaziri, G. Weihs, A. Zeilinger, Entanglement of the orbital angular momentum states of photons, *Nature 412* (2001) 313–316.
[28] T.G. Mason, K. Ganesan, J.H. van Zanten, D. Wirtz, S.C. Kuo, Particle tracking microrheology of complex fluids, *Phys. Rev. Lett. 79* (1997) 3282–3285.
[29] N. Murazawa, S. Juodkazis, V. Jarutis, Y. Tanamura, H. Misawa, Viscosity measurement using a rotating laser-trapped microsphere of liquid crystal, *Europhys. Lett. 73* (2006) 800–805.
[30] T.A. Nieminen, N.R. Heckenberg, H. Rubinsztein-Dunlop, Optical measurement of microscopic torques, *J. Mod. Opt. 48* (2001) 405–413.
[31] M.J. Padgett, Rotation of Particles in Optical Tweezers, in this book, 2008.
[32] S.J. Parkin, G. Knöner, T.A. Nieminen, N.R. Heckenberg, H. Rubinsztein-Dunlop, Measurement of the total optical angular momentum transfer in optical tweezers, *Opt. Exp. 14* (2006) 6963–6970.
[33] S. Parkin, G. Knoener, W. Singer, T.A. Nieminen, N.R. Heckenberg, H. Rubinsztein-Dunlop, in: *Laser Manipulation of Cells and Tissues*, in: *Methods in Cell Biology*, Academic Press, Elsevier, 2007, pp. 526–561 (Chapter 19).
[34] S.J. Parkin, T.A. Nieminen, N.R. Heckenberg, H. Rubinsztein-Dunlop, Optical measurement of torque exerted on an elongated object by a noncircular laser beam, *Phys. Rev. A 70* (2004) 023816.
[35] E.J. Peterman, F. Gittes, C.F. Schmidt, Laser-induced heating in optical traps, *Biophys. J. 84* (2003) 1308–1316.
[36] D.E. Soper, Classical Field Theory, John Wiley & Sons, 1976.
[37] L. Starrs, P. Bartlett, Colloidal dynamics in polmer solutions: Optical two-point microrheology measurements, *Faraday Discuss. 123* (2002) 323–333.
[38] K. Svoboda, S.M. Block, Biological applications of optical forces, *Annu. Rev. Biophys. Biophys. Chem. 23* (1994) 247–285.
[39] K. Svoboda, C.F. Schmidt, B.J. Schnapp, S.M. Block, Direct observation of kinesin stepping by optical trapping interferometry, *Nature 365* (1993) 721–727.
[40] J.M. Tiffany, The viscosity of human tears, *Int. Ophthalmol. 15* (1991) 371–376.
[41] S.F. Tolić-Nørrelykke, E. Schäffer, J. Howard, F.S. Pavone, F. Jülicher, H. Flyvbjerg, Calibration of optical tweezers with positional detection in the back focal plane, *Rev. Sci. Instrum. 77* (2006) 103101.
[42] L. Tskhovrebova, J. Trinick, J.A. Sleep, R.M. Simmons, Elasticity and unfolding of single molecules of the giant muscle protein titin, *Nature 387* (1997) 308–312.
[43] P.A. Valberg, D.F. Albertini, Cytoplasmic motions, rheology, and structure probed by a novel magnetic particle method, *J. Cell Biol. 101* (1985) 130–140.
[44] T.A. Waigh, Microrheology of complex fluids, *Rep. Prog. Phys. 68* (2005) 685–742.
[45] H. Yin, M.D. Wang, K. Svoboda, R. Landick, S.M. Block, J. Gelles, Transcription against an applied force, *Science 270* (1995) 1653–1657.

Chapter 11

Orbital Angular Momentum in Quantum Communication and Information

Sonja Franke-Arnold[1] *and John Jeffers*[2]

[1] *University of Glasgow, Scotland*
[2] *University of Strathclyde, Glasgow, Scotland*

Light has a long tradition in serving as a model representation for quantum communication systems. Its polarization provides two orthogonal states, for example, horizontal, and vertical, which can be used to encode one bit of information in a quantum system (say, horizontal represents 0 and vertical represents 1). The system does not necessarily need to be horizontally or vertically polarized; it could be left or right circularly polarized or linearly polarized at some other angle. This allows the system to be in a superposition of the orthogonal bit states and therefore to represent what is known as a *qubit*. Various alternative physical qubit systems are competing in the run toward other feasible quantum information applications, including cold trapped ions, Bose condensates, and quantum dots. Polarized light, however, is particularly suitable for proof of principle investigations, as its generation, manipulation, and detection is comparatively simple, fast, and inexpensive. The field is advancing rapidly, and quantum key distribution has already entered the public domain [1].

In 1992 Les Allen and colleagues showed that light can carry orbital angular momentum (OAM), and, in particular, that an azimuthal phase dependence $e^{i\ell\phi}$ of a light beam corresponds to ℓ units of orbital angular momentum [2]. Such phase dependence is characteristic of either Laguerre–Gaussian or Bessel modes. Each of these mode families provides an infinite-dimensional orthogonal basis set, and with hindsight it seems obvious to employ OAM modes as representations of higher-dimensional

STRUCTURED LIGHT AND ITS APPLICATIONS

qubits called *qunits* (sometimes *qudits*). It took about a decade until the advent of the first quantum mechanical studies of orbital angular momentum, and Zeilinger's group stated in their landmark experiment on OAM entanglement [3] that "... this approach provides a practical route to entanglement that involves many orthogonal quantum states ... and could be of considerable importance in the field of quantum information, enabling, for example, more efficient use of communication channels in quantum cryptography." The superposition principle is the first main advantage of the quantum world over the classical in information processing. Entanglement, where the individual parts of a multi-component quantum system do not have independent properties, is another. An entangled state of a two-component system (formally known as a bipartite system) cannot be written as a simple product of states of the individual systems. If the two entangled components of the system are spatially separated, something which is not difficult to arrange for light, measurement results taken on one system will be correlated with measurement results on the other. This bipartite entanglement is exploited heavily in experiments using OAM, as is described later in the chapter.

For some applications, higher-dimensional systems are found to be superior. The most immediate advantage is given by the availability of a larger "alphabet," consisting of the various OAM states. In principle, Laguerre–Gaussian (LG) beams provide infinitely many orthogonal OAM states, although in realistic experiments the number of distinguishable OAM modes is currently limited to about 10. This still offers a considerable increase in data capacity. Higher-dimensional quantum systems are also known to improve the level of security in quantum cryptography under the presence of noise [4], and are required by some quantum protocols [5] and quantum computation schemes [6]. Maybe most intriguingly, there are tasks that can be solved more efficiently using higher-dimensional systems, e.g., the Byzantine agreement problem [7] or quantum coin tossing [8].

There are two routes to increasing the dimensionality of a system: using several entangled qubit systems or increasing the dimensionality of the fundamental system, as shown in Figure 11.1. There have been various attempts employing the first approach, which were based on interferometric combinations of qubits. Here the dimension of the system is provided by the individual paths of the interferometers. Early experiments included the entanglement of spatial modes [9], of polarization states [10] and notably of time and energy in time-binning experiments [11]. Multiple nested and/or concatenated interferometers provide many experimental challenges however, especially with regard to path length stability, and so this approach seems limited in scope. To date, OAM provides the only noninterferometric way of encoding qunits onto photons. Since 2001, the quantum mechanical study of OAM states has developed rapidly, and several groups worldwide are involved with the production of entangled OAM

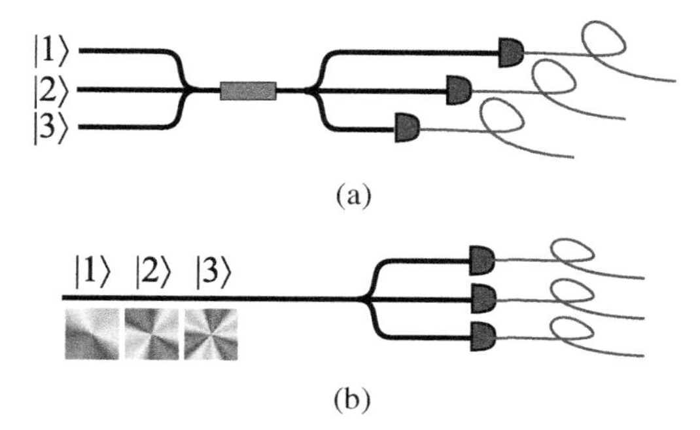

Figure 11.1 The data capacity of quantum communication can be increased by combining many two-bit systems as in time-binning (a), or by using a higher-dimensional base system like the OAM of light (b). See color insert.

states, the encoding, manipulation and decoding of OAM information, and the investigation of first quantum protocols. Most of these efforts are concentrating on the OAM of single photons, however, it is worth noting that even classical OAM modes offer advantages in quantum communication and exhibit an intrinsic security against eavesdropping as will be shown in later in the first section.

In this chapter, we will describe the main features of the field. In the following section, we will summarize the production and detection of quantum states in general and OAM states in particular. Later, we will look at the generation of superposition states of the OAM and briefly outline concepts for storing quantum information encoded in OAM. The final section describes some of the many protocols that require higher-dimensional spaces. Finally we will summarize and provide an outlook for future developments.

11.1 SENDING AND RECEIVING QUANTUM INFORMATION

All quantum physics experiments have the same basic form. Some kind of state preparation device takes a quantum system and places it in an appropriate state. This state then evolves, and subsequently the system is measured by a measuring device.

Formally the state of the system is normally described by a density operator, $\hat{\rho}$, which encodes not only the quantum information, but any classical probabilistic information as well. The ideal preparation device will attempt to produce pure states for which the information encoded in the state is in the complex probability amplitudes. For pure states, we can simply write $\hat{\rho} = |\psi\rangle\langle\psi|$, and the density operator is a simple projector. Otherwise, the state is mixed and can be written as a probabilistic sum of such terms

$$\hat{\rho} = \sum_j P_j |\psi_j\rangle\langle\psi_j|. \tag{1}$$

The quantum states generated in optics are always mixed to some degree, so in this chapter $|\psi_j\rangle\langle\psi_j|$ identifies a particular OAM projector, normally informally called a state, and P_j denotes the probability with which this state is produced.

Measurements are described by probability operators, which correspond to measurement results; each possible measurement result j has a corresponding probability operator $\hat{\pi}_j$ [12]. If measurement result j corresponds to measuring the system in a pure state $|\psi_j\rangle$, then the appropriate probability operator can be written as

$$\hat{\pi}_j \propto |\psi_j\rangle\langle\psi_j|. \tag{2}$$

Otherwise, if the measurement corresponds to a mixed state,

$$\hat{\pi}_j \propto \sum_k W_{jk} |\psi_k\rangle\langle\psi_k|, \tag{3}$$

where W_{jk} are the weights of the particular states $|\psi_k\rangle$ in measurement result j. The proportionality constant is determined by the fact that the probability of all possible measurement results must sum to unity, which means that the sum of the probability operators (sometimes known as a *probability operator measure* or a positive operator-valued measure [12]) is $\sum_j \hat{\pi}_j = \hat{1}$, the identity operator for the system. The standard probability postulate of quantum mechanics can easily be cast in terms of this formalizm. The probability that measurement result j is obtained is $P(j) = \mathrm{Tr}(\hat{\rho}\hat{\pi}_j)$, where the trace is taken over the full set of system space states.

Evolution between preparation and measurement can take one of two forms. It may be unitary, and reversible, or nonunitary, and therefore cause decoherence. Unitary evolution is designed into many quantum information systems, presenting one of the main reasons for the advantages of quantum information processing over classical. Unitary evolution allows the manipulation and transmutation of superposition states. Nonunitary evolution typically corresponds to a loss of quantum information from the system and therefore is normally to be avoided. One of the main benefits of quantum optical systems is that there is very little uncontrolled interaction of the physical quantum system (photons) with any external environment during propagation, especially in free space. Photons therefore make excellent "flying" qubits and, as we shall see, qunits.

11.1.1 Generation of Entangled OAM States

Quantum entanglement lies at the heart of many applications in quantum communication and information. Some quantum cryptographic protocols in particular rely upon the nonlocal nature of entangled particles, where an operation performed on the sender particle appears to set instantaneously the state of the entangled particle at the receiver end. Spontaneous parametric down-conversion is by now an established source for entangled photons, generating correlations in arrival times, momentum, polarization, and images [13]. In 2001, entanglement of OAM was demonstrated for the first time in spontaneous parametric down-conversion [3]. In such systems, measuring the OAM of one of the down-converted photons instantaneously defines the OAM of the remote partner photon. The entanglement is associated with the conservation of OAM at the single-photon level in the down-conversion process. It is fair to say that this experiment kick-started the newly emerging area of quantum information with OAM-carrying light. In order to demonstrate the conservation of OAM, an argon ion laser with an OAM of $m_{\text{pump}}\hbar$ per photon was used to pump a BBO type I crystal, where $m_{\text{pump}} = -1, 0, 1$. The down-converted photons were then tested for their OAM. This was achieved using a hologram containing a suitable fork dislocation followed by on-axis intensity detection. In order to identify a photon in the $m = \ell$ mode, it was passed through a $-\ell$ hologram. In the first diffraction order, the photon is converted into the fundamental Gaussian $m = 0$ mode, the only mode with on-axis intensity, which can efficiently couple into a single mode fiber and cause a detector click. This measurement has corresponding probability operators

$$\hat{\pi}_{\ell} = |\ell\rangle\langle\ell| \tag{4}$$

when the detector fires, and

$$\hat{\pi}_{\bar{\ell}} = \hat{1} - |\ell\rangle\langle\ell| \tag{5}$$

when it does not, where $\hat{1}$ is the identity operator for the angular momentum space. With this technique, the OAM states $m = -2, -1, 0, 1, 2$ could be distinguished, and coincidences between the two emitted photons were recorded. Coincidences were predominantly observed for cases where the OAM of the two down-converted photons summed to that of the pump beam, deducing a two-photon wavefunction characteristic for entanglement

$$\begin{aligned}|\Psi_{\text{2photon}}\rangle = c_{00}|0\rangle|0\rangle + c_{11}\big(|+1\rangle|-1\rangle + |-1\rangle|+1\rangle\big) \\ + c_{22}\big(|+2\rangle|-2\rangle + |-2\rangle|+2\rangle\big) + \cdots\end{aligned}$$

for a pump beam with zero OAM. While these experiments confirmed conservation of OAM for each generated photon pair, both individual down-converted beams are incoherent and the OAM of a classical beam is the average of the cumulated OAM. Mathematically, this is evident from the mixed density operator of one of the down-converted modes

$$\begin{aligned}\rho_A &= \mathrm{Tr}_B\big(|\Psi_{2\mathrm{photon}}\rangle\langle\Psi_{2\mathrm{photon}}|\big)\\ &= |c_{00}|^2|0\rangle\langle 0| + |c_{11}|^2\big(|1\rangle\langle 1| + |-1\rangle\langle -1|\big)\\ &\quad + |c_{22}|^2\big(|2\rangle\langle 2| + |-2\rangle\langle -2|\big) + \cdots,\end{aligned}$$

where A and B denote the two different beams. In order to demonstrate that the down-converted photons are indeed entangled, it is necessary to show that the two-photon state is not just a mixture of OAM states but a coherent superposition. If one photon is measured in a superposition of two or more OAM states, its partner photon will be projected into the corresponding superposition. Different possibilities to generate superpositions of OAM states will be outlined in the next section. A superposition of two OAM states is characterized by an off-axis phase singularity associated with a localized vanishing intensity. This intensity zero will occur at the position where the amplitudes of the two LG modes are equal and their phases are opposite. Instead, a mixture of OAM states would not contain an intensity zero, because in this case the intensities rather than the amplitudes are added, resulting in finite intensities throughout the beam profile. Experimentally such superpositions of $m = 0$ and $m = 2$ were realized by shifting the phase dislocation of an $\ell = 2$ hologram away from the beam center. The resulting mode of the partner photon was detected by scanning the corresponding detector across the beam profile, and an intensity adequate for a superposition state was observed.

The entanglement in OAM can be derived from the phase matching condition during the nonlinear process [14]. For a pump beam propagating along z, momentum conservation in the transverse plane restricts the x and y wave vector components of the down-converted photons according to $\delta(k_x^{(A)} + k_x^{(B)})\delta(k_y^{(A)} + k_y^{(B)})$. The corresponding position representation can be obtained by Fourier transform,

$$\begin{aligned}\delta(x_A - x_B)\delta(y_A - y_B) &= \frac{\delta(r_A - r_B)}{r_A}\delta(\phi_A - \phi_B)\\ &= \frac{\delta(r_A - r_B)}{2\pi r_A}\sum_{m=-\infty}^{+\infty} e^{im\phi_A}e^{-im\phi_B}. \end{aligned} \tag{6}$$

Here r and ϕ are the radial and azimuthal coordinates of the emitted photons, and from this simplified argument it is clear that phase-matching forces the entangled photons to be correlated in their radial position and into OAM eigenmodes with OAM m and $-m$ that sum to zero.

11.1.2 Detection of OAM States at the Single Photon Level

The OAM of a light beam can be visualized by interference with a plane wave reference beam, generating an m-lobed intensity pattern. These methods are described elsewhere in this book. Most applications in quantum information, however, are based upon single photons, and therefore a method is required to detect the OAM of single photons. The method employed in [3] obviously works on the single photon level but is restricted to a two-valued response, as can be seen from equations (4) and (5). It can determine whether or not the photon is in a particular OAM state (or in a particular superposition of OAM states). More generally, it is desirable to determine the OAM of a photon by using some sort of multichannel analyzer. As the OAM basis is in principle infinite, any real experiment will need to restrict the number of OAM states that can be distinguished. In 2002, an interferometric mode analyzer was developed [15] that can sort photons into even and odd OAM states, or more generally in OAM classes mod($2m$). This method relies on the rotational symmetry of the OAM modes: rotating a mode carrying an OAM of $m\hbar$ by an angle α generates a phase dependence $\exp(im(\phi+\alpha))$ corresponding to a phase shift of $m\alpha$. The m-dependent phase shift can be utilized to generate constructive and destructive interference. If $m\alpha$ is a multiple of 2π, the rotated mode is identical to the original mode; if $m\alpha$ is an odd multiple of π, the rotated mode is out of phase. The first stage of the OAM sorter consists of a Mach–Zehnder interferometer, rotating the mode in one arm by $\alpha=\pi$ with respect to the other. By correctly adjusting the path lengths, modes with even m will exit through one channel, and modes with odd m through the other, as shown in Figure 11.2. The even modes can be sorted in the same way by imposing a rotation of $\alpha=\pi/2$ into OAM classes "$m \bmod 4 = 0$" and "$m \bmod 4 = 2$." The odd OAM modes can either be converted into even modes by passage through an $\ell=1$ hologram before entering the second stage interferometer, or an additional m-independent phase shift of $\Phi_c=-\pi/2$ can be added in one of the interferometer arms, changing the phase dependence to $\exp i(m\phi+m\alpha+\Phi_c)$. In the experiment, a rotation by α is implemented by inserting two Dove prisms, rotated with respect to each other by $\alpha/2$. Each Dove prism flips the transverse mode profile, and their combination thus produces an m-dependent rotation. An m-independent phase shift is generated simply by a suitably inclined glass plate. The mixed probability operator corresponding to the first level of sorting is

$$\hat{\pi}_{\text{even}} \propto |0\rangle\langle 0| + |2\rangle\langle 2| + |-2\rangle\langle -2| + \cdots, \tag{7}$$

$$\hat{\pi}_{\text{odd}} \propto |1\rangle\langle 1| + |-1\rangle\langle -1| + |3\rangle\langle 3| + |-3\rangle\langle -3| + \cdots. \tag{8}$$

By combining rotations of modes within interferometers and addition of OAM by holograms, one can achieve an unmixed set of probability operators (4), each corre-

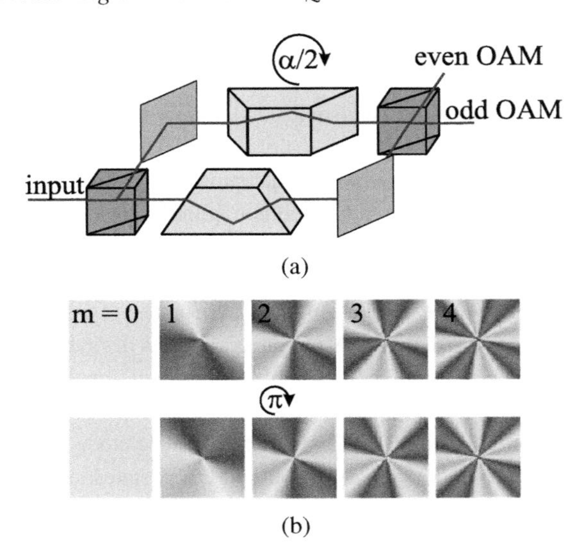

Figure 11.2 (a) The first stage of the OAM channel analyzer [15] is sorting photons in even and odd OAM classes. Two Dove prisms at an angle $\pi/2$ rotate the mode profile by $\alpha = \pi$. (b) Even OAM modes with twofold symmetry remain the same under rotation by π, whereas odd OAM modes become exactly out of phase. See color insert.

sponding to a particular value of OAM, provided that the amount of angular momentum in the prepared state is limited.

There are advantages and disadvantages to each of the two main methods of determining the OAM of photons. The hologram method is the easier of the two experimentally, but can only determine if a photon is of a particular angular momentum—otherwise the photon is lost. This has implications for data transmission rates. The interferometric method is in principle lossless, but precise determination of the OAM using this method may rely on several interferometers, all of which need to be stabilized. This is a considerable experimental challenge.

A different suggestion to determine the OAM of photons on the quantum level uses the optical frequency shift [16], which is an extension of the optical frequency shift measurement proposed in [17]. In this case, a rotating Dove prism imparts an m-dependent frequency shift on the light beam so that photons of different OAM will be detected at different frequencies. The probability operators corresponding to the measurement correspond to different frequencies of single photon states, but are effectively the same as equation (4). To our knowledge, this method has not yet been implemented at the single photon level.

11.1.3 Intrinsic Security

Although most quantum information applications rely on single photons, the multidimensionality provided by the OAM may also prove useful in classical communication systems. In this case, the main advantage lies in the increased "alphabet" and the associated higher data capacity. The transfer of classical information encoded as eight pure OAM states was demonstrated in [18]. Both the transmitter and receiver unit were based on computer controlled holograms implemented using spatial light modulators. The chosen "alphabet" consisted of the OAM values $\ell = -16, -12, -8, -4, 4, 8, 12, 16$, where the separation in ℓ allowed for a smaller overlap of the beam profiles and hence a better identification. The residual $\ell = 0$ beam was used for alignment. The different "letters" were generated from precalculated holograms written on the spatial light modulator in the transmitter unit. In the receiver unit, the analyzing hologram was designed to diffract the transmitted light beam into 9 directions, each with a different helicity. The beams were arranged on a 3×3 grid and imaged with a CCD camera from which the transmitted OAM could be discerned by a high on-axis intensity.

It is worth noting that information encoded as OAM provides some degree of intrinsic security. Any attempt to eavesdrop by intersecting the transmitted information away from the beam axis will be subject to an angular restriction and misalignment. A restriction in the angular position $\Delta\Phi$ causes a spread in OAM $\Delta\ell$, as angle and orbital angular momentum are linked in an uncertainty relation [14]. The distribution of ℓ states is determined by the Fourier transform of the azimuthal dependence of the aperture function. This means that an eavesdropper can only achieve an error-free measurement of the transmitted data by having access to the information over the full 2π angular range. Further potential corruption of data can result from a misalignment of the receiver [19]. A lateral misalignment corresponds to a measurement of ℓ along an axis parallel to the beam axis, introducing an additional extrinsic angular momentum that is in measurements indistinguishable from the original intrinsic OAM, and causes a shift of the central ℓ value in the OAM distribution. An angular misalignment means that in the new measurement basis a pure OAM is projected onto a superposition of OAM values centered around the original OAM. In combination, misalignment broadens the OAM distribution and displaces the mean value, posing huge practical difficulties on error-free eavesdropping [18]. Of course, the same potential data corruption applies for all detection of information encoded as OAM, be it an eavesdropper or the intended recipient. Although demonstrated with classical light beams, it will still hold for single photons in the quantum regime. Optical communication via OAM states is therefore, in practice, only feasible along the line of sight between transmitter and receiver.

In this section, we have concentrated on entangled OAM states and detection at the single photon level. More generally, any system that implements quantum communication or processing via OAM will require the generation of and transformation between particular OAM states—techniques that have been well established since the first OAM experiments in the 1990s. The most convenient tool for changing the OAM of a light beam is via holograms that allow a controlled modification of the phase structure of a beam. Flexibility can be added by using spatial light modulators (SLMs), which are computer-controlled refractive elements, to implement such holograms. In the next section, we will address different methods to generate superpositions of OAM states.

11.2 EXPLORING THE OAM STATE SPACE

The possibility of forming superpositions of states is one of the central features of quantum mechanics. Many applications in quantum information rely on the increased processing possibilities that result from working with superpositions, and much recent research deals with the engineering of specific OAM states. For optical spin angular momentum, i.e. polarization, superpositions of horizontal and vertical polarized light, $c_{\leftrightarrow}|\leftrightarrow\rangle + c_{\updownarrow}|\updownarrow\rangle$, are given by light polarized either along a rotated axis for real amplitudes c, or in light that is, in various degrees, elliptically polarized. As OAM occupies an infinite-dimensional Hilbert space, there is a far wider range of possible superposition states. Superpositions of OAM states can be prepared by various methods. In each of these, a pure m state is passed through optical components that manipulate the phase of the light beam.

11.2.1 Superpositions of OAM States

Probably the simplest possibility is to superimpose an LG-mode with the fundamental Gaussian mode, $c_0|0\rangle + c_m|m\rangle$, reminiscent of polarization superposition states. A superposition of this kind will have a phase singularity at a position shifted away from the center, at a radius where the intensity of the LG-mode matches that of the Gaussian, and at an angle where the modes are exactly out of phase. A very straightforward way to realize such superpositions is in an interferometer [20], where a hologram placed into one of the arms can convert an incoming Gaussian mode into the desired LG mode, and amplitudes and phases of the two arms can be adjusted by attenuating elements and phase plates. For realistic experiments, however, the interferometric preparation of superposition states may not be suitable, mainly due to the necessity to keep the phase in the different arms stable throughout the measurements.

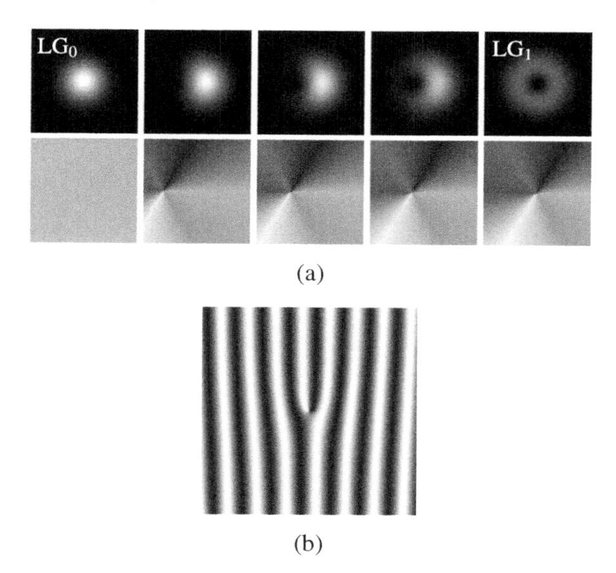

Figure 11.3 (a) Intensities (*top*) and phases (*bottom*) of various superpositions of the fundamental mode, LG_0, and a mode carrying one unit of OAM, LG_1. (b) Hologram generating the LG_1 mode. See color insert.

A more controllable scheme is offered by passing a pure m state through a hologram containing a fork dislocation, where the center of the hologram is displaced from the beam center by a variable distance d, as shown in Figure 11.3. For an $\ell = 1$ hologram and an incoming $\ell = 0$ mode, any superposition of $c_0|0\rangle + c_1|1\rangle$ can be produced [3,20]. The two extreme cases are given for perfect alignment, $d = 0$, when the beam will be projected onto the mode of the hologram, $|1\rangle$, while for an infinitely large displacement the mode will be left unchanged in $|0\rangle$. A displaced hologram of higher optical charge will produce superpositions of more modes, however, a controlled generation of any given superposition state is no longer possible.

In 2002, Molina-Terriza and colleagues devised the first scheme that allowed the engineering of OAM states beyond two dimensions [16], and enabled manipulation, including addition and removal of specific vector state projections. They suggested employing specific light distributions that yield the desired OAM superposition. For a given light distribution, the superposition can be found by calculating the projection into the spiral harmonics $\exp(im\phi)$. From the projection into the mth harmonic, one can compute the weight P_m of the mode with OAM $\ell = m$. In the proposal, the light fields are "vortex pancakes," single-charged vortices nested in a Gaussian host [21]. By adequately positioning N vortices, any superposition $\sum_{\ell=0}^{N} c_\ell|\ell\rangle$ can be generated.

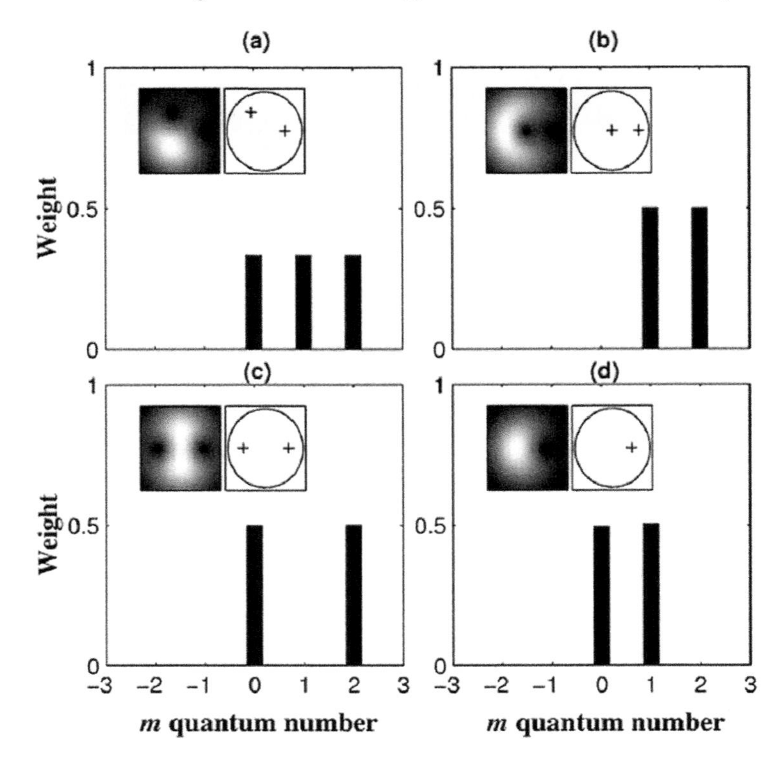

Figure 11.4 Controlled preparation of specific 3-dimensional photon states with an $\ell = 2$ "vortex pancake." (a) Equidistributed states, (b)–(d) equal populations in two out of the three OAM states $m = 0, 1, 2$. Reprinted from [16].

To some extent, this work can be understood as an extension of the shifted hologram [3,20], where the control parameter is the distance of the beam axis from the core of a single vortex. Here, instead, the light field is made up of several vortices distributed around the beam center, and the control parameter is their separation. Note that a vortex necklace made up of N vortices prepares a superposition state of maximal $(N + 1)$ OAM states. The case $N = 1$ describes the shifted hologram approach, generating superposition states of $|0\rangle$ and $|1\rangle$. For small numbers of vortices, the mode weighting can be analyzed analytically, and the required vortex separation for specific states can be determined, as shown in Figure 11.4. In general, light fields are decomposed into a series of infinitely many spiral harmonics, corresponding to superpositions of infinitely many OAM states. In principle, any infinite superposition state could be produced by a suitable vortex pancake, determined by solving the inverse problem. However, a general technique for this has not yet been developed.

11.2.2 Generating Entangled Superposition States

Many quantum communication protocols rely on maximally entangled states. The reason for this is that a maximally entangled state, when traced over one of its spatially separated modes, leaves a maximally mixed state of the remaining modes. This property guarantees that there is no quantum information in the remaining modes alone. Unfortunately, experimental setups usually generate nonmaximally entangled states. OAM entangled photons are conventionally produced via spontaneous parametric down-conversion. The OAM is conserved [3] (strictly speaking, only for collinearly generated photons, but to a very good degree also for emission by an angle of a few degrees off the pump beam). The probability to produce a photon pair with certain OAM depends on the overlap of the pump mode and the down-converted photon modes at the $\chi^{(2)}$ crystal. For a Gaussian pump beam, higher-order LG beams are produced with decreasing probability, and the entangled state will be

$$c_{00}|0,0\rangle + c_{11}|1,1\rangle + c_{22}|2,2\rangle + \cdots, \tag{9}$$

where $c_{00} > c_{11} > c_{22} \cdots$ and the kets give the absolute values of ℓ of photon 1 and 2, respectively. This has been observed experimentally [3], and the relative amplitudes have been shown to rely on the beam waists, positions, and the crystal length [16,22, 23].

Generation of engineered entangled OAM states by shaping the pump beam of a parametric downconversion system has been suggested [22]. If the light distribution of the pump beam is in the shape of a vortex necklace, the weighting of its OAM contributions can be controlled [16]. Phase-matching within the nonlinear crystal allows only the generation of downconverted beams with OAM that obey angular momentum conservation, i.e., the mth spiral harmonic of the pump beam generates photons with $m_1 + m_2 = m$. By manipulating the vortex positions of the pump beam, both phase and amplitude of the superposition states of the downconverted photons can be controlled. With this technique, particular entangled states could be generated on demand, e.g., maximally entangled states or other states required for optimal quantum protocols. While the theoretical work [22] presents a powerful tool for quantum information processing in arbitrary Hilbert dimensions, it remains to be seen how well this approach works in practical setups where higher-order radial contributions LG_p^ℓ and matching of beam waists could play an important role.

In the method described in the previous paragraph [22], classical information is imprinted on the pump beam in order to generate specific, e.g., maximally entangled, OAM beams. However, it is also possible to "distill" specific quantum states by manipulating the quantum information imprinted on the quantum state. In order

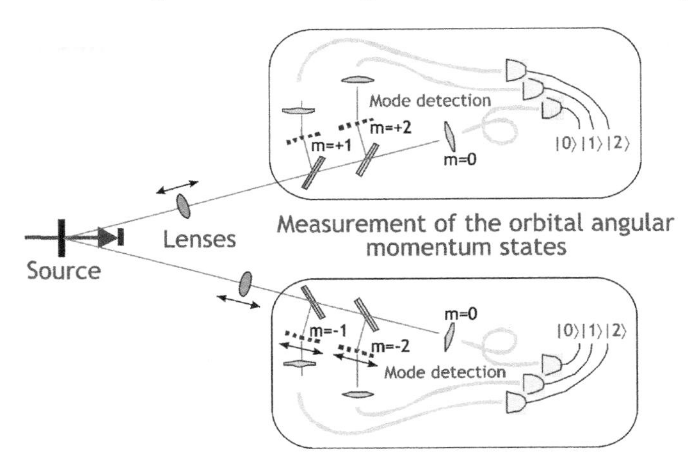

Figure 11.5 Experimental setup for entanglement concentration. Shifting the two lenses modifies the coupling efficiency of different OAM modes, allowing the postselection of maximally entangled states. Reprinted from [24].

to concentrate the entanglement of downconversion states and ideally generate maximally entangled states, the amplitudes c_{ii} have to be modified. Vaziri and colleagues have implemented "spatial filtering" to do this for OAM states of $|\ell| = 0, 1, 2$ [24]. The waist size of LG modes grows with $|\ell|$ so that efficient coupling into single mode fiber requires different lens settings. In reverse, by changing the lens setting in the setup shown in Figure 11.5, the coupling efficiency and therefore the detection efficiency of different $|\ell|$ modes can be varied, and by optimizing the coupling efficiency of the mode with the lower emission probability, the detection efficiencies can be assimilated. It is important to note that a lens acting as a local filter on one photon is simultaneously acting (in coincidence detection or in postselection) as a remote filter on its partner photon. By favoring coupling of $\ell = 2$ over $\ell = 1$ in one arm and $\ell = 1$ over $\ell = 0$ in the other, a state very close to the three-dimensional maximally entangled state $|\psi\rangle = \frac{1}{\sqrt{3}}(|0,0\rangle + |1,1\rangle + |2,2\rangle)$ was generated.

11.2.3 Storing OAM Information

While photons offer the most convenient system for generating engineered superpositions of OAM states, and spontaneous parametric downconversion provides a well-established method to generate entangled states, ultimately the optical quantum information may need to be processed and stored. This is particularly important as the entangled down-conversion photon pairs are produced at random times. It may be possible to store optical quantum information in cavities, and maybe even process it by controlled modification of the cavity mirrors; however, to the best of our knowledge, this has not been investigated in any detail. Instead, optical OAM information could be transferred onto atoms or Bose–Einstein condensates, and much recent work has been undertaken in this field.

A diffraction grating was written into a cloud of cold atoms, so that an incoming Gaussian beam was diffracted into beams carrying OAM [25]. Alternatively, entanglement of OAM states has been demonstrated between a collective atomic excitation and a single photon [26]. A weak Gaussian write pulse was directed at a sample of cold Rubidium atoms, inducing the generation of an anti-Stokes photon. Measurement of the OAM of the anti-Stokes photon projects the atomic sample into the correlated OAM state. One could assume that the atomic sample "memorizes" the phase difference between the write pulse and the emitted anti-Stokes photon and stores this information as atomic excitation. The information can be extracted from the atomic sample by the reverse process: A Gaussian read pulse excites the atoms, inducing the generation of a Stokes photon, which picks up the OAM from the atomic excitation and can be measured to carry the correct amount of OAM. In all of these cases, the OAM of the atoms is in the form of a mechanical angular momentum. An interesting possibility to control the atomic OAM was presented in [27], where the OAM states were manipulated by Larmor precession in a quadrupole magnetic field.

While these processes show possible methods of extracting OAM-carrying light from atoms, the mechanism of how OAM is transferred between atoms and light is not well understood, and the comparatively fast decoherence of the typical thermal atomic samples renders them unsuitable for processing or storing quantum information. Several schemes have been devised that transfer OAM from light beams onto atoms, namely Bose condensates. In these theoretical ideas, vortices are generated within a BEC via stimulated Raman processes with light that carries OAM [28]. The first experiment demonstrating coherent transfer of OAM between light and atoms was achieved by Phillips group at NIST. In 2006, they transferred optical OAM onto a sodium Bose condensate via two-photon stimulated Raman scattering [29]. The condensate is exposed to counterpropagating LG_0^1 and LG_0^1 beams. An atom, initially at rest, absorbs a photon with OAM and re-emits a stimulated photon without OAM, thus acquiring a linear momentum of $2\hbar k$ and an OAM of $\hbar$, causing the condensate to rotate. The OAM of the condensate was tested interferometrically by generating one condensate fraction at $2\hbar k$ with an OAM of $\hbar$ and one at $-2\hbar k$ with an OAM of $-\hbar$, which are then interfered. The resulting interference pattern is characteristic for an average OAM of zero, indicating a coherent process. This was confirmed by the fact that superpositions of different OAM states could be produced. Once OAM information is stored as vortices in a BEC it may be recovered with a "vortex sorter," which extracts the weights of the various spiral harmonics by utilizing symmetries of the differently charged vortices [30] similar to the optical OAM sorter [15].

In summary, various techniques are being developed that enable the coherent transfer of OAM from light (as a carrier of quantum information) onto atoms. In combination with well-established atom optical techniques, this introduces new and exciting possibilities to quantum processing.

11.3 QUANTUM PROTOCOLS

11.3.1 Advantages of Higher Dimensions

Higher-dimensional spaces and the qunits that are used as the alphabets increase the information density per physical system sent between sender and receiver. This information density increase may improve quantum information processing efficiency. Using qunits can also improve security, for example in quantum key distribution under the presence of noise [4]. In 3-D quantum cryptography schemes, the noise level may be as high as 22.5%, whereas for the 2-D BB84 [31] and Ekert [32] entangled state schemes the maximum noise level is 14.6%. Some quantum communication protocols require higher-dimensional quantum systems [5,33], and their security may be guaranteed by higher-dimensional Bell-type inequalities [34].

In this section, we will review the main protocols based upon angular momentum states. Vaziri and colleagues [35] have shown that it is possible to engineer bipartite 3-D maximally entangled states suitable for quantum communication. The preparation of the states is performed using type I downconversion in BBO to produce pairs of entangled photons. By passing each photon through a suitable combination of an $\ell = 1$ and an $\ell = -1$ hologram, superpositions of the form $c_0|0\rangle + c_1|1\rangle + c_2|2\rangle$ can be produced, and coincidences measured with an arrangement similar to that shown in Figure11.5. Ren and colleagues [36] have used a similar scheme to produce hyper-entangled (entangled in OAM, polarization and energy-time) photons. Again these states have three OAM basis states. Thus, the higher-alphabet protocols are experimentally realizable now.

One optical primitive that can be useful for 3-D optical quantum information schemes is the OAM analog of the polarizing beam splitter. This device has been proposed by Zou and Mathis using two symmetric 3-D splitters and a pair of Dove prisms [37]. It can be used for both preparing and measuring OAM states. Given a 1-photon state ℓ $(= 0, 1, 2)$ in spatial input mode i $(= 0, 1, 2)$, the device makes the transformation

$$|\ell\rangle_i \rightarrow |\ell\rangle_{\ell \ominus i}, \tag{10}$$

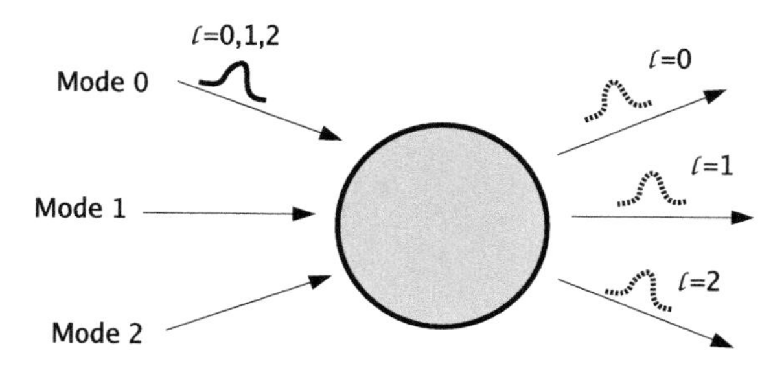

Figure 11.6 Schematic OAM beam splitter.

where the symbol $\ominus$ denotes subtraction modulo 3. Schematically, the device works as shown in Figure 11.6. Any photon of OAM ℓ input into port 0 of the beam splitter appears in output ℓ. The obvious analogous transformations occur for the other modes. In addition to using the beam splitter for preparation and measurement, it can also be used in a purification protocol [38].

Another method of OAM-state preparation and measurement was proposed by Molina-Terriza and colleagues [39]. This method uses the previously described hologrammatic technique, extended and combined with state tomography to identify which state has been prepared.

11.3.2 Communication Schemes

Most qunit-based quantum communication protocols are refinements of Ekert's 2-D entanglement-based protocol [32]. In this scheme, Alice and Bob are each sent one particle of an entangled pair. They each choose at random to measure one observable of a complementary pair on their particle. Afterward, they do not communicate their results, but merely which observable they decided to measure. The entanglement in the system guarantees that when they measure the same observable, their results are perfectly correlated, and any eavesdropping can be detected as a decrease in this correlation. The classic photon-based scheme uses down-converted photons entangled in polarization, as shown in Figure 11.7. If both Alice and Bob measure linear polarization with the same axes, their results are correlated. If they choose axes oriented at $\pi/4$, their results should be uncorrelated, and they throw them away. By comparing a small amount of the supposedly correlated data, they can detect eavesdropping, and they can use the remainder to establish a secret key. The classic BB84 scheme [31] shown in Figure 11.8 does not use entanglement; rather Alice chooses at random one of the four possible polarization states to send to Bob, who makes the same random

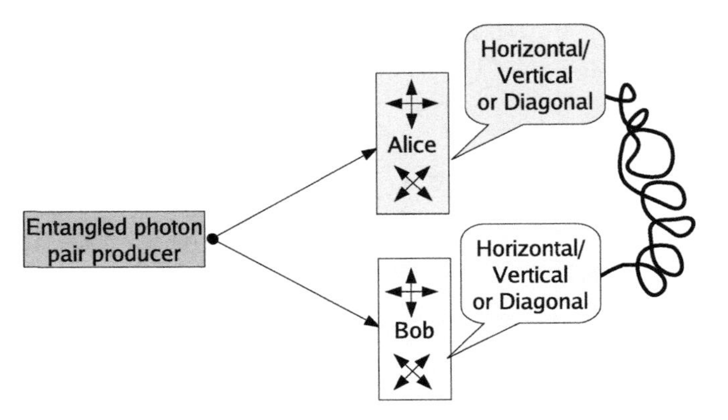

Figure 11.7 Schematic entangled state protocol for quantum key distribution.

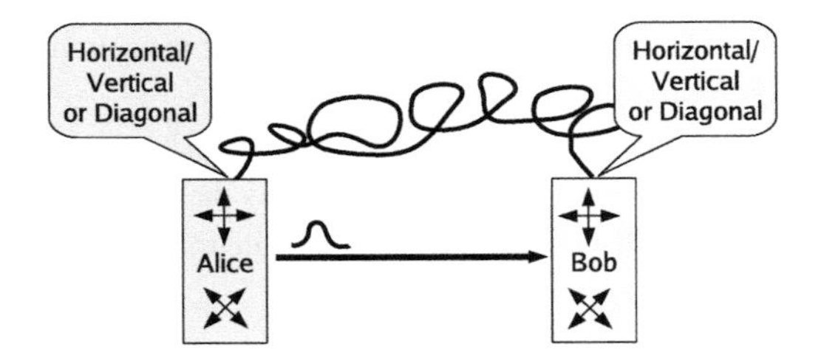

Figure 11.8 Schematic BB84 protocol for quantum key distribution.

measurement choice as in the Ekert scheme. The remainder of the protocol is identical to the entangled state version.

Bourennane and colleagues considered a simple extension of the BB84 protocol to higher dimensions [33]. Similarly, Bechmann-Pasquinucci and Peres [4] utilized an extension of the BB84-type schemes using a 3-state system. No bipartite entanglement properties of twin-beam angular momentum were required. Implementation of such schemes would merely require the production of pure single-beam superposition states of at least three different angular momentum values, which could be readily achieved by postselection in a twin-beam experiment. The postselecting measurement would provide the random selection of one of the states sent. Indeed, for 2-D systems this type of postselection is the basis of entangled state quantum cryptography [32].

An entanglement-based 3-D scheme has been realized by Gröblacher and colleagues [40], who managed to send a key with almost maximally entangled states, as shown in Figure 11.9. In this protocol, Alice and Bob each randomly select which angular momentum measurement they will make at the detectors by the insertion of

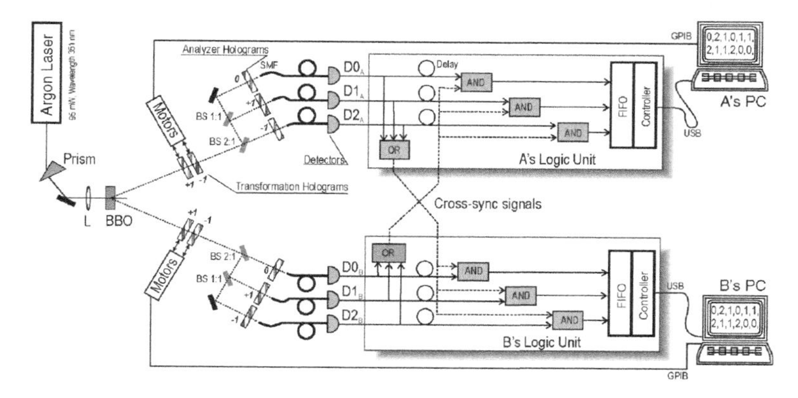

Figure 11.9 Key-distribution protocol of Gröblacher and colleagues, from [40].

transformation holograms. When they select the same hologram, they make the same measurement and a key is produced. Some of the data produced when they select different holograms can be used to violate a Bell-type inequality, and so acts as a security check, eliminating the possibility of an eavesdropper.

A variant on the quantum key distribution theme is the scheme of Karimipour and colleagues [5], which generalizes an earlier-proposed quantum secret sharing protocol [41]. Alice and Bob each require a generalized two-particle Bell-state (reference) of the n-level system, and each also receives one half of another entangled pair. They both perform an entanglement-swapping operation between the generalized Bell-states and their half of the entangled pair. Entanglement swapping is one of the more intriguing possibilities in quantum physics, whereby it is possible to entangle two particles that have never interacted in the past. The two particles must initially be part of entangled pairs, so that a joint measurement on one of each of the particles of the entangled pair leaves the remaining two particles unentangled. In the Karimipour scheme, after the operation has been performed, one of Alice's particles which initially formed the Bell state is entangled with one of Bob's. By simple communication of experimental details, the parties can then establish a secret key.

Another slightly unusual variant is the bit commitment protocol realized by Langford and colleagues [42]. In bit commitment, Alice sends a bit of information to Bob, over which she maintains control after he receives it in that she decides when the information is revealed to Bob. However, once Bob has received the bit, Alice cannot change its value—a quantum betting system. Langford and colleagues used a hologram-based measurement scheme to record the full density matrix for qutrit states. They then used this measurement to show that, in principle, entangled qutrits provide a more secure way to undertake bit commitment.

Another qutrit protocol is the quantum coin flipping scheme of Ambainis, Spekkens and Rudolph [8,43], realized for superpositions of three different OAM states by Zeilinger's group [44]. Coin flipping is a means of generating random bits shared by two parties. If these bits are generated by Alice and sent to Bob, then Alice might have control over which bit is generated, unbeknown to Bob. Quantum coin flipping protocols seek to minimize any possible bias that Alice can exert. Here Alice needs to manufacture four possible states of a 3-state system: $|0\rangle + |1\rangle$, $|0\rangle - |1\rangle$, $|0\rangle + |2\rangle$ and $|0\rangle - |2\rangle$. She sends one of them to Bob, and then tells him which state to measure. If Bob's measurement indicates the state Alice claims to have sent, then by simple classical communication they can generate a bit value. If Alice cheats by sending different states, she is limited to 75% control over the bit that she sends. For qubit schemes, Alice has greater control—over 90%. The four states that Alice sends could be produced using postselection in a twin-beam experiment. This would not allow Alice to have any control over the bit value; however, she could in principle find other ways of generating "cheat" states.

The main feature of quantum communication protocols is that there is increased security available using qunits rather than qubits. In quantum computing, the main criterion is efficiency and not security. There are schemes to perform quantum computation using qunits as the basic logical states [6,45–49], which would allow the use of fewer physical qunits than qubits to perform the same computation. These may prove useful in the long term if physical computational resources are limited. However, even qubit all-optical quantum computing is still far from developed, and little of the information-processing associated with the qunit computational schemes has at present been implemented in OAM-based schemes.

11.4 CONCLUSIONS AND OUTLOOK

Three basic elements are required for any quantum information system, the reliable production and identification of different quantum states, and the evolution that provides the required processing. Quantum optical technology is relatively advanced, having been developed for several decades before the beginning of the explosion of interest in quantum information. Advances in OAM state production and detection have therefore been rapid. Furthermore, the use of photons to encode quantum information is robust. They do not interact with one another unless such an interaction is specifically engineered. Photons also do not decohere rapidly; normally photons are either lost, in which case they are not detected, or they appear at the detectors as ordered. Evolution is thus relatively easy to control and engineer with standard optical elements.

In this chapter, we have described the use of optical OAM in quantum information systems. Its fundamental advantage, the possibility of using qunits rather than qubits, has been the main driver for developments in the field. The infinite-dimensional state space offers the immediate benefit of a larger "alphabet" and thus increased data-capacity, and the possibility to explore higher-dimensional protocols. There have been several proof of principle experiments in recent years, and this is likely to continue. This means that encoding quantum information in OAM is anticipated to have a significant future.

ACKNOWLEDGMENTS

The authors would like to thank the UK Engineering and Physical Sciences Research Council for financial support. Sonja Franke-Arnold is a Dorothy Hodgkin Research Fellow of the Royal Society.

REFERENCES

[1] A. Poppe, A. Fedrizzi, R. Ursin, H.R. Böhm, T. Lorünser, O. Maurhardt, M. Peev, M. Suda, C. Kurtsiefer, H. Weinfurter, T. Jennewein, A. Zeilinger, Practical quantum key distribution with polarization entangled photons, *Opt. Exp. 12* (2004) 3865.

[2] L. Allen, M.W. Beijersbergen, R.J.C. Spreeuw, J.P. Woerdman, Orbital angular-momentum of light and the transformation of Laguerre–Gaussian laser modes, *Phys. Rev. A 45* (1992) 8185.

[3] A. Mair, A. Vaziri, G. Weihs, A. Zeilinger, Entanglement of the orbital angular momentum states of photons, *Nature (London) 412* (2001) 313.

[4] H. Bechmann-Pasquinucci, A. Peres, Quantum cryptography with 3-state systems, *Phys. Rev. Lett. 85* (2000) 3313.

[5] V. Karimipour, S. Bagherinezhad, A. Bahraminasab, Entanglement swapping of generalized cat states and secret sharing, *Phys. Rev. A 65* (2002) 042320.

[6] S.D. Bartlett, H. de Guise, B.C. Sanders, Quantum encodings in spin systems and harmonic oscillators, in: R.G. Clark (Ed.), *Proceedings of IQC'01*, Rinton, Princeton, NJ, 2001, pp. 344–347, *Phys. Rev. A 65* (2002) 052316.

[7] M. Fitzi, N. Gisin, U. Maurer, Quantum solution to the Byzantine agreement problem, *Phys. Rev. Lett. 87* (2001) 217901.

[8] A. Ambainis, A new protocol and lower bounds for quantum coin flipping, *Proc. STOC 01* (2001) 134, quant-ph/0204022.

[9] M. Żukowski, A. Zeilinger, M.A. Horne, Realizable higher-dimensional two-particle entanglements via multiport beam splitters, *Phys. Rev. A 55* (1996) 2564.

[10] R.B.M. Clarke, V.M. Kendon, A. Chefles, S.M. Barnett, E. Riis, M. Sasaki, Experimental realization of optimal detection strategies for overcomplete states, *Phys. Rev. A 64* (2001) 012303.

[11] R.T. Thew, A. Acin, H. Zbinden, N. Gisin, Experimental realization of entangled qutrits for quantum communication, *Quantum Inf. Proc. 4* (2004), quant-ph/0307122.

[12] C.W. Helstrom, Quantum Detection and Estimation Theory, Academic Press, New York, 1976.

[13] See e.g., M.I. Kolobov (Ed.), *Quantum Imaging*, Springer-Verlag, Berlin, 2007.

[14] S. Franke-Arnold, S. Barnett, E. Yao, J. Leach, J. Courtial, M. Padgett, Uncertainty principle for angular position and angular momentum, *New J. Phys. 6* (2004) 103.

[15] J. Leach, M.J. Padgett, S.M. Barnett, S. Franke-Arnold, J. Courtial, Measuring the orbital angular momentum of a single photon, *Phys. Rev. Lett. 88* (2002) 257901; H. Wei, X. Xue, J. Leach, M.J. Padgett, S.M. Barnett, S. Franke-Arnold, J. Courtial, Simplified measurement of the orbital angular momentum of single photons, *Opt. Commun. 223* (2003) 117.

[16] G. Molina-Terriza, J.P. Torres, L. Torner, Management of the angular momentum of light: Preparation of photons in multidimensional vector states of angular momentum, *Phys. Rev. Lett. 88* (2002) 013601.

[17] G. Nienhuis, Doppler effect induced by rotating lenses, *Opt. Commun. 132* (1996) 8; J. Courtial, D.A. Robertson, K. Dholakia, L. Allen, M.P. Padgett, Rotational frequency shift of a light beam, *Phys. Rev. Lett. 81* (1998) 4828.

[18] G. Gibson, J. Courtial, M.J. Padgett, M. Vasnetsov, V. Pas'ko, S.M. Barnett, S. Franke-Arnold, Free-space information transfer using light beams carrying orbital angular momentum, *Opt. Exp. 12* (2004) 5448.

[19] M.V. Vasnetsov, V.A. Pas'ko, M.S. Soskin, Analysis of orbital angular momentum of a misaligned optical beam, *New J. Phys. 7* (2005) 46.

[20] A. Vaziri, G. Weihs, A. Zeilinger, Superpositions of the orbital angular momentum for applications in quantum experiments, *J. Opt. B 4* (2002) S47.

[21] G. Indebetouw, Optical vortices and their propagation, *J. Mod. Opt. 40* (1993) 73.

[22] J.P. Torres, Y. Deyanova, L. Torner, G. Molina-Terriza, Preparation of engineered two-photon entangled states for multidimensional quantum information, *Phys. Rev. A 67* (2003) 052313.

[23] H.H. Arnaut, G.A. Barbosa, Quantum cryptography with 3-state systems, *Phys. Rev. Lett. 85* (2000) 286.

[24] A. Vaziri, J.-W. Pan, Th. Jennewein, G. Weihs, A. Zeilinger, Concentration of higher dimensional entanglement: Qutrits of photon orbital angular momentum, *Phys. Rev. Lett. 91* (2003) 227902.

[25] S. Barreiro, J.W.R. Tabosa, Generation of light carrying orbital angular momentum via induced coherence grating in cold atoms, *Phys. Rev. Lett. 90* (2003) 133001.

[26] R. Inoue, N. Kanai, T. Yonehara, Y. Miyamoto, M. Koashi, M. Kozuma, Entanglement of orbital angular momentum states between an ensemble of cold atoms and a photon, *Phys. Rev. A 74* (2006) 053809.

[27] D. Akamatsu, M. Kozuma, Coherent transfer of orbital angular momentum from an atomic system to a light field, *Phys. Rev. A 67* (2003) 023803.

[28] Z. Dutton, J. Ruostekoski, Transfer and storage of vortex states in light and matter waves, *Phys. Rev. Lett. 93* (2004) 193602; K.T. Kapale, J.P. Dowling, Vortex phase qubit: Generating arbitrary, counterrotating, coherent superpositions in Bose–Einstein condensates via optical angular momentum beams, *Phys. Rev. Lett. 95* (2005) 173601.

[29] M.F. Andersen, C. Ryu, P. Cladé, V. Natarajan, A. Vaziri, K. Helmerson, W.D. Phillips, Quantized rotation of atoms from photons with orbital angular momentum, *Phys. Rev. Lett. 97* (2006) 170406.

[30] G. Whyte, J. Veitch, P. Öhberg, J. Courtial, Vortex sorter for Bose–Einstein condensates, *Phys. Rev. A 70* (2004) 011603(R).

[31] C.H. Bennett, G. Brassard, Quantum cryptography: Public key distribution and coin tossing, in: *Proceedings of the IEEE International Conference on Computers, Systems, and Signal Processing*, Bangalore, India, IEEE, New York, 1984, p. 175.

[32] A. Ekert, Quantum cryptography based on Bell's theorem, *Phys. Rev. Lett. 67* (1991) 661.

[33] M. Bourennane, A. Karlsson, G. Björk, Quantum key distribution using multilevel encoding, *Phys. Rev. A 64* (2001) 012306.

[34] S. Massar, S. Pironio, J. Roland, B. Gisin, Bell inequalities resistant to detector inefficiency, *Phys. Rev. A 66* (2002) 052112.

[35] A. Vaziri, G. Weihs, A. Zeilinger, Experimental two-photon, three-dimensional entanglement for quantum communication , *Phys. Rev. Lett. 89* (2002) 240401.

[36] X.-F. Ren, G.-P. Guo, J. Li, C.-F. Li, G.-C. Guo, Engineering of multi-dimensional entangled states of photon pairs using hyper-entanglement, *Chin. Phys. Lett. 23* (2006) 552.

[37] X.B. Zou, W. Mathis, Scheme for optical implementation of orbital angular momentum beam splitter of a light beam and its application in quantum information processing, *Phys. Rev. A 71* (2005) 042324.

[38] M.A. Martín-Delgado, M. Navasués, Single-step distillation protocol with generalized beam splitters, *Phys. Rev. A 68* (2003) 012322.

[39] G. Molina-Terriza, A. Vaziri, J. Rehacek, Z. Hradil, A. Zeilinger, Triggered qutrits for quantum communication protocols, *Phys. Rev. Lett. 92* (2004) 167903.

[40] S. Gröblacher, Th. Jennewein, A. Vaziri, G. Weihs, A. Zeilinger, Experimental quantum cryptography with qutrits, *New J. Phys. 8* (2006) 75.

[41] A. Cabello, Quantum key distribution without alternative measurements, *Phys. Rev. A 61* (2000) 052312.

[42] N.K. Langford, R.B. Dalton, M.D. Harvey, J.L. O'Brien, G.J. Pryde, A. Gilchrist, S.D. Bartlett, A.G. White, Measuring entangled qutrits and their use for quantum bit commitment, *Phys. Rev. Lett. 93* (2004) 053601.

[43] R.W. Spekkens, T. Rudolph, Degrees of concealment and bindingness in quantum bit commitment protocols, *Phys. Rev. A 65* (2002) 012310; R.W. Spekkens, T. Rudolph, Quantum protocol for cheat-sensitive weak coin flipping, *Phys. Rev. Lett. 89* (2002) 227901.

[44] G. Molina-Terriza, A. Vaziri, R. Ursin, A. Zeilinger, Experimental quantum coin tossing, *Phys. Rev. Lett. 94* (2005) 040501.

[45] S.D. Bartlett, H. de Guise, B.C. Sanders, Quantum encodings in spin systems and harmonic oscillators, *Phys. Rev. A 65* (2002) 052316.

[46] M.A. Nielsen, M.J. Bremner, J.M. Dodd, A.M. Childs, C.M. Dawson, Universal simulation of Hamiltonian dynamics for quantum systems with finite-dimensional state spaces, *Phys. Rev. A 66* (2002) 022317.

[47] J. Daboul, X. Wang, B.C. Sanders, Quantum gates on hybrid qudits, *J. Phys. A: Math. Gen. 36* (2003) 2525.

[48] G.K. Brennen, S.S. Bullock, D.P. O'Leary, Efficient circuits for exact-universal computation with qudits, *Quantum Inf. Comput. 6* (2005) 436.

[49] D.P. O'Leary, G.K. Brennen, S.S. Bullock, Parallelism for quantum computation with qudits, *Phys. Rev. A 74* (2006) 032334.

Chapter 12

Optical Manipulation of Ultracold Atoms

G. Juzeliūnas[1] *and P. Öhberg*[2]

[1]*Institute of Theoretical Physics and Astronomy of Vilnius University, Lithuania*
[2]*Heriot-Watt University, United Kingdom*

12.1 BACKGROUND

The concept of using light to manipulate ensembles of, or indeed individual, atoms, goes back a long time. From Maxwell's theory of electromagnetism, it became clear that light carries a momentum that can be transferred to particles [1]. Classically, a light beam will induce forces on a dipole. These forces depend upon the shape of the light beam—both the intensity and the phase. With the advent of the laser in the 1960s, it became possible to address in an unprecedented way the mechanical forces on atoms where the internal level structure of the atoms was exploited. This opened a path toward laser cooling and trapping atoms, where, in particular, the quantum mechanical nature of the atoms needed to be taken into account [2–4].

Optical manipulation of quantum objects has come a long way since the early attempts a century ago to manipulate the dynamics of thermal gases. In this chapter, we will first briefly review the mechanisms for trapping ensembles of ultracold atoms. These techniques will then be applied to neutral atoms that form a Bose–Einstein condensate. The optical trap will form the basis for manipulating the cold atoms where we rely on the coherent nature of the ultracold sample of atoms and the intensity of the light.

Subsequently, we will discuss a situation where the phase and the intensity of the incident light both play a crucial role. Here, we will consider a different kind of op-

STRUCTURED LIGHT AND ITS APPLICATIONS

tical manipulation where the laser fields are applied to induce vector and scalar potentials acting on atoms. The induced potentials have a geometric nature and depend exclusively upon the relative intensity and relative phase of the laser beams involved rather than on their absolute intensity. The approach relies on the ability to prepare the atoms in superpositions of the internal energy states of the atom. Interestingly, this technique provides a way to optically induce an effective magnetic field acting on electrically neutral atoms. This happens if the applied laser fields have a nontrivial topology, e.g., if they carry an orbital angular momentum along the propagation direction [5–7].

12.2 OPTICAL FORCES AND ATOM TRAPS

It was known since the time of Maxwell that light exerts a force on classical dipoles. The resulting force does indeed depend on the gradient of the amplitude and phase of the light [8],

$$m\ddot{\mathbf{R}} = \mathbf{d} \cdot (\nabla \mathbf{E}) = \mathbf{d} \cdot (\nabla \xi + \xi \nabla \theta) e^{i(\omega t + \theta)}, \tag{1}$$

where m is the mass of the dipole, $\mathbf{d}$ is the dipole moment, and $\mathbf{E}(\mathbf{R}, t) = \epsilon \xi(\mathbf{R}) e^{i(\omega t + \theta(\mathbf{R}))}$ is the electric field with the corresponding amplitude ξ, polarization ϵ, phase θ, and frequency ω.

An atom can be considered as a prototype dipole. For this purpose, let us restrict ourselves to two energy levels, as shown in Figure 12.1. Our goal is to describe the atom quantum mechanically, but allow the light field to be classical. The Hamiltonian for the atom is then

$$H = \frac{\mathbf{P}^2}{2m} + \hat{H}_0 - \mathbf{d} \cdot \mathbf{E}(\mathbf{R}, t), \tag{2}$$

where $\mathbf{P}^2/2m$ is the kinetic energy associated with the center of mass motion of the atom, $\hat{H}_0$ is the Hamiltonian for the unperturbed internal motion, and $\mathbf{d} \cdot \mathbf{E}(\mathbf{R}, t)$ is the interaction between the atom and the light field, which is based upon the dipole approximation. With the Hamiltonian from equation (2) and the Ehrenfest theorem, we obtain the expression for the force

$$F = m\ddot{\mathbf{r}} = \langle \nabla (\mathbf{d} \cdot \mathbf{E}) \rangle = \langle \mathbf{d} \cdot \epsilon \rangle \nabla \xi(\mathbf{r}, t), \tag{3}$$

where $\mathbf{r} = \langle \mathbf{R} \rangle$ and $\xi(\mathbf{r}, t) = \xi(\mathbf{r}) e^{i(\omega t + \theta(\mathbf{r}))}$. On the right side above, we have assumed that the force is uniform across the atomic wave packet. For a thorough discussion, we refer to references [9,10,2,8].

The two-level atom driven by a laser has been studied extensively [9–12,8]. In order to obtain an expression for the force acting on the atom, we need to calculate

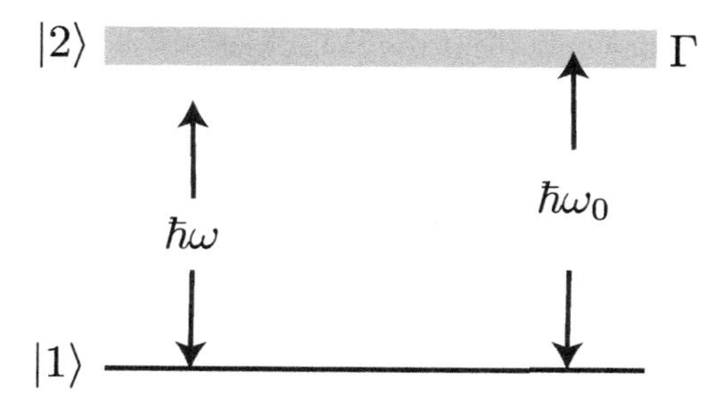

Figure 12.1 The atom is described as a two level system with energy difference $\hbar\omega_0$ between stat $|1\rangle$ and $|2\rangle$, and the frequency of the driving laser by ω. The decay rate of the excited state $|2\rangle$ is given by Γ.

the response of the atom to the light, i.e., the susceptibility or polarization, $\langle \mathbf{d} \cdot \boldsymbol{\epsilon} \rangle$. For this, we assume a monochromatic field of the form

$$\xi(\mathbf{r}, t) = \frac{1}{2} E(\mathbf{r}) e^{i(\theta(\mathbf{r}) + \omega t)}, \tag{4}$$

where $E(\mathbf{r})$ is the amplitude, ω is the laser frequency, and θ is a space-dependent phase factor. From the Schrödinger equation, we obtain the two coupled equations for the probability amplitudes C_1 and C_2 for the atom to be in state $|1\rangle$ and $|2\rangle$, respectively. By choosing a rotating frame according to

$$C_1 = D_1 e^{i\frac{1}{2}(\delta t + \theta)}, \tag{5}$$

$$C_2 = D_2 e^{-i\frac{1}{2}(\delta t + \theta)}, \tag{6}$$

we obtain the equations [9]

$$i\dot{D}_1 = \frac{1}{2}(\delta + \dot{\theta}) D_1 - \frac{\Omega}{2} D_2, \tag{7}$$

$$i\dot{D}_2 = -\frac{1}{2}(\delta + \dot{\theta}) D_2 - \frac{\Omega}{2} D_1. \tag{8}$$

In deriving the equations for D_1 and D_2, we have introduced the detuning $\delta = \omega - \omega_0$ and used the rotating wave approximation where rapidly oscillating terms are neglected [12]. The dipole moment d for the transition between state 1 and 2 is given by $d = \langle 1|\mathbf{d} \cdot \boldsymbol{\epsilon}|2\rangle$, and the Rabi frequency is defined by

$$\Omega = \frac{dE(t)}{\hbar}. \tag{9}$$

It is convenient to introduce the density matrix at this stage, which is defined as $\rho_{nm} = C_n C_m^*$ or $\sigma_{nm} = D_n D_m^*$ with

$$\rho_{11} = \sigma_{11}, \tag{10}$$

$$\rho_{22} = \sigma_{22}, \tag{11}$$

$$\rho_{12} = \sigma_{12} e^{i(\theta+\omega t)}, \tag{12}$$

$$\rho_{21} = \sigma_{21} e^{-i(\theta+\omega t)}. \tag{13}$$

From equations (7) and (8), we can see that the matrix elements of the density matrix obey

$$\dot{\sigma}_{11} = -\frac{i}{2}\Omega(\sigma_{12} - \sigma_{21}) + \Gamma\sigma_{22}, \tag{14}$$

$$\dot{\sigma}_{22} = \frac{i}{2}\Omega(\sigma_{12} - \sigma_{21}) - \Gamma\sigma_{22}, \tag{15}$$

$$\dot{\sigma}_{12} = -i(\delta + \dot{\theta})\sigma_{12} + \frac{i}{2}\Omega(\sigma_{22} - \sigma_{11}) - \frac{1}{2}\Gamma\sigma_{12}, \tag{16}$$

where we have introduced the spontaneous emission rate Γ to incorporate decay processes [12].

The density matrix now allows us to calculate the expectation value for the dipole moment, which is given by

$$\langle \mathbf{d} \cdot \epsilon \rangle = d(\rho_{12} + \rho_{21}) = d\big(\sigma_{12} e^{i(\theta+\omega t)} + \sigma_{21} e^{-i(\theta+\omega t)}\big). \tag{17}$$

With this expression and again utilizing the Rotating Wave Approximation, we get from equations (3) and (4) the force

$$\mathbf{F} = \frac{d}{2}\big(\sigma_{12} + \sigma_{21} - i(\sigma_{12} - \sigma_{21})\big) = \frac{\hbar}{2}(U\nabla\Omega + V\Omega\nabla\theta), \tag{18}$$

where we have introduced the notation $U = \sigma_{12} + \sigma_{21}$ and $V = i(\sigma_{12} - \sigma_{21})$, and used the fact that $\dot{\theta} = \nabla\theta(\mathbf{r}) \cdot \dot{\mathbf{r}}$. If the atomic motion is slow, such that the phase of the atomic state, $\dot{\theta}$, does not change much during the lifetime $1/\Gamma$ of the excited state, we can restrict ourselves to the steady-state solution of the density matrix and put the time derivatives of the left side equal to zero in equations (14)–(16). The solutions for the corresponding U and V are then

$$U = \frac{\delta}{\Omega}\frac{s}{s+1} \tag{19}$$

$$V = \frac{\Gamma}{2\Omega}\frac{s}{s+1}, \tag{20}$$

where s is the saturation parameter

$$s = \frac{\Omega^2/2}{(\delta + \dot{\theta})^2 + \Gamma^2/4}. \tag{21}$$

The force acting upon the atom now consists of two parts—the dipole force and the radiation force, respectively,

$$\mathbf{F} = \mathbf{F}_{\rm dip} + \mathbf{F}_{\rm pr}, \tag{22}$$

with

$$\mathbf{F}_{\rm dip} = -\frac{\hbar(\delta + \dot{\theta})}{2} \frac{\nabla s}{s+1}, \tag{23}$$

$$\mathbf{F}_{\rm pr} = -\frac{\hbar\Gamma}{2} \frac{s}{s+1} \nabla\theta. \tag{24}$$

In the case of plane waves, the latter radiation force $\mathbf{F}_{\rm pr}$, often referred to as the radiation pressure, is proportional to the wave vector $\mathbf{k} = \nabla\theta$. For trapping purposes, on the other hand, the former dipole force $\mathbf{F}_{\rm dip}$ is more important. The force $\mathbf{F}_{\rm dip}$ is determined by the intensity of the laser field. If $s \ll 1$ and $|\delta| \gg \Gamma, \Omega$, we get the corresponding potential using $\mathbf{F}_{\rm dip} = \nabla W$

$$W = \frac{\hbar\Omega^2}{4\delta} = \frac{d^2E^2}{4\delta\hbar}. \tag{25}$$

From this expression, we can see that if the intensity of the light is inhomogeneous, we obtain a nonzero force whose direction depends upon the sign of the detuning. For a focused Gaussian beam, this means that the atoms are attracted to the high intensity if the laser is red-detuned ($\delta < 0$), i.e., the atoms are the high field seekers. On the other hand, if the laser is blue-detuned ($\delta > 0$), the atoms are the low field seekers and are repelled from the center of the beam.

12.3 THE QUANTUM GAS: BOSE–EINSTEIN CONDENSATES

During recent decades, experimental techniques for trapping and cooling atoms have developed enormously. Experimentalist reached a major goal in 1995 when they were able to trap and cool atomic gases of ^{87}Rb [13], ^{23}Na [14] and ^{7}Li [15] to temperatures low enough to see striking effects of the quantum nature of these gases. The atomic Bose–Einstein condensate (BEC) was born. This literally opened the floodgates in terms of experimental and theoretical activity. One of the main advantages with ultracold atomic quantum gases of either bosons or fermions is the unprecedented possibility to change and manipulate the physical parameters such as density of the cloud, geometry, or even the interaction strength between the atoms [16]. In addition, the underlying theory describing these gases is remarkably accurate, which have resulted in a very fruitful coexistence between theory and experiments. In this brief introduction to quantum gases, we will give an overview of the basic concepts. In particular, we will concentrate on the theoretical tools we possess and need in order to describe these systems.

12.3.1 Bose–Einstein Condensation in a Cloud of Atoms

There are two types of particles in Nature: *Bosons* which have an integer spin, and *fermions* which have a half integer spin associated with them. Bosons are governed by a symmetric multiparticle wave function and are allowed to all occupy the same quantum state. Fermions, on the other hand, obey the Pauli exclusion principle, which tells us that there cannot be two or more fermions in the same quantum state. The original idea of Bose–Einstein condensation dates back to 1924, when S.N. Bose and A. Einstein were working on a statistical description of light [17,18]. They were able to show that there is a phase transition in a gas of noninteracting particles, where a "condensation" of particles into the lowest state takes place as a consequence of quantum statistical effects. Much later, this phenomenon drew renewed interest in the context of superfluidity when in 1938 F. London predicted that the origin of superfluidity was in Bose–Einstein condensation [19].

Experimental techniques to trap and cool atoms where developed much later. Experimentalists made big advances in the late 1970s when they developed new techniques that used laser cooling and magnetic trapping. The obvious candidate had so far been hydrogen, since it is a light atom with consequently a relatively high critical temperature [20]

$$T_c \sim \frac{\hbar^2 \rho^{2/3}}{m k_B}, \tag{26}$$

where k_B is the Boltzmann constant, m is the mass of the atom, and ρ is the density. Highly sophisticated methods were developed for cooling hydrogen [21], which fortunately paved the way for future experiments. However, it turned out to be surprisingly difficult to reach the quantum regime, which requires high densities in combination with low temperatures. In the 1980s, another candidate(s) entered the scene. Neutral alkali atoms turned out to be well suited for laser cooling and trapping. This is because alkali atoms have suitable level structures and optical transitions that can be addressed with available lasers. Eventually, using a combination of laser cooling and trapping, weak magnetic trapping based upon the Zeeman shift, and evaporative cooling, the experimental groups of Cornell and Wieman at Boulder, Colorado, and Wolfgang Ketterle at MIT succeeded in reaching the required high densities and low temperatures for Bose–Einstein condensation. In these two experiments, ^{87}Rb and ^{23}Na were used.

With these experiments, a completely new physical system had been created and now had to be understood. The atomic cloud was trapped, which meant that the condensation did not only take place in momentum space, as it had been traditionally looked at in homogeneous systems, but also in coordinate space. This was new and has resulted in numerous remarkable experiments by groups all over the world where the Bose–Einstein condensate phenomenon is observed in a direct way simply by looking at the density of the cloud and its dynamics.

12.3.2 The Condensate and Its Description

A Bose–Einstein condensate can be understood as a macroscopically occupied single quantum state. We will now look at the weakly interacting gas. The phase transition describing the onset of BEC can be considered using an ideal gas, where the critical temperature is readily derived (see any undergraduate textbook on statistical mechanics [20]). In the following, we will assume zero temperature. This is indeed a legitimate approximation. Present cooling techniques allow the experimentalist to go far below the critical temperature. This is typically in the micro Kelvin regime, where any contribution from the remaining thermal component can be neglected in most cases, as shown in Figure 12.2.

For a dilute gas, only two-body collisions take place. In addition, we obviously have a cold gas, hence we consider only s-wave scattering as the mechanism for the interaction. The interaction potential is therefore of the form [20]

$$V_{\text{int}}(\mathbf{r}-\mathbf{r}') = \frac{4\pi\hbar^2 a}{m}\delta(\mathbf{r}-\mathbf{r}'), \tag{27}$$

where the interaction is described by the single parameter a, called the s-wave scattering length. This is a result of the cold collisions in the gas. For a derivation of equation (27), we have to solve the two-body scattering problem in the limit of zero momentum.

With these assumptions, we get a Hamiltonian of the form

$$\hat{H} = \int d\mathbf{r}\left\{\hat{\Psi}^{\dagger}(\mathbf{r})\left[-\frac{\hbar^2}{2m}\nabla^2 + V_{\text{ext}}\right]\hat{\Psi}(\mathbf{r}) + \frac{g}{2}\hat{\Psi}^{\dagger}(\mathbf{r})\hat{\Psi}^{\dagger}(\mathbf{r})\hat{\Psi}(\mathbf{r})\hat{\Psi}(\mathbf{r})\right\}, \tag{28}$$

where $g = 4\pi\hbar^2 a/m$. As such, this is rather intractable and we have to succumb to approximations. The field operators $\Psi(\mathbf{r}, t)$ and $\Psi^{\dagger}(\mathbf{r}, t)$ destroy and creates, respectively, a particle at $\mathbf{r}$ at time t, and obey the usual bosonic commutation rules

$$\left[\hat{\Psi}(\mathbf{r}), \hat{\Psi}^{\dagger}(\mathbf{r}')\right] = \delta(\mathbf{r}-\mathbf{r}') \tag{29}$$

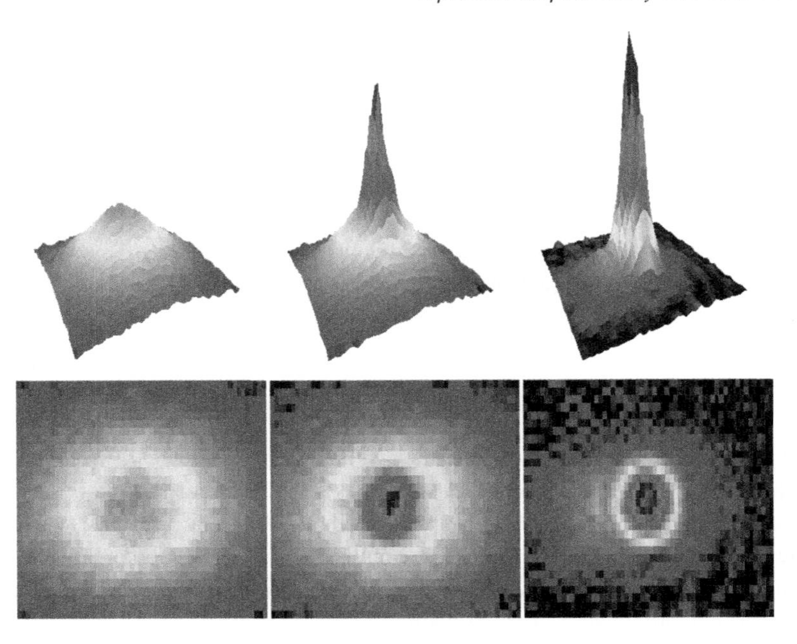

Figure 12.2 The onset of BEC is seen as a sharp peak in the density in the center of the trap. In the figures, the temperature is lowered from left to right. To the far right, we see a pure condensate with a negligible thermal component. The pictures are from the BEC experiment at University of Strathclyde, Glasgow, UK [22]. See color insert.

and

$$\left[\hat{\Psi}(\mathbf{r},t),\hat{\Psi}(\mathbf{r}',t)\right]=\left[\hat{\Psi}^{\dagger}(\mathbf{r},t),\hat{\Psi}^{\dagger}(\mathbf{r}',t)\right]=0. \tag{30}$$

Using these commutation rules, we obtain the Heisenberg equation of motion for the field operator

$$i\hbar\frac{\partial}{\partial t}\hat{\Psi}=[\hat{H},\hat{\Psi}]=-\frac{\hbar^2}{2m}\nabla^2\hat{\Psi}+V_{\text{ext}}(r)\hat{\Psi}+g\hat{\Psi}^{\dagger}\hat{\Psi}\hat{\Psi}. \tag{31}$$

We now split the field operator into the operator for a lowest mode and a part representing the fluctuations and thermal excitations,

$$\hat{\Psi}(\mathbf{r})=\hat{\Psi}_0(\mathbf{r})+\delta\hat{\Psi}(\mathbf{r}). \tag{32}$$

At zero temperature, we can as a first approximation neglect the term standing for the fluctuations $\delta\hat{\Psi}(\mathbf{r})$. In the presence of a condensate, the lowest mode is macroscopically populated, so we can write

$$\hat{\Psi}(\mathbf{r})=\Psi(\mathbf{r})\hat{a}_0\approx\Psi(\mathbf{r})\sqrt{N}. \tag{33}$$

Here we have replaced the annihilation operator $\hat{a}_0$ by $\sqrt{N}$, which is often referred to the Bogoliubov approximation (see, for instance, [4]). This is a legitimate approximation provided the number of atoms N in the condensate is sufficiently large. In other words, we have replaced the field operator by its average

$$\hat{\Psi}(\mathbf{r}) \approx \langle \hat{\Psi}(\mathbf{r}) \rangle = \Psi(\mathbf{r})\sqrt{N}. \tag{34}$$

The resulting equation of motion for the condensate "wavefunction" $\Psi(\mathbf{r})$ then becomes

$$i\hbar \frac{\partial}{\partial t}\Psi(\mathbf{r},t) = \left[-\frac{\hbar^2}{2m}\nabla^2 + V_{\text{ext}}(\mathbf{r}) + g\left|\Psi(\mathbf{r},t)\right|^2\right]\Psi(\mathbf{r},t). \tag{35}$$

This is the celebrated Gross–Pitaevskii equation [4,3]. It is the true workhorse when describing the dynamics of a Bose–Einstein condensate. The Gross–Pitaevskii equation is based on mean field theory, in which each atom feels the presence of all the other atoms through the effective potential. The potential is proportional to the density of the cloud, providing the nonlinear behavior of the condensate. The Gross–Pitaevskii equation is a very useful tool and has been used extensively to describe the properties of Bose–Einstein condensates.

The time-independent version of equation (35) is readily achieved by taking the Ansatz $\Psi(\mathbf{r},t) = \varphi(\mathbf{r})e^{-i\mu t\hbar}$, where μ is the chemical potential

$$\mu\varphi(\mathbf{r}) = \left[-\frac{\hbar^2}{2m}\nabla^2 + V_{\text{ext}}(\mathbf{r}) + g\left|\varphi(\mathbf{r})\right|^2\right]\varphi(\mathbf{r}). \tag{36}$$

A typical density distribution is shown in Figure 12.4, where the atoms are trapped in a harmonic external potential. Due to the interactions between the atoms, the density is not Gaussian, but is closer to an inverse parabola. This can be understood in terms of the Thomas–Fermi approximation [23] which neglects the kinetic energy term in the time-independent Gross–Pitaevskii equation (36) and gives

$$\left|\varphi(\mathbf{r})\right|^2 = \left[\mu - V_{\text{ext}}(\mathbf{r})\right]/g. \tag{37}$$

For harmonic traps, the Thomas–Fermi approximation does indeed give the shape of an inverse parabola for the atomic density shown in Figure 12.4. The approximation works well for trapping frequencies $\omega_z \ll \mu/\hbar$, and only at the edge of the cloud it inevitably breaks down.

12.3.3 Phase Imprinting the Quantum Gas

The trapping of a quantum gas can be achieved by a far detuned laser beam, as shown in the previous section, where the absorption of the light is avoided. With this technique, it is possible to shape the density of the Bose–Einstein condensate if we can shape the intensity of the light beam. The optical trap does not, however, affect the phase of the quantum gas, which is well defined for every atom in the coherent Bose–Einstein condensate. In order to be able to shape not only the density but also the phase of the Bose–Einstein condensate, we can use the so-called *phase imprinting technique*, which relies on a dynamic process, in contrast to the static optical trap.

The method of phase imprinting consists of passing a short off-resonant laser pulse through an appropriately designed absorption plate or spatial light modulator, which alters the intensity profile of the light beam [24,25]. The shaped light pulse is then allowed to propagate through the Bose–Einstein condensate. In the following, we will illustrate the mechanism by looking at a one-dimensional cloud.

The trapped Bose–Einstein condensate can be considered dynamically one-dimensional if the radial trapping frequency is larger than the corresponding chemical potential, $\omega_r \ll \mu/\hbar$, and the longitudinal confinement of atoms in the trapping potential is much weaker than that in the transverse direction. The Gross–Pitaevskii equation then takes the form

$$i\hbar\frac{\partial}{\partial t}\Psi(z,t) = \left[-\frac{\hbar^2}{2m}\frac{\partial^2}{\partial z^2} + V(z) + W(z,t) + g_{1\mathrm{D}}\left|\Psi(z,t)\right|^2\right]\Psi(z,t), \tag{38}$$

where $W(z,t)$ describes the interaction with the external laser, i.e., the dipole potential generated by the far-detuned laser pulse. The static trapping potential is given by the potential $V(z)$ which is typically harmonic in z. The one-dimensional dynamics is ensured with a renormalized mean field strength,

$$g_{1\mathrm{D}} = g\frac{m\omega_r}{2\pi\hbar}, \tag{39}$$

where ω_r is the transverse trapping frequency. It is important to remember that the dynamics can indeed be one-dimensional, but the collisions are three-dimensional. A true one-dimensional scenario can be achieved by a combination of strong transversal confinement and a low density of the bosonic gas. This phenomenon has been studied extensively both theoretically and experimentally, and is related to a phenomenon where the bosonic gas gets fermionic properties [26].

If the duration of the far off-resonant laser pulse is short compared to the correlation time,

$$t_{\mathrm{corr}} = \frac{\hbar}{\mu}, \tag{40}$$

the condensate does not have time to react, so the dominating term in the right side of equation (38) is given by $W(z,t)$. Consequently, we can write the solution of equation

(38) after the pulse has passed through the condensate as

$$\Psi(z) = e^{-i\int dt' \, W(z,t')}\Psi_0(z, t = 0), \tag{41}$$

where $\Psi_0(z, t = 0)$ is the initial state of the condensate. If the incident pulse is sufficiently short, we can extend the integration over time to infinity, hence the acquired phase is given by

$$\phi(z) = \Delta t W(z), \tag{42}$$

where Δt is a measure of the width of the pulse in the time domain. The potential $W(z)$ is, as in the case of the optical trap, given by equation (25), so the acquired phase $\phi(z)$ depends on the intensity of the pulse and its duration. In other words, the phase imprinting relies on the timing and shaping of the light beam intensity.

The phase imprinting technique offers a versatile tool for preparing a Bose–Einstein condensate in some chosen state. Generally the prepared state is not an eigenstate of the trapped quantum gas. Consequently, the phase imprinting can be exploited to induce coherent dynamics of the Bose–Einstein condensate. This effect was used when creating dark solitons in Bose–Einstein condensates [24,25]. A *soliton* is a topological excitation, or kink in the wave function, which propagates in the atomic cloud without losing its shape. The soliton is also a particular solution of the nonlinear Gross–Pitaevskii equation. Admittedly, the imprinted state is not the exact soliton solution. But it is close. The phase imprinting procedure therefore does indeed produce dark solitons, i.e., a density notch, in the case of repulsive interactions between the atoms. We are, of course, not restricted to only soliton dynamics. By choosing, for instance, a phase that has a quadratic dependence on position, we can induce focusing or defocusing. Similarly, a linear dependence in position will induce a momentum kick to the gas, which can be used as a method to coherently split a condensate. This is shown in Figures 12.3 and 12.5.

Interestingly, there is a close analogy between optics and the phase imprinting on a Bose–Einstein condensate that acts as a phase plate for the atoms. This is most clearly seen in the focusing dynamics of the cloud of atoms when a parabolically shaped phase is imprinted onto the atoms. Such a phase acts as a lens for the atoms. The focused matter wave, i.e., the condensate, does not, however, focus to a single point later in time due to the interactions between the atoms, but the overall dynamics closely resemble the focusing of light [27,28].

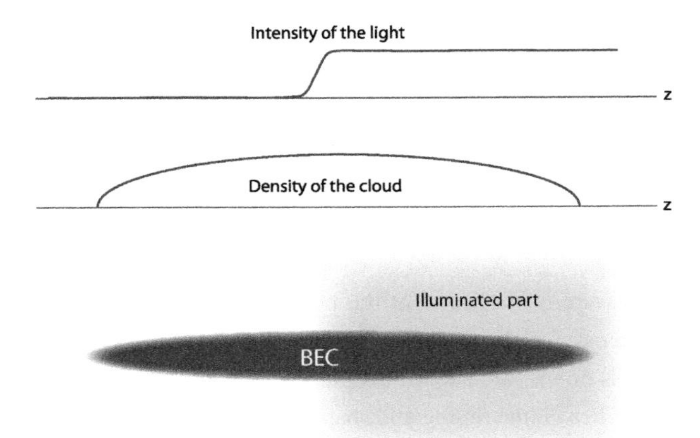

Figure 12.3 The phase imprinting technique can for instance be used for engineering a Bose–Einstein condensate with a sharp phase slip. The resulting dynamics will give rise to dark solitons in the case of repulsive interaction between the atoms.

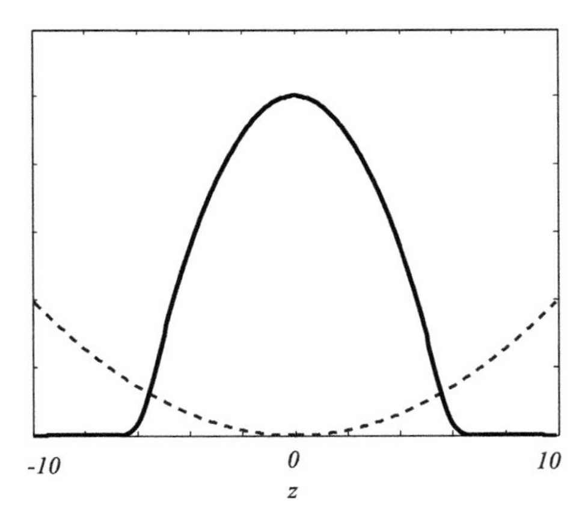

Figure 12.4 The density of the Bose–Einstein condensate has a parabolic shape for $\hbar\omega_z \ll \mu$. The dashed curve indicates the harmonic external trap, $\frac{1}{2}m\omega_z^2 z^2$.

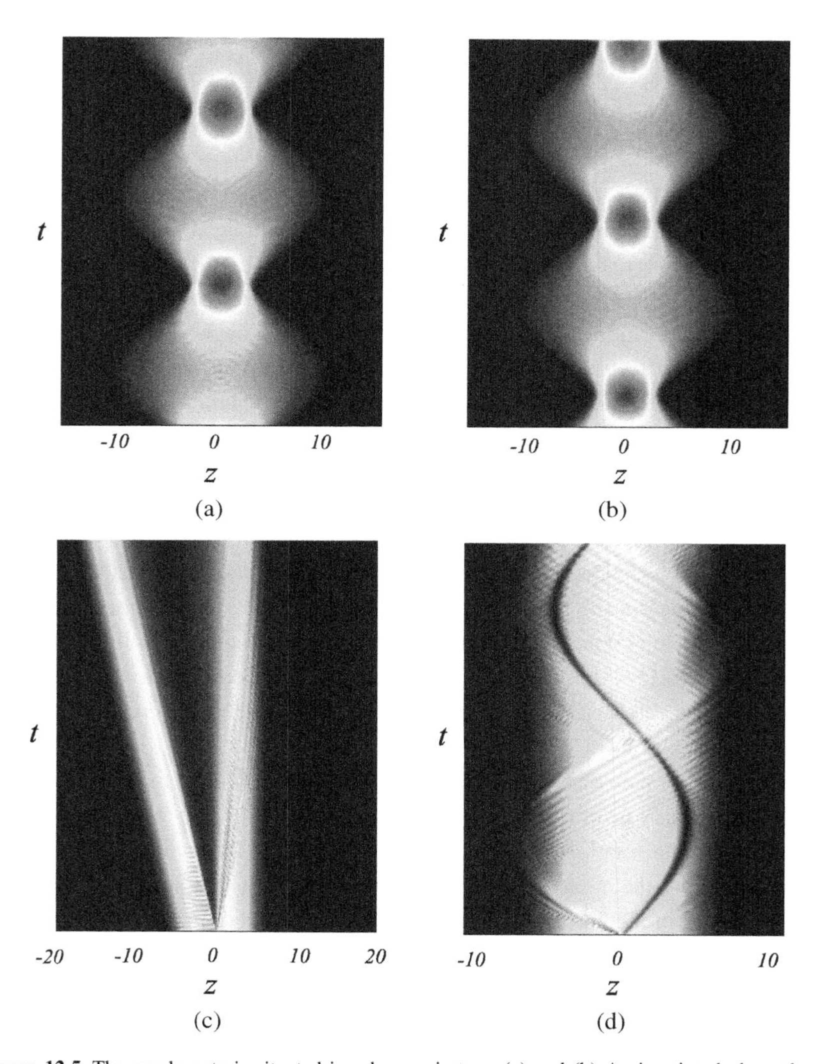

Figure 12.5 The condensate is situated in a harmonic trap. (a) and (b) An imprinted phase that is quadratic in z induces defocusing or focusing, depending upon the sign of the phase gradient. (c) If the phase is chosen such that it is zero for $z > 0$ and linear in z for $z < 0$, the result is a splitting of the cloud where part of the cloud is separated and the remaining part stays stationary during a time much smaller than $1/\omega_z$. (d) With a sharp phase slip imprinted, the result is a dark soliton that oscillates in the cloud (see Figure 12.3). See color insert.

12.4 LIGHT-INDUCED GAUGE POTENTIALS FOR COLD ATOMS

12.4.1 Background

Up to now we have shown that the shaped intensity of the light beam can be used for trapping and phase imprinting a Bose–Einstein condensate. The phase of the light has not played a significant role so far. In this section, we will consider a situation where both the intensity and the phase of the incident laser fields are important in the optical manipulation of atoms. Specifically, we will show how two or more laser fields can induce effective vector and trapping potentials for the atomic center of mass motion. This will give us a new tool for manipulating the neutral atoms because of an induced effective magnetic field. For this, the laser beams should act on the atoms in an Electromagnetically Induced Transparency (EIT) [29–33] configuration. The induced gauge potential has a geometric nature and depends upon the relative intensity and the relative phase of the incident laser fields rather than on their absolute intensities and phases. The technique provides a way to optically induce an effective magnetic field acting on electrically neutral atoms. This happens if the applied laser fields have a nontrivial topology, e.g., if they carry an orbital angular momentum along the propagation direction [5–7,34,35]. The appearance of the effective vector potential is a manifestation of the Mead–Berry connection [36,37] which is encountered in many different areas of physics [38–43].

In passing, we note that the usual way to produce an effective magnetic field in a cloud of electrically neutral atoms is to rotate the system such that the vector potential will appear in the rotating frame of references [44–46]. This would correspond to a situation where the atoms feel a uniform magnetic field. Yet stirring an ultracold cloud of atoms in a controlled manner is a rather demanding task. There have also been suggestions to take advantage of a discrete periodic structure of an optical lattice to introduce asymmetric atomic transitions between the lattice sites [47–50]. Using this approach, one can induce a nonvanishing phase for the atoms moving along a closed path on the lattice, i.e., one can simulate a magnetic flux [51,47–50]. However, such a way of creating the effective magnetic field is inapplicable to an atomic gas that does not constitute a lattice. The light-induced gauge potentials are free from all these drawbacks [5–7,34,52,53]. Furthermore, using these techniques it is possible to induce not only the usual (Abelian) gauge potentials [5–7,34,42,43,53], but also non-Abelian gauge potentials [52,54], whose Cartesian components do not commute. This will be considered in detail in subsequent sections.

12.4.2 General Formalism for the Adiabatic Motion of Atoms in Light Fields

We will start by adapting the general theory of the adiabatic dynamics [36–40] to the center of mass motion of atoms in stationary laser fields. For this we consider atoms with multiple internal states. The full atomic Hamiltonian is

$$\hat{H} = \frac{\hat{p}^2}{2m} + \hat{H}_0(\mathbf{r}) + \hat{V}(\mathbf{r}), \tag{43}$$

where $\hat{\mathbf{p}} \equiv -i\hbar\nabla$ is the momentum operator for an atom positioned at $\mathbf{r}$, and m is the atomic mass. Here the Hamiltonian $\hat{H}_0(\mathbf{r})$ describes the electronic degrees of freedom of the atom, and $\hat{V}(\mathbf{r})$ represents an external trapping potential. Note that the atomic Hamiltonian $\hat{H}_0(\mathbf{r})$ accommodates effects due to external light fields in addition to the internal dynamics.

For a fixed position $\mathbf{r}$, the atomic Hamiltonian $\hat{H}_0(\mathbf{r})$ can be diagonalized to give a set of, say, N dressed states $|\chi_n(\mathbf{r})\rangle$ of the atom coupled with the light fields. The dressed states are characterized by eigenvalues $\varepsilon_n(\mathbf{r})$, with $n = 1, 2, \dots, N$. The full quantum state of the atom describing both internal and motional degrees of freedom can then be expanded in terms of the dressed states as

$$|\Phi\rangle = \sum_{n=1}^{N} \Psi_n(\mathbf{r}) \big|\chi_n(\mathbf{r})\big\rangle, \tag{44}$$

where $\Psi_n(\mathbf{r}) \equiv \Psi_n$ is a wave-function for the center of mass motion of the atom in the internal state n. Substituting equation (44) into the Schrödinger equation $i\hbar\partial|\Phi\rangle/\partial t = \hat{H}|\Phi\rangle$, one arrives at a set of coupled equations for the components Ψ_n. Introducing the N-dimensional column vector $\Psi = (\Psi_1, \Psi_2, \dots, \Psi_N)^T$, it is convenient to represent these equations in a matrix form,

$$i\hbar\frac{\partial}{\partial t}\Psi = \left[\frac{1}{2m}(-i\hbar\nabla - \mathbf{A})^2 + U\right]\Psi, \tag{45}$$

where $\mathbf{A}$ and U are $N \times N$ matrices with the following elements

$$\mathbf{A}_{n,m} = i\hbar\big\langle\chi_n(\mathbf{r})\big|\nabla\chi_m(\mathbf{r})\big\rangle, \tag{46}$$

$$U_{n,m} = \varepsilon_n(\mathbf{r})\delta_{n,m} + \big\langle\chi_n(\mathbf{r})\big|\hat{V}(\mathbf{r})\big|\chi_m(\mathbf{r})\big\rangle. \tag{47}$$

The latter matrix U includes contributions from both the internal atomic energies and also the external trapping potential. The former matrix $\mathbf{A}$ is the gauge potential that appears due to the position dependence of the atomic dressed states. If the off-diagonal elements of the matrices $\mathbf{A}$ and U are much smaller than the difference in the atomic energies $U_{nn} - U_{mm}$, the *adiabatic approximation* can be applied by neglecting the off-diagonal contributions. This leads to a separation of the dynamics in different dressed states. Atoms in any one of the dressed states evolve according to a separate Hamiltonian in which the gauge potential $\mathbf{A}$ reduces to the 1×1 matrix, i.e., the gauge potential becomes Abelian. The adiabatic approximation fails if there are degenerate (or nearly degenerate) dressed states, so that the off-diagonal (nonadiabatic) couplings between the degenerate dressed states can no longer be ignored. In that case, the gauge potentials no longer reduce to the 1×1 matrices. They are non-Abelian provided their Cartesian components do not commute.

Let us assume that the first q atomic dressed states are degenerate (or nearly degenerate) and that these levels are well separated from the remaining $N - q$ levels. Neglecting transitions to the remaining states, one can project the full Hamiltonian onto this subspace. As a result, one arrives at the closed Schrödinger equation for the reduced column vector $\tilde{\Psi} = (\Psi_1, \ldots, \Psi_q)^\top$

$$i\hbar \frac{\partial}{\partial t}\tilde{\Psi} = \left[\frac{1}{2m}(-i\hbar\nabla - \mathbf{A})^2 + U + \Phi\right]\tilde{\Psi}, \tag{48}$$

where $\mathbf{A}$ and U are the truncated $q \times q$ matrices. The projection of the term $\mathbf{A}^2$ to the q-dimensional subspace cannot entirely be expressed in terms of a truncated matrix $\mathbf{A}$. This gives rise to a geometric scalar potential Φ, which is again a $q \times q$ matrix,

$$\begin{aligned}\Phi_{n,m} &= \frac{1}{2m}\sum_{l=q+1}^{N} \mathbf{A}_{n,l} \cdot \mathbf{A}_{l,m} \\ &= \frac{\hbar^2}{2m}\left(\langle\nabla\chi_n|\nabla\chi_m\rangle + \sum_{k=1}^{q}\langle\chi_n|\nabla\chi_k\rangle\langle\chi_k|\nabla\chi_m\rangle\right),\end{aligned} \tag{49}$$

with $n, m \in (1, \ldots, q)$. The reduced $q \times q$ matrix $\mathbf{A}$ is the Mead–Berry connection [36,37], also known as the effective *vector potential*. It is related to a curvature (an effective "magnetic" field) $\mathbf{B}$ as

$$B_i = \frac{1}{2}\epsilon_{ikl}F_{kl}, \qquad F_{kl} = \partial_k A_l - \partial_l A_k - \frac{i}{\hbar}[A_k, A_l]. \tag{50}$$

Note that the term $\frac{1}{2}\varepsilon_{ikl}[A_k, A_l] = (\mathbf{A} \times \mathbf{A})_i$ does not vanish in general because the vector components of $\mathbf{A}$ do not necessarily commute. In fact this term reflects the non-Abelian character of the gauge potentials.

In the next section, we will consider a situation where two laser beams are coupled to the atoms in the so-called *Λ configuration*. In this scheme, there is a single nondegenerate electronic state (known as a *dark state*). Thus, the atomic center of mass undergoes the adiabatic motion influenced by the (Abelian) vector and trapping potentials. Later in the chapter, we will analyze a tripod scheme of laser–atom interactions that provides two degenerate dark states. In that case, one has non-Abelian light-induced gauge potentials.

12.5 LIGHT-INDUCED GAUGE POTENTIALS FOR THE Λ SCHEME

12.5.1 General

We will now consider an ensemble of cold three-level atoms in the Λ configuration with two ground states $|1\rangle$ and $|2\rangle$ and an electronically excited state $|0\rangle$, as shown in Figure 12.6. For example, the states $|1\rangle$ and $|2\rangle$ can be different hyperfine ground states of an atom. The atoms interact with two resonant laser beams in the EIT configuration, as shown in Figure 12.6. The first beam has a frequency of ω_1 and a wave-vector of $\mathbf{k}_1$, and it induces the atomic transitions $|1\rangle \to |0\rangle$ with Rabi frequency $\Omega_1 \equiv \mu_{01}E_1/2$, where E_1 is the electric field strength and μ_{01} is the dipole moment for the transition from the ground state $|1\rangle$ to the excited state $|0\rangle$. The second beam is characterized by the frequency ω_2 and wave-vector $\mathbf{k}_2$. It causes the transition $|2\rangle \to |0\rangle$ with a Rabi frequency $\Omega_2 \equiv \mu_{02}E_2/2$.

When adopting the rotating wave approximation, the Hamiltonian for the electronic degrees of freedom of an atom interacting with the two beams becomes

$$\hat{H}_0(\mathbf{r}) = \epsilon_{21}|2\rangle\langle 2| + \epsilon_{01}|0\rangle\langle 0| - \hbar\big(\Omega_1|0\rangle\langle 1| + \Omega_2|0\rangle\langle 2| + \text{H.c.}\big), \qquad (51)$$

where ϵ_{21} and ϵ_{01} are, respectively, the energies of the detuning from the two- and single-photon resonances. Note that the spatial dependence of the Hamiltonian $\hat{H}_0(\mathbf{r})$ emerges through the spatial dependence of the Rabi frequencies $\Omega_1 \equiv \Omega_1(\mathbf{r})$ and $\Omega_2 \equiv \Omega_2(\mathbf{r})$.

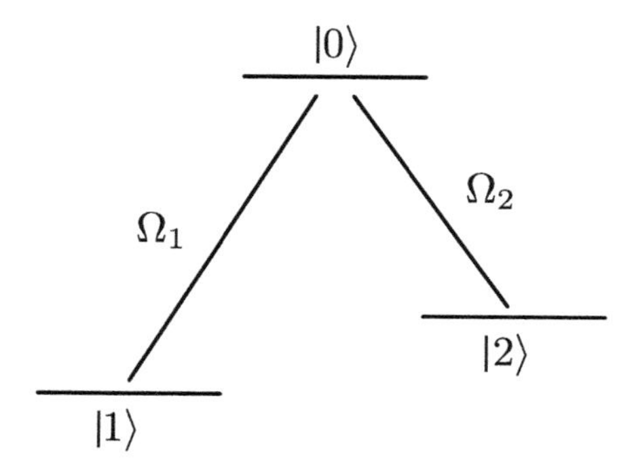

Figure 12.6 The EIT Λ configuration with two laser beams Ω_1 and Ω_2 coupling the levels.

Neglecting the two-photon detuning ($\epsilon_{21} = 0$), the Hamiltonian (51) has the eigenstate

$$|D\rangle = \frac{1}{\sqrt{1 + |\zeta|^2}}\big(|1\rangle - \zeta|2\rangle\big), \tag{52}$$

representing a coherent superposition of both ground states, where

$$\zeta = \frac{\Omega_1}{\Omega_2} \tag{53}$$

is the ratio of the amplitudes of the laser fields. It is characterized by a zero eigenenergy: $\hat{H}_0(\mathbf{r})|D\rangle = 0$. Since the state $|D\rangle$ has no contribution from the excited electronic state $|0\rangle$ and is not coupled to that state, it is immune to absorption and spontaneous emission. Therefore, the state $|D\rangle$ is called the dark state [29–33]. We are interested in a situation where the atoms are kept in their dark state $|D\rangle \equiv |D(\mathbf{r})\rangle$, so that the full atomic state-vector is

$$|\Phi\rangle = \Psi_D(\mathbf{r})\big|D(\mathbf{r})\big\rangle, \tag{54}$$

where Ψ_D is the wave-function for the center of mass motion of the dark-state atoms. If an atom is in the dark state $|D\rangle$, the laser beams induce the absorption paths $|2\rangle \to |0\rangle$ and $|1\rangle \to |0\rangle$, which interfere destructively, resulting in the Electromagnetically Induced Transparency [29–33]. In such a situation, the transitions to the upper atomic level $|0\rangle$ are suppressed. That is why the dark state has no contribution by the excited electronic state $|0\rangle$.

Suppose once again that the laser fields are tuned to the two-photon resonance: $\epsilon_{21} = 0$. The remaining two photon mismatch (if any) can be accommodated within the trapping potential

$$\hat{V}(\mathbf{r}) = V_1(\mathbf{r})|1\rangle\langle 1| + V_2(\mathbf{r})|2\rangle\langle 2| + V_0(\mathbf{r})|0\rangle\langle 0|, \tag{55}$$

where $V_j(\mathbf{r})$ is the trapping potential for an atom in the electronic state j, with $j = 0, 1, 2$. Applying the treatment presented in the previous section, the center of mass dynamics of the dark-state atoms is described by the equation of motion

$$i\hbar\frac{\partial}{\partial t}\Psi_D = \left[\frac{1}{2m}(-i\hbar\nabla - \mathbf{A})^2 + V_{\text{eff}}\right]\Psi_D, \tag{56}$$

where $\mathbf{A}$ and V_{eff} are the effective vector and trapping potentials, respectively,

$$\mathbf{A} = i\hbar\langle D|\nabla D\rangle, \tag{57}$$

$$V_{\text{eff}} = V + \phi, \tag{58}$$

with

$$V = \frac{V_1(\mathbf{r}) + |\zeta|^2 V_2(\mathbf{r})}{1 + |\zeta|^2}, \tag{59}$$

$$\phi = \frac{\hbar^2}{2M}\big(\langle D|\nabla D\rangle^2 + \langle \nabla D|\nabla D\rangle\big). \tag{60}$$

Since $V_1(\mathbf{r})$ and $V_2(\mathbf{r})$ are the trapping potentials for an atom in the electronic states 1 and 2, the potential V represents the external trapping potential for an atom in the dark state.

In this way, the effective trapping potential V_{eff} is composed of the external trapping potential V and the geometric scalar potential ϕ. The former V is determined by the shape of the trapping potentials $V_1(\mathbf{r})$ and $V_2(\mathbf{r})$, as well as the intensity ratio $|\zeta|^2$. The latter geometric potential ϕ is determined exclusively by the spatial dependence of the dark state $|D\rangle$ emerging through the spatial dependence of the ratio of the Rabi frequencies $\zeta = \Omega_1/\Omega_2$. Note that the effective vector potential $\mathbf{A}$ has a geometric nature as well because it also originates from the spatial dependence of the dark state.

12.5.2 Adiabatic Condition

The separation between the energies of the dark state and the remaining dressed atomic states of the Λ system is characterized by the total Rabi frequency $\Omega = \sqrt{\Omega_1^2 + \Omega_2^2}$. Assuming that the laser fields are tuned to the one- and two-photon resonances ($\epsilon_{01}, \epsilon_{21} \ll \hbar\Omega$), the adiabatic approach holds if the off-diagonal matrix elements in equation (45) are much smaller than the total Rabi frequency Ω. This leads to the following condition

$$F \ll \Omega, \tag{61}$$

where the velocity-dependent term

$$F = \frac{1}{1 + |\zeta|^2}|\nabla\zeta \cdot \mathbf{v}| \tag{62}$$

reflects the two-photon Doppler detuning. Note that the condition (61) does not accommodate effects due to the decay of the excited atoms. The dissipative effects can be included by replacing the energy of the one-photon detuning ϵ_{01} by $\epsilon_{01} - i\hbar\gamma_0$, where γ_0 is the excited-state decay rate. In such a situation, the dark state can be shown to acquire a finite lifetime

$$\tau_D \sim \gamma_0^{-1}\Omega^2/F^2, \tag{63}$$

which should be large compared to other characteristic times of the system.

The condition (61) implies that the inverse Rabi frequency Ω^{-1} should be smaller than the time an atom travels a characteristic length over which the amplitude or the phase of the ratio $\zeta = \Omega_1/\Omega_2$ changes considerably. The latter length exceeds the optical wavelength, and the Rabi frequency can be of the order of 10^7 to 10^8 s^{-1} [55]. Consequently, the adiabatic condition (61) should hold for atomic velocities up to the order of tens of meters per second, i.e., up to extremely large velocities in the context of ultra-cold atomic gases. The allowed atomic velocities become lower if the spontaneous decay of the excited atoms is taken into account. According to equation (63), the atomic dark state acquires then a finite lifetime τ_D, which is equal to γ_0^{-1} times the ratio Ω^2/F^2. The atomic decay rate γ_3 is typically of the order 10^7 s^{-1}. Therefore, in order to achieve long-lived dark states, the atomic velocity should not be too large. For instance, if the atomic velocities are of the order of a centimeter per second (a typical speed of sound in an atomic BEC), the atoms should survive in their dark states up to a few seconds. This is comparable to the typical lifetime of an atomic BEC.

12.5.3 Effective Vector and Trapping Potentials

Let us express the ratio of Rabi frequencies ζ in terms of amplitude and phase as

$$\zeta = \frac{\Omega_1}{\Omega_2} = |\zeta| e^{iS}. \tag{64}$$

Using expression (52) for the dark state, the effective vector potential takes the form

$$\mathbf{A} = -\hbar \frac{|\zeta|^2}{1+|\zeta|^2} \nabla S. \tag{65}$$

The effective magnetic field is consequently

$$\mathbf{B} = \hbar \frac{\nabla S \times \nabla |\zeta|^2}{(1+|\zeta|^2)^2}, \tag{66}$$

and the geometric scalar potential reads

$$\phi = \frac{\hbar^2}{2M} \frac{(\nabla|\zeta|)^2 + |\zeta|^2 (\nabla S)^2}{(1+|\zeta|^2)^2}. \tag{67}$$

One can easily recognize that the gauge potential $\mathbf{A}$ yields a nonvanishing effective magnetic field $\mathbf{B} = \nabla \times \mathbf{A}$ only if the gradients of the relative intensity and the relative phase are both nonzero and not parallel to each other. Therefore the effective magnetic field cannot be induced using the plane waves for the Λ scheme [42,43]. However, plane waves can indeed be used in a more complicated tripod setup [52,54], which we will consider in the next section.

Equation (66) has a very intuitive interpretation. Here $\nabla[|\zeta|^2/(1+|\zeta|^2)]$ is a vector that connects the "center of mass" of the two light beams, and ∇S is proportional to the vector of the relative momentum of the two light beams. Thus, a nonvanishing **B** requires a *relative orbital angular momentum* of the two light beams. We will see that this is the case for light beams with a vortex [5–7,34] or if one uses two counter-propagating light beams of finite diameter with an axis offset [53].

12.5.4 Co-Propagating Beams with Orbital Angular Momentum

Let us suppose that the incident laser beams can carry an orbital angular momentum along the propagation axis z, as shown in Figure 12.7. In this case, the spatial distribution of the beams is [56,57]

$$\Omega_1 = \Omega_1^{(0)} e^{i(k_1 z + l_1 \phi)} \tag{68}$$

and

$$\Omega_2 = \Omega_2^{(0)} e^{i(k_2 z + l_2 \phi)}, \tag{69}$$

where $\Omega_1^{(0)}$ and $\Omega_2^{(0)}$ are slowly varying amplitudes, $\hbar l_1$ and $\hbar l_2$ are the corresponding orbital angular momenta per photon along the propagation axis z, and ϕ is the azimuthal angle. The phase of the ratio $\zeta = \Omega_1/\Omega_2$ then reads $S = l\phi$. Therefore, the effective vector potential and the magnetic field take the form

$$\mathbf{A} = -\frac{\hbar l}{\rho} \frac{|\zeta|^2}{1+|\zeta|^2} \hat{\mathbf{e}}_\phi, \tag{70}$$

$$\mathbf{B} = \frac{\hbar l}{\rho} \frac{1}{(1+|\zeta|^2)^2} \hat{\mathbf{e}}_\phi \times \nabla |\zeta|^2, \tag{71}$$

where $l = l_1 - l_2$ is the difference in the winding numbers of the laser beams, $\mathbf{e}_\phi$ is the unit vector in the azimuthal direction, and ρ is the cylindrical radius. Note that although both beams are generally allowed to have nonzero orbital angular momentum by equations (68) and (69), it is desirable for the angular momentum to be zero for one of these beams. In fact, if both l_1 and l_2 were nonzero, the amplitudes Ω_1 and Ω_2 would simultaneously go to zero at the origin where $\rho = 0$. In that situation, the

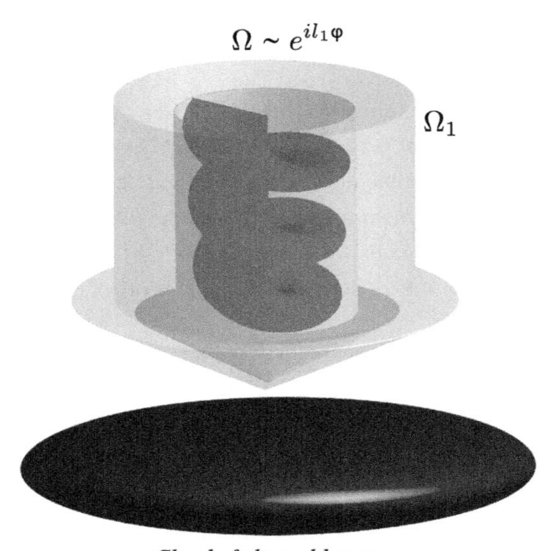

Figure 12.7 At least one of the two coupling beams in the EIT configuration should have an orbital angular momentum. See color insert.

total Rabi frequency $\Omega = \sqrt{\Omega_1^2 + \Omega_2^2}$ would also vanish, leading to the violation of the adiabatic condition (61) at $\rho = 0$.

If the beams are cylindrically symmetric, the intensity ratio $|\zeta|^2$ depends upon the cylindrical radius ρ only. In that case, the effective magnetic field is directed along the z-axis:

$$\mathbf{B} = -\hat{\mathbf{e}}_z \frac{\hbar l}{\rho} \frac{1}{(1+|\zeta|^2)^2} \frac{\partial}{\partial \rho} |\zeta|^2. \tag{72}$$

It is evident that the effective magnetic field is nonzero only if the ratio $\zeta = \Omega_1/\Omega_2$ contains a nonzero phase ($l = l_1 - l_2 \neq 0$) and the amplitude $|\zeta|$ has a radial dependence ($\partial|\zeta|/\partial\rho \neq 0$).

The light-induced magnetic field affects the atomic motion in the xy-plane. This might lead to a number of phenomena, such as the de Haas–van Alphen effect in the cloud of atomic fermions [5], or the light-induced Meissner effect [34] in the atomic Bose–Einstein condensates. Furthermore, the light-induced potentials alter the expansion dynamics of the atomic cloud [35]. A more detailed analysis of the light-induced gauge potentials for this geometry is presented in references [5–7,34].

12.5.5 Counterpropagating Beams with Shifted Transverse Profiles

We will now consider a different scenario [53], where we will use two counterpropagating light beams of finite diameter with an axis offset for which $\Omega_1 = \Omega_1^{(0)} e^{ik_1 y}$ and $\Omega_2 = \Omega_2^{(0)} e^{-ik_2 y}$, where $\Omega_1^{(0)}$ and $\Omega_2^{(0)}$ are real amplitudes with shifted transverse profiles, as shown in Figure 12.8. The beams possess a required relative orbital angular momentum similarly to two point particles with constant momenta passing each other at some finite distance, hence, an effective magnetic field will be generated.

The phase of the ratio $\zeta = \Omega_1/\Omega_2$ is now given by

$$S = ky, \quad k = k_1 + k_2, \tag{73}$$

so that $\nabla S = k\hat{\mathbf{e}}_y$ where $\hat{\mathbf{e}}_y$ is a unit Cartesian vector. The spatial dependence of the intensity ratio $|\zeta|^2 = |\Omega_1/\Omega_2|^2$ is determined by the spatial profiles of both $|\Omega_1|^2$ and $|\Omega_2|^2$. Since the light beams counterpropagate along the y-axis, their intensities depend weakly on the propagation distance y. Furthermore, we shall disregard the z-dependence of the intensity ratio $|\zeta|^2$. This is legitimate, for instance, if the atomic motion is confined to the xy-plane due to a strong trapping potential in the z-direction. Using equation (66), one arrives at the following strength of the light-induced effective magnetic field

$$\mathbf{B} = -\hat{\mathbf{e}}_z \hbar k \frac{\partial}{\partial x} \frac{|\zeta|^2}{(1+|\zeta|^2)}. \tag{74}$$

The effective magnetic field $\mathbf{B}$ is oriented along the z-axis. Its magnitude B depends generally upon the x-coordinate, yet it has a weak y-dependence as long as the paraxial approximation holds.

One possible application of this technique is to study quantum Hall phenomena and thus the possibility to enter the lowest Landau level (LLL) regime for the trapped atoms. In doing so, we have to estimate the maximum strength of the magnetic field.

Figure 12.8 Two counterpropagating and overlapping laser beams interact with a cloud of cold atoms.

For this we determine the minimum area needed for a magnetic flux to correspond to a single flux quantum $2\pi\hbar$. From equation (74), we recognize that this area is given by the product λx_{eff}, where x_{eff} is the effective separation between the two beam centers and $\lambda = 4\pi/k$. To reach the LLL in a two-dimensional gas, the atomic density therefore has to be smaller than one atom per λx_{eff}.

The above analysis holds as long as the atoms move sufficiently slow to remain in their dark states. This is the case if the adiabatic condition given by equation (62) holds. In the present situation, the adiabatic condition takes the form

$$F^2 = \frac{1}{(1+|\zeta|^2)^2}\left[\left(v_x \frac{\partial}{\partial x}|\zeta|\right)^2 + \left(|\zeta| k v_y\right)^2\right] \ll \Omega^2. \tag{75}$$

Let us assume that both beams are characterized by Gaussian profiles with the same amplitude Ω_0 and width σ

$$|\Omega_j| = \Omega_0 \exp\left(-\frac{(x-x_j)^2}{\sigma^2}\right), \quad j = 1, 2. \tag{76}$$

In the paraxial approximation, the Gaussian beams have the width $\sigma \equiv \sigma(y) = \sigma_0[1 + (\lambda y/\pi\sigma_0^2)]^{1/2}$, where $\sigma_0 \equiv \sigma(0)$ is the beam waist and λ is the wavelength. Since $k_1 \approx k_2 \approx k/2$, we have $\lambda \approx 4\pi/k$ for both beams. We are interested mostly in distances $|y|$ much less than the confocal parameter of the beams $b = 2\pi\sigma_0^2/\lambda \approx k\sigma_0^2/2$. For such distances, $|y| \ll b$, the width $\sigma(y)$ is close to the beam waist: $\sigma(y) \approx \sigma_0$.

Suppose the beams are centered at $x_1 = x_0 + \Delta x/2$ and $x_2 = x_0 - \Delta x/2$. The intensity ratio then reads: $|\zeta|^2 \equiv |\Omega_1/\Omega_2|^2 = \exp[(x-x_0)/a]$, where $a \equiv a(y) = \sigma^2/4\Delta x$ is the relative width of the two beams. Thus we have

$$\mathbf{B} = -\hbar k \frac{1}{4a\cosh^2((x-x_0)/2a)} \mathbf{e}_z, \tag{77}$$

$$V_{\text{eff}}(\mathbf{r}) = V(\mathbf{r}) + \frac{\hbar^2 k^2}{2m} \frac{(1 + 1/4a^2k^2)}{4\cosh^2((x-x_0)/2a)}, \tag{78}$$

where $V(\mathbf{r})$ is the external trapping potential. It is evident that both $\mathbf{B}$ and $V_{\text{eff}}(\mathbf{r})$ are maximum at the central point $x = x_0$ and decrease quadratically for $|x - x_0| \ll a$. The term *quadratic* in the displacement $x - x_0$ can be cancelled out in the effective trapping potential (78) by taking an external potential $V(\mathbf{r})$ containing the appropriate quadratic term. The frequency of the external potential fulfilling such a condition is

$$\omega_{\text{ext}} = \frac{\hbar k}{4am}\sqrt{1 + 1/4a^2k^2}. \tag{79}$$

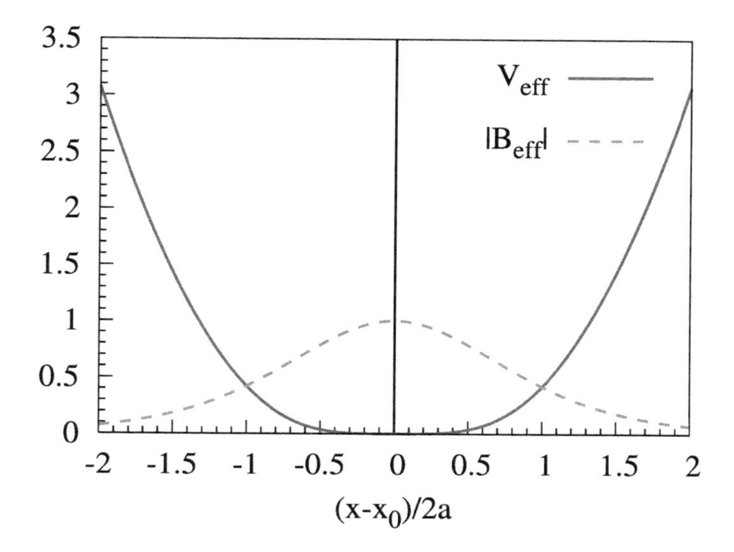

Figure 12.9 Effective trapping potential V_{eff} and effective magnetic field B_{eff} produced by counter-propagating Gaussian beams. The external harmonic potential V_{ext} cancels the quadratic term in the overall potential V_{eff}. The effective magnetic field is plotted in the units of $B(0) \equiv \hbar k/4a$, whereas the effective trapping potential is plotted in the units of $\hbar\omega_{\text{rec}}(1 + 1/4a^2k^2)$, with $\omega_{\text{rec}} = \hbar k^2/2m$.

With this, the overall effective trapping potential becomes constant up to terms of the fourth order in $x - x_0$. In the vicinity of the central point ($|x - x_0| \ll a$), the magnetic field strength is $B \approx \hbar k/4a$. The corresponding magnetic length and cyclotron frequency are $\ell_B \approx \sqrt{\hbar/B} = 2\sqrt{a/k}$ and $\omega_c = B/m \approx \hbar k/4am$. The magnetic length ℓ_B is much smaller than the relative width of the two beams, $\ell_B \ll a$, provided the latter is much larger than the optical wave length, $ak \gg 1$. In that case, many magnetic lengths fit within the interval $|x - x_0| < a$ across the beams. Furthermore, the cyclotron frequency is then approximately equal to the frequency of the external trap: $\omega_c \approx \omega_{\text{ext}}$, both of them being much less than the recoil frequency.

Figure 12.9 shows the effective trapping potential and effective magnetic field calculated using equations (77) and (78), where the external harmonic potential V_{ext} with frequency ω_{ext} (equation (79)) is added to cancel the quadratic term in the overall potential $V(\mathbf{r})$. The magnetic field is seen to be close to its maximum value in the area of constant potential, where $|x - x_0| \ll a$. For larger distances, the effective trapping potential forms a barrier, so the atoms can be trapped in the region of large magnetic field.

12.6 LIGHT-INDUCED GAUGE FIELDS FOR A TRIPOD SCHEME

12.6.1 General

Let us now consider a more complex tripod scheme [52,58] of the atom-light coupling shown in Figure 12.10, in which there is an additional third laser driving the transitions between an extra ground state 3 and the excited state 0. Assuming exact single- and two-photon resonances, the Hamiltonian of the tripod system reads in interaction representation

$$\hat{H}_0 = -\hbar\big(\Omega_1|0\rangle\langle 1| + \Omega_2|0\rangle\langle 2| + \Omega_3|0\rangle\langle 3|\big) + \text{H.c.} \tag{80}$$

The Hamiltonian $\hat{H}_0$ has two eigen-states $|D_j\rangle$ $(j = 1, 2)$ characterized by zero eigen-energies $\hat{H}_0|D_j\rangle = 0$. The eigen-states $|D_j\rangle$ are the dark states containing no excited-state contribution, as one can see in equations (82) and (83).

Parameterizing the Rabi-frequencies Ω_μ with angle and phase variables according to

$$\begin{aligned}\Omega_1 &= \Omega\sin\theta\cos\phi\, e^{iS_1},\\ \Omega_2 &= \Omega\sin\theta\sin\phi\, e^{iS_2},\\ \Omega_3 &= \Omega\cos\theta e^{iS_3},\end{aligned} \tag{81}$$

where $\Omega = \sqrt{|\Omega_1|^2 + |\Omega_2|^2 + |\Omega_3|^2}$, the adiabatic dark states read

$$|D_1\rangle = \sin\phi e^{iS_{31}}|1\rangle - \cos\phi e^{iS_{32}}|2\rangle, \tag{82}$$

$$|D_2\rangle = \cos\theta\cos\phi e^{iS_{31}}|1\rangle + \cos\theta\sin\phi e^{iS_{32}}|2\rangle - \sin\theta|3\rangle, \tag{83}$$

with $S_{ij} = S_i - S_j$ being the relative phases. As in the Λ scheme, the dark states are eigen-states of the Hamiltonian $\hat{H}_0$ with zero eigen-energy: $\hat{H}_0|D_j\rangle = 0$. The dark

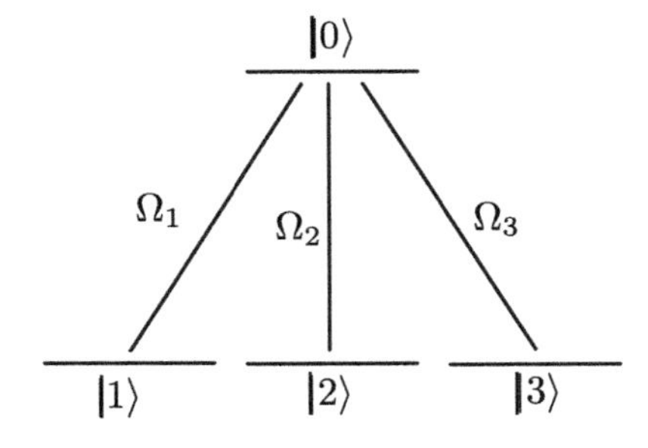

Figure 12.10 The tripod configuration.

states depend upon the atomic position through the spatial dependence of the Rabi-frequencies Ω_j. This leads to the appearance of the gauge potentials $\mathbf{A}$ and Φ considered below.

We are interested in a situation where the atoms are kept in their dark states. This can be done neglecting transitions from the dark states to the bright state $|B\rangle \sim \Omega_1^*|1\rangle + \Omega_2^*|2\rangle + \Omega_3^*|3\rangle$. The latter is coupled to the excited state $|0\rangle$ with the Rabi frequency Ω, so the two states $|B\rangle$ and $|0\rangle$ split into a doublet separated from the dark states by the energies $\pm\Omega$. The adiabatic approximation is justified if Ω is sufficiently large compared to the two-photon detuning due to the laser mismatch and/or Doppler shift. In that case, the internal state of an atom does indeed evolve within the dark state manifold. The atomic state-vector $|\Phi\rangle$ can then be expanded in terms of the dark states according to $|\Phi\rangle = \sum_{j=1}^{2} \Psi_j(\mathbf{r})|D_j(\mathbf{r})\rangle$, where $\Psi_j(\mathbf{r})$ is the wave-function for the center of mass motion of the atom in the jth dark state. Adapting the general treatment used is the section on light-induced gauge potentials for cold atoms, the atomic center of mass motion is described by a two-component wave-function

$$\Psi = \begin{pmatrix} \Psi_1 \\ \Psi_2 \end{pmatrix}, \tag{84}$$

which obeys the Schrödinger equation

$$i\hbar\frac{\partial}{\partial t}\Psi = \left[\frac{1}{2m}(-i\hbar\nabla - \mathbf{A})^2 + V + \Phi\right]\Psi, \tag{85}$$

where the potentials $\mathbf{A}$, Φ, and V are 2×2 matrices. The former $\mathbf{A}$ and Φ are light-induced gauge potentials emerging due to the spatial dependence of the atomic dark states [52]

$$\begin{aligned}
\mathbf{A}_{11} &= \hbar\left(\cos^2\phi\nabla S_{23} + \sin^2\phi\nabla S_{13}\right),\\
\mathbf{A}_{12} &= \hbar\cos\theta\left(\frac{1}{2}\sin(2\phi)\nabla S_{12} - i\nabla\phi\right),\\
\mathbf{A}_{22} &= \hbar\cos^2\theta\left(\cos^2\phi\nabla S_{13} + \sin^2\phi\nabla S_{23}\right),
\end{aligned} \tag{86}$$

and

$$\Phi_{11} = \frac{\hbar^2}{2m}\sin^2\theta\left(\frac{1}{4}\sin^2(2\phi)(\nabla S_{12})^2 + (\nabla\phi)^2\right),$$

$$\Phi_{12} = \frac{\hbar^2}{2m} \sin\theta \left(\frac{1}{2} \sin(2\phi) \nabla S_{12} - i \nabla \phi \right)$$
$$\left(\frac{1}{2} \sin(2\theta) \left(\cos^2 \phi \nabla S_{13} + \sin^2 \phi \nabla S_{23} \right) - i \nabla \theta \right),$$
$$\Phi_{22} = \frac{\hbar^2}{2m} \left(\frac{1}{4} \sin^2(2\theta) \left(\cos^2 \phi \nabla S_{13} + \sin^2 \phi \nabla S_{23} \right)^2 + (\nabla \theta)^2 \right). \tag{87}$$

Since the level scheme considered in Figure 12.10 corresponds to that of Alkali atoms where $|1\rangle$, $|2\rangle$, and $|3\rangle$ are Zeeman components of hyperfine levels, it is natural to assume that the external trapping potential is diagonal in these states and has the form $V = V_1(\mathbf{r})|1\rangle\langle 1| + V_2(\mathbf{r})|2\rangle\langle 2| + V_3(\mathbf{r})|3\rangle\langle 3|$. This still takes into account the fact that magnetic, magneto-optical, or optical dipole forces can be different or various Zeeman states. According to equation (47), the external potential in the adiabatic basis is then given by a 2×2 matrix with elements $V_{jk} = \langle D_j|V|D_k\rangle$. Using the expressions for the dark states (82) and (83), we arrive at the following matrix elements of the external potential [52]

$$V_{11} = V_2 \cos^2 \phi + V_1 \sin^2 \phi,$$
$$V_{12} = \frac{1}{2}(V_1 - V_2) \cos\theta \sin(2\phi),$$
$$V_{22} = \left(V_1 \cos^2 \phi + V_2 \sin^2 \phi \right) \cos^2 \theta + V_3 \sin^2 \theta. \tag{88}$$

At this point, it is instructive to consider some specific examples.

12.6.2 The Case where $S_{12} = 0$

Let us first assume that the laser fields that couple the levels $|1\rangle$ and $|2\rangle$ are copropagating and have the same frequency and the same orbital angular momentum (if any). In this case, their relative phase is fixed and can be put to zero $S_{12} = 0$. This leads to $S_{13} = S_{23} \equiv S$, and the expressions for the vector potential simplify to

$$\mathbf{A} = \hbar \begin{pmatrix} \nabla S & -i \cos\theta \nabla \phi \\ i \cos\theta \nabla \phi & \cos^2 \theta \nabla S \end{pmatrix}. \tag{89}$$

The components of the 2×2 matrix of the effective magnetic field can be easily evaluated to be

$$\mathbf{B}_{11} = 0,$$
$$\mathbf{B}_{12} = i\hbar \sin\theta e^{-iS} \nabla\theta \times \nabla\phi - \hbar \cos\theta e^{-iS} \nabla S \times \nabla\phi \left(1 + \cos^2 \theta \right),$$
$$\mathbf{B}_{22} = -2\hbar \cos\theta \sin\theta \nabla\theta \times \nabla S. \tag{90}$$

One recognizes that a large magnetic field requires large gradients of the relative intensities of the fields, parametrized by the angles ϕ and θ and a large gradient of the relative phase S. Gradients of ϕ and θ on the order of the wavenumber k can be achieved by using standing-wave fields. Large gradients of S can be obtained from a running wave Ω_3 orthogonal to the other two or by a vortex beam with large orbital angular momentum. In this case, magnetic fluxes as large as 1 Dirac flux quantum per atom can be reached.

We now construct a specific field configuration that leads to a magnetic monopole. For this we will consider two copropagating and circularly polarized fields $\Omega_{1,2}$ with opposite orbital angular momenta $\pm\hbar$ along the propagation axis z, whereas the third field Ω_3 propagates in x direction and is linearly polarized along the y-axis [52]:

$$\Omega_{1,2} = \Omega_0 \frac{\rho}{R} \mathrm{e}^{i(kz \mp \varphi)}, \qquad \Omega_3 = \Omega_0 \frac{z}{R} \mathrm{e}^{ik'x}. \tag{91}$$

Here ρ is the cylindrical radius, and φ is the azimuthal angle. It should be noted that these fields have a vanishing divergence and obey the Helmholtz equation. The total intensity of the laser fields (91) vanishes at the origin which is a singular point.

The vector potential associated with the fields can be calculated using equation (86):

$$\mathbf{A} = -\hbar \frac{\cos\vartheta}{r \sin\vartheta} \hat{\mathrm{e}}_\varphi \begin{pmatrix} 0 & 1 \\ 1 & 0 \end{pmatrix} + \frac{\hbar}{2} (k\hat{\mathrm{e}}_z - k'\hat{\mathrm{e}}_x)$$

$$\left[(1+\cos^2\vartheta) \begin{pmatrix} 1 & 0 \\ 0 & 1 \end{pmatrix} + (1-\cos^2\vartheta) \begin{pmatrix} 1 & 0 \\ 0 & -1 \end{pmatrix} \right]. \tag{92}$$

The first term proportional to σ_x corresponds to a magnetic monopole of the unit strength at the origin. This is easily seen by calculating the magnetic field

$$\mathbf{B} = \frac{\hbar}{r^2} \hat{\mathrm{e}}_r \begin{pmatrix} 0 & 1 \\ 1 & 0 \end{pmatrix} + \cdots. \tag{93}$$

The dots indicate nonmonopole field contributions proportional to the Pauli matrices σ_z and σ_y, and to the unity matrix.

12.7 ULTRA-RELATIVISTIC BEHAVIOR OF COLD ATOMS IN LIGHT-INDUCED GAUGE POTENTIALS

12.7.1 Introduction

In this section, we will show how cold atoms can acquire properties of ultra-relativistic fermions [54] if they are manipulated properly by light fields acting upon the atoms in the tripod configuration (see also references [51,59] for similar effects with atoms in optical lattices). Specifically, we demonstrate that by choosing certain light fields the vector potential can be made proportional to an operator of spin $1/2$. For small momenta, the atomic motion becomes then equivalent to the ultra-relativistic motion of two-component Dirac fermions, as is the case also for electrons in graphene near the Fermi surface [60–68]. In this section, we will discuss an experimental setup for observing such a quasi-relativistic behavior for the cold atoms. Furthermore, we will show that the atoms can experience negative refraction and focusing by Veselago-type lenses [69,70].

It is important to realize that the velocity of the quasi-relativistic atoms is of the order of a centimeter per second. This is ten orders of magnitude smaller than the speed of light in a vacuum $c \approx 3 \times 10^8$ m/s. For comparison, the velocity of the Dirac-type electrons in graphene is only two orders of magnitude smaller than c [63]. Thus, the ultra-relativistic behavior of cold atoms manifests itself at extremely small velocities.

12.7.2 Formulation

To demonstrate the ultra-relativistic behavior of cold atoms [54], we will consider the tripod scheme where the first two lasers have the same intensities and counterpropagate in the x-direction, while the third one propagates in the negative y-direction, as shown in Figure 12.11. Specifically, we have $\Omega_1 = \Omega \sin\theta e^{-i\kappa x}/\sqrt{2}$, $\Omega_2 = \Omega \sin\theta e^{i\kappa x}/\sqrt{2}$, and $\Omega_3 = \Omega \cos\theta e^{-i\kappa y}$, where $\Omega = \sqrt{|\Omega_1|^2 + |\Omega_2|^2 + |\Omega_3|^2}$ is the total Rabi frequency and θ is the mixing angle defining the relative intensity of the third laser field.

The potentials $\mathbf{A}$, Φ, and V have been considered in the previous section for arbitrary light fields acting upon tripod atoms. In the present configuration of the light fields, the potentials take the form [54]

$$\mathbf{A} = \hbar\kappa \begin{pmatrix} \mathbf{e}_y & -\mathbf{e}_x \cos\theta \\ -\mathbf{e}_x \cos\theta & \mathbf{e}_y \cos^2\theta \end{pmatrix}, \tag{94}$$

$$\Phi = \begin{pmatrix} \hbar^2\kappa^2 \sin^2\theta/2m & 0 \\ 0 & \hbar^2\kappa^2 \sin^2(2\theta)/8m \end{pmatrix}, \tag{95}$$

$$V = \begin{pmatrix} V_1 & 0 \\ 0 & V_1 \cos^2\theta + V_3 \sin^2\theta \end{pmatrix}, \tag{96}$$

with $\mathbf{e}_x$ and $\mathbf{e}_y$ being unit Cartesian vectors. Here the external trapping potential is assumed to be the same for the first two atomic states, $V_1 = V_2$.

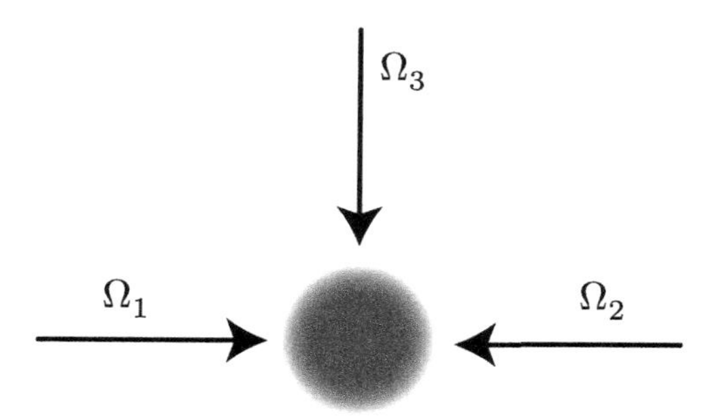

Figure 12.11 The three laser beams incident on the cloud of atoms in the tripod configuration.

In what follows, we take $V_3 - V_1 = \hbar^2\kappa^2 \sin^2(\theta)/2m$. This can be achieved by detuning the third laser from the two-photon resonance by the frequency $\Delta\omega_3 = \hbar\kappa^2 \sin^2\theta/2m$. Thus, the overall trapping potential simplifies to $V + \Phi = V_1 \mathrm{I}$ (up to a constant), where I is the unit matrix. In other words, both dark states are affected by the same trapping potential $V_1 \equiv V_1(\mathbf{r})$.

Furthermore, we take the mixing angle $\theta = \theta_0$ to be such that $\sin^2\theta_0 = 2\cos\theta_0$, giving $\cos\theta_0 = \sqrt{2} - 1$. In that case, the vector potential can be represented in a symmetric way in terms of the Pauli matrices σ_x and σ_z,

$$\mathbf{A} = \hbar\kappa'(-\mathbf{e}_x\sigma_x + \mathbf{e}_y\sigma_z) + \hbar\kappa_0\mathbf{e}_y\mathrm{I}, \tag{97}$$

where $\kappa' = \kappa\cos\theta_0 \approx 0.414\kappa$ and $\kappa_0 = \kappa(1 - \cos\theta_0)$. Although the vector potential is constant, it cannot be eliminated via a gauge transformation, because the Cartesian components A_x and A_y do not commute. Thus, the light-induced potential $\mathbf{A}$ is non-Abelian. Such a non-Abelian gauge potential can also be induced in optical lattices using other techniques [50].

12.7.3 Quasi-Relativistic Behavior of Cold Atoms

It is convenient to introduce the new dark states:

$$|D_1'\rangle = \frac{1}{\sqrt{2}}\big(|D_1\rangle + i|D_2\rangle\big)e^{i\kappa_0 y}, \tag{98}$$

$$|D_2'\rangle = \frac{i}{\sqrt{2}}\big(|D_1\rangle - i|D_2\rangle\big)e^{i\kappa_0 y}. \tag{99}$$

The transformed two component wave-function is related to the original one according to $\Psi' = \exp(-i\kappa_0 y)\exp(-i\frac{\pi}{4}\sigma_x)\Psi$, where σ_x is the Pauli spin matrix. The exponential factor $\exp(-i\kappa_0 y)$ induces a shift in the origin of the momentum $\mathbf{k} \to \mathbf{k} - \kappa_0\mathbf{e}_y$.

With the new set of dark states, we get the vector potential $\mathbf{A}' = -\hbar\kappa'\sigma_{\perp}$, where $\sigma_{\perp} = \mathbf{e}_x\sigma_x + \mathbf{e}_y\sigma_y$ is the spin $\frac{1}{2}$ operator in the xy-plane. The transformed equation of the atomic motion takes the form

$$i\hbar\frac{\partial}{\partial t}\Psi' = \left[\frac{1}{2m}(-i\hbar\nabla + \hbar\kappa'\boldsymbol{\sigma}_{\perp})^2 + V_1\right]\Psi'. \tag{100}$$

In this way, the vector potential governing the atomic motion is proportional to the spin operator $\sigma_{\perp}$.

If the trapping potential V_1 is constant, we can consider plane-wave solutions,

$$\Psi'(\mathbf{r}, t) = \Psi_{\mathbf{k}} e^{i\mathbf{k}\cdot\mathbf{r} - i\omega_{\mathbf{k}}t}, \quad \Psi_{\mathbf{k}} = \begin{pmatrix} \Psi_{1\mathbf{k}} \\ \Psi_{2\mathbf{k}} \end{pmatrix}, \tag{101}$$

where $\omega_{\mathbf{k}}$ is the eigen-frequency. The $\mathbf{k}$-dependent spinor $\Psi_{\mathbf{k}}$ obeys the stationary Schrödinger equation $H_{\mathbf{k}}\Psi_{\mathbf{k}} = \hbar\omega_{\mathbf{k}}\Psi_{\mathbf{k}}$, with the following $\mathbf{k}$-dependent Hamiltonian

$$H_{\mathbf{k}} = \frac{\hbar^2}{2m}(\mathbf{k} + \kappa'\boldsymbol{\sigma}_{\perp})^2 + V_1. \tag{102}$$

For small wave-vectors ($k \ll \kappa'$), the atomic Hamiltonian reduces to the Hamiltonian for the 2D relativistic motion of a two-component massless particle of the Dirac type,

$$H_{\mathbf{k}} = \hbar v_0 \mathbf{k}\cdot\boldsymbol{\sigma}_{\perp} + V_1 + m v_0^2, \tag{103}$$

where $v_0 = \hbar\kappa'/m$ is the velocity of such a quasi-relativistic particle. The velocity v_0 represents the recoil velocity corresponding to the wave-vector κ' and is typically of the order of a centimeter per second.

The Hamiltonian $H_{\mathbf{k}}$ commutes with the 2D chirality operator $\sigma_{\mathbf{k}} = \mathbf{k}\cdot\boldsymbol{\sigma}_{\perp}/k$. The latter is characterized by the eigenstates

$$\Psi_{\mathbf{k}}^{\pm} = \frac{1}{\sqrt{2}}\begin{pmatrix} 1 \\ \pm\frac{k_x + ik_y}{k} \end{pmatrix}, \tag{104}$$

for which $\sigma_{\mathbf{k}}\Psi_{\mathbf{k}}^{\pm} = \pm\Psi_{\mathbf{k}}^{\pm}$. The eigenstates (104) are also eigenstates of the Hamiltonian $H_{\mathbf{k}}$ with eigen-frequencies $\omega_{\mathbf{k}} \equiv \omega_{\mathbf{k}}^{\pm}$. In the following, the atomic motion is assumed to be confined in the xy-plane. The dispersion is then given by

$$\hbar\omega_{\mathbf{k}}^{\pm} = \hbar v_0\left(k^2/2\kappa' \pm k\right) + V_1 + m v_0^2, \tag{105}$$

where the upper (lower) sign corresponds to the upper (lower) dispersion branch. The atomic motion in different dispersion branches is characterized by opposite chirality if the direction $\mathbf{k}/k$ is fixed. For small wave-vectors ($k \ll \kappa'$), the dispersion simplifies to $\hbar\omega_{\mathbf{k}}^{\pm} = \pm\hbar v_0 k + V_1 + m v_0^2$, where the upper (lower) sign corresponds to a linear cone with a positive (negative) group velocity, $v_g^{\pm} = \pm v_0$. Exactly the same dispersion is featured for electrons near the Fermi level in graphene [60–64].

12.7.4 Proposed Experiment

To observe the quasi-relativistic behavior of cold atoms, the following experimental situation has been proposed [54]. Suppose that initially an atom (or a dilute atomic cloud) is in the internal state $|3\rangle$ with a translational motion described by a wave-packet with a central wave-vector $\mathbf{k}_{\text{in}}$ and a wave-vector spread $\Delta k \ll k_{\text{in}}$. The full initial state-vector is then given by $|\Psi_{\text{in}}\rangle = \psi(\mathbf{r})e^{i\mathbf{k}_{\text{in}}\cdot\mathbf{r}}|3\rangle$, where the envelope function $\psi(\mathbf{r})$ varies slowly within the wavelength $\lambda_{\text{in}} = 2\pi/k_{\text{in}}$. The cold atoms can be set in motion using various techniques, e.g., by means of the two-photon scattering that induces a recoil momentum $\hbar\mathbf{k}_{in} = \hbar\mathbf{k}_{\text{2phot}}$ to the atoms, where $\mathbf{k}_{\text{2phot}}$ is a wave-vector of the two-photon mismatch.

Initially, all three lasers are off. Subsequently, the lasers are switched on in a counterintuitive manner, switching the lasers 1 and 2 on first, followed by the laser 3. At the beginning of this stage, the internal state $|3\rangle$ coincides with the dark state $|D_2\rangle$, so the full initial state-vector can be rewritten as $|\Psi_{\text{in}}\rangle = \psi(\mathbf{r})e^{i\mathbf{k}_{\text{in}}\cdot\mathbf{r}}|D_2\rangle$. If the laser 3 is switched on sufficiently slowly, the atom remains in the dark state $|D_2\rangle$ during the whole switch-on stage. Yet the duration of the switching on should be sufficiently short to prevent the dynamics of the atomic center of mass at this stage. Immediately after the lasers reach their steady state, the multicomponent wave-function reads:

$$\Psi = \begin{pmatrix} 0 \\ 1 \end{pmatrix} \psi(\mathbf{r})e^{i\mathbf{k}_{\text{in}}\cdot\mathbf{r}}. \tag{106}$$

Expressing $|D_2\rangle$ as a function of $|D'_{1,2}\rangle$, the transformed multicomponent wave-function takes the form

$$\Psi' = \frac{1}{\sqrt{2}} \begin{pmatrix} -i \\ 1 \end{pmatrix} \psi(\mathbf{r})e^{i\mathbf{k}_c\cdot\mathbf{r}}, \tag{107}$$

where $\mathbf{k} = \mathbf{k}_{\text{in}} - \kappa_0\mathbf{e}_y$ is the central wave-vector.

Let us now consider the subsequent atomic dynamics in the laser fields. As mentioned earlier in this section, to have ultra-relativistically behaving atoms, the wave-number k should be small ($k \ll \kappa$) so that $\mathbf{k}$ is a small addition to $\mathbf{k}_{\text{in}} = \kappa_0\mathbf{e}_y + \mathbf{k}$. Furthermore, the wave-vector spread $\Delta k \ll k$, i.e., the width of the atomic wave-packet, is much larger than the central wavelength. Hence, the dynamics is sensitive to the direction of the central wave-vector $\mathbf{k}$. To illustrate this we will consider two specific cases.

(i) If $\mathbf{k} = \pm k\mathbf{e}_y$, the wave-function (107) can be represented as:

$$\Psi' = -i\Psi_{\mathbf{k}}^{\pm}\psi(\mathbf{r})e^{\pm iky}, \quad \Psi_{\mathbf{k}}^{\pm} = \frac{1}{\sqrt{2}} \begin{pmatrix} 1 \\ i \end{pmatrix}. \tag{108}$$

The upper (lower) sign in $\mathbf{k} = \pm k\mathbf{e}_y$ corresponds to a situation where the atom is characterized by a positive (negative) chirality, hence being in the upper (lower) dispersion branch. In both cases, the atomic wave-packet propagates along the y axis with the velocity $\mathbf{v}_0 = \mathbf{e}_y \hbar\kappa'/m$.

(ii) If the wave-vector is along the x-axis ($\mathbf{k} = k\mathbf{e}_x$), the multicomponent wave-function (107) takes the form

$$\Psi' = \left(c_+ \Psi_\mathbf{k}^+ + c_- \Psi_\mathbf{k}^-\right)\psi(\mathbf{r})e^{i\mathbf{k}\cdot\mathbf{r}}, \quad \Psi_\mathbf{k}^\pm = \frac{1}{\sqrt{2}}\begin{pmatrix} 1 \\ \pm 1 \end{pmatrix}, \tag{109}$$

where $c_\pm = (-i \pm 1)/2$. In this case, the initial wave-packet splits into two with equal weights ($|c_\pm^2| = 1/2$) and the same wave-vector $\mathbf{k}$. The two wave-packets are characterized by the different chiralities and thus move in opposite directions. The wave-packet with a positive chirality (plus sign in $\Psi_\mathbf{k}^\pm$) belongs to the upper dispersion branch and moves along the x-axis with a velocity $\mathbf{v}_0 = \mathbf{e}_x \hbar\kappa'/m$. On the other hand, the wave-packet characterized by a negative chirality (minus sign in $\Psi_\mathbf{k}^\pm$) moves with a velocity $\mathbf{v}_0 = -\mathbf{e}_x \hbar\kappa'/m$.

Suppose the time is sufficiently small ($v_0 t < d$) so the wave-packets of the width d are not yet spatially separated. The internal atomic state will then undergo temporal oscillations between the dark states $|D_2\rangle$ and $|D_1\rangle$, with a frequency equal to $\omega_\mathbf{k}^+ - \omega_\mathbf{k}^- = 2v_0 k$. Such an internal dynamics can be detected by switching the laser 3 off at a certain time. This transforms the dark state $|D_2\rangle$ to the physical state $|3\rangle$. Subsequently, one can measure the population of the state $|3\rangle$ for various delay times and various wave-vectors $\mathbf{k}$. The chiral nature of the atomic motion will manifest itself in the oscillations of the population of the atomic state $|3\rangle$ if $\mathbf{k}$ is along the x-axis, and the absence of such oscillations if $\mathbf{k}$ is along the y-axis.

Furthermore, as a consequence of the constructed Hamiltonian (103), the quasi-relativistic atoms can show negative refraction at a potential barrier and thus exhibit focusing by Veselago-type lenses [69,70]. Consider incident atoms that are in the upper dispersion branch and propagate along the y-axis with a wave-vector $\mathbf{k} = k\mathbf{e}_y$. Let us place a potential barrier of a height $2\hbar v_0 k$ at an angle of incidence α, as shown in Figure 12.12. Inside the barrier, the atoms are transferred to the lower dispersion branch with $\mathbf{k}_t = -k[\cos(2\alpha)\mathbf{e}_y + \sin(2\alpha)\mathbf{e}_x]$. This would lead to the negative refraction of cold atoms at the barrier as shown in Figure 12.12. Thus, the potential barrier can act as a flat lens that refocuses the atomic wave-packet.

In this way, we have shown how the atomic motion can be equivalent to the dynamics of ultra-relativistic (massless) two-component Dirac fermions. As a result, the ultracold atoms can experience negative refraction and focusing by Veselago-type lenses. In addition, the chiral nature of the atomic motion is manifested through dy-

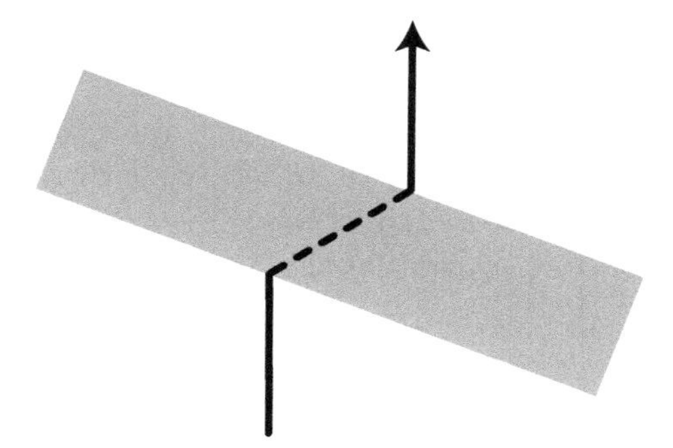

Figure 12.12 Negative refraction of cold atoms at a potential barrier. The incoming and outgoing atoms are in the upper dispersion branch with a wave-vector $\mathbf{k}$ (solid line), whereas the atoms inside the barrier are in a lower dispersion branch with a wave-vector $\mathbf{k}_t$ (dashed line).

namics of the population of the internal atomic states, which is highly sensitive to the direction of the center of mass motion.

12.8 FINAL REMARKS

In this chapter, we have considered different types of manipulation of cold atoms by light fields. We have reviewed the mechanisms for trapping ensembles of ultracold atoms. The optical trap formed the basis for manipulating the cold atoms where we relied upon on the coherent nature of the ultracold sample of atoms and the intensity of the light.

Subsequently, we have discussed a situation where the phase and the intensity of the incident light both play a crucial role, i.e., we have considered a different kind of optical manipulation where the laser fields are applied to induce vector and scalar potentials for the atoms. The induced potentials have a geometric nature and depend exclusively upon the relative intensity and relative phase of the laser beams involved rather on their absolute intensity. The approach relies on the ability to prepare the atoms in superpositions of the internal energy states of the atom.

The technique provides a way to optically produce an effective magnetic field acting upon electrically neutral atoms. This happens if the applied laser fields have a nontrivial topology (e.g., if they carry an orbital angular momentum along the propagation direction [5–7,34,35]) or if the atom-light system contains more than one degenerate dark state. The latter situation appears in the tripod configuration of the light-atom system exhibiting two degenerate dark states. Consequently, the light-induced potentials are 2×2 matrices, whose Cartesian components generally do not commute [52,

54], i.e., the potentials are non-Abelian. In such a situation, nontrivial light-induced gauge potentials can be produced, even using plane waves.

Finally, we noted that the tripod scheme of the light-matter coupling can have other important applications, e.g., it can be used to produce solitons in atomic Bose–Einstein condensates [71]. Using this method, it is possible to circumvent the restriction set by the diffraction limit inherent to conventional methods, such as the phase imprinting [24,25].

REFERENCES

[1] J.D. Jackson, Classical Electrodynamics, Wiley, New York, 1998.

[2] P. Meystre, Atom Optics, American Institute of Physics, 2001.

[3] C.J. Pethick, H. Smith, Bose–Einstein Condensation in Dilute Gases, Cambridge Univ. Press, Cambridge, 2001.

[4] L.P. Pitaevskii, S. Stringari, Bose–Einstein Condensation, Clarendon Press, Oxford, 2003.

[5] G. Juzeliūnas, P. Öhberg, Slow light in degenerate Fermi gases, *Phys. Rev. Lett. 93* (2004) 033602.

[6] G. Juzeliūnas, P. Öhberg, J. Ruseckas, A. Klein, Effective magnetic fields in degenerate atomic gases induced by light beams with orbital angular momenta, *Phys. Rev. A 71* (2005) 053614.

[7] G. Juzeliūnas, J. Ruseckas, P. Öhberg, Effective magnetic fields induced by eit in ultra-cold atomic gases, *J. Phys. B 38* (2005) 4171.

[8] S. Stenholm, The semiclassical theory of laser cooling, *Rev. Mod. Phys. 58* (1986) 699.

[9] R.J. Cook, Atomic motion in resonant radiation: An application of Ehrenfest's theorem, *Phys. Rev. A 20* (1979) 224.

[10] C. Cohen-Tannoudji, J. Dupont-Roc, G. Grynberg, Atom–Photon Interactions, Wiley, New York, 1998.

[11] S.M. Barnett, P.M. Radmore, Methods in Theoretical Quantum Optics, Clarendon Press, Oxford, 1997.

[12] R. Loudon, The Quantum Theory of Light, Oxford Univ. Press, 1979.

[13] M.H. Anderson, J.R. Ensher, M.R. Matthews, C.E. Wieman, E.A. Cornell, Observation of Bose–Einstein condensation in a dilute atomic vapor, *Science 269* (1995) 198.

[14] K.B. Davis, M.-O. Mewes, M.R. Andrews, N.J. van Druten, D.S. Durfee, D.M. Kurn, W. Ketterle, Bose–Einstein condensation in a gas of sodium atoms, *Phys. Rev. Lett. 75* (1995) 3969.

[15] C.C. Bradley, C.A. Sackett, J.J. Tollett, R.G. Hulet, Evidence of Bose–Einstein condensation in an atomic gas with attractive interactions, *Phys. Rev. Lett. 75* (1995) 1687.

[16] J. Stenger, S. Inouye, M. Andrews, H.-J. Miesner, D. Stamper-Kurn, W. Ketterle, Strongly enhanced inelastic collisions in a Bose–Einstein condensate near Feshbach resonances, *Phys. Rev. Lett. 82* (1999) 2422.

[17] S.N. Bose, Plancks Gesetz und Lichtquantenhypothese, *Z. Phys. 26* (1924) 178.

[18] A. Einstein, Quantentheorie des einatomigen Idealen Gases, *Sitzungsber. Kgl. Preuss. Akad. Wiss. 3* (1924) 261.

[19] F. London, The λ-phenomenon of liquid helium and the Bose–Einstein degeneracy, *Nature 141* (1938) 643.

[20] K. Huang, Statistical Mechanics, Wiley, New York, 1987.

[21] D. Kleppner, in: M. Inguscio, S. Stringari, C. Wieman (Eds.), *Bose–Einstein Condensation in Atomic Gases: Enrico Fermi International School of Physics*, IOS Press, US, 1999.

[22] A.S. Arnold, C.S. Garvie, E. Riis, Large magnetic storage ring for Bose–Einstein condensates, *Phys. Rev. A 73* (2006) 041606.

[23] G. Baym, C.J. Pethick, Ground-state properties of magnetically trapped Bose-condensed rubidium gas, *Phys. Rev. Lett. 76* (1996) 6.

[24] S. Burger, K. Bongs, S. Dettmer, W. Ertmer, K. Sengstock, A. Sanpera, G.V. Shlyapnikov, M. Lewenstein, Dark solitons in Bose–Einstein condensates, *Phys. Rev. Lett. 83* (1999) 5198.

[25] J. Denschlag, J.E. Simsarian, D.L. Feder, C.W. Clark, L.A. Collins, J. Cubizolles, L. Deng, E.W. Hagley, K. Helmerson, W.P. Reinhardt, S.L. Rolston, B.I. Schneider, W.D. Phillips, Generating solitons by phase engineering of a Bose–Einstein condensate, *Science 287* (2000) 97.

[26] M.D. Girardeau, E.M. Wright, Breakdown of time-dependent mean-field theory for a one-dimensional condensate of impenetrable bosons, *Phys. Rev. Lett. 84* (2000) 5239.

[27] G. Whyte, P. Öhberg, J. Courtial, Transverse laser modes in Bose–Einstein condensates, *Phys. Rev. A 69* (2004) 053610.

[28] D.R. Murray, P. Öhberg, Matter wave focusing, *J. Phys. B 38* (2005) 1227.

[29] E. Arimondo, Coherent population trapping in laser spectroscopy, *Prog. Opt. 35* (1996) 259.

[30] S.E. Harris, Electromagnetically induced transparency, *Phys. Today 50* (1997) 36.

[31] A.B. Matsko, O. Kocharovskaja, Y. Rostovtsev, G.R. Welch, A.S. Zibrov, M.O. Scully, Slow, ultraslow, stored, and frozen light, *Adv. At. Mol. Opt. Phys. 46* (2001) 191.

[32] M.D. Lukin, Trapping and manipulating photon states in atomic ensembles, *Rev. Mod. Phys. 75* (2003) 457.

[33] M. Fleischhauer, A. Imamoglu, J.P. Maragos, Electromagnetically induced transparency: Optics in coherent media, *Rev. Mod. Phys. 77* (2005) 633.

[34] G. Juzeliūnas, P. Öhberg, Creation of effective magnetic fields using electromagnetically induced transparency, *Opt. Spectroscopy 99* (2005) 357.

[35] P. Öhberg, G. Juzeliūnas, J. Ruseckas, M. Fleischhauer, Filled landau levels in neutral quantum gases, *Phys. Rev. A 72* (2005) 053632.

[36] M.V. Berry, Quantal phase factors accompanying adiabatic changes, *Proc. R. Soc. A 392* (1984) 45.

[37] C.A. Mead, The geometric phase in molecular systems, *Rev. Mod. Phys. 64* (1992) 51.

[38] F. Wilczek, A. Zee, Appearance of gauge structure in simple dynamical systems, *Phys. Rev. Lett. 52* (1984) 2111.

[39] J. Moody, A. Shapere, F. Wilczek, Realizations of magnetic-monopole gauge fields: Diatoms and spin precession, *Phys. Rev. Lett. 56* (1986) 893.

[40] R. Jackiw, Berry's phase-topological ideas from atomic, molecular and optical physics, *Comments At. Mol. Phys. 21* (1988) 71.

[41] C.-P. Sun, M.-L. Ge, Generalizing Born–Oppenheimer approximations and observable effects of an induced gauge field, *Phys. Rev. D 41* (1990) 1349.

[42] R. Dum, M. Olshanii, Gauge structures in atom–laser interaction: Bloch oscillations in a dark lattice, *Phys. Rev. Lett. 76* (1996) 1788.

[43] P.M. Visser, G. Nienhuis, Geometric potentials for subrecoil dynamics, *Phys. Rev. A 57* (1998) 4581.

[44] V. Bretin, S. Stock, Y. Seurin, J. Dalibard, Fast rotation of a Bose–Einstein condensate, *Phys. Rev. Lett. 92* (2004) 050403.

[45] V. Schweikhard, I. Coddington, P. Engels, V.P. Mogendorff, E.A. Cornell, Rapidly rotating Bose–Einstein condensates in and near the lowest landau level, *Phys. Rev. Lett. 92* (2004) 040404.
[46] M.A. Baranov, K. Osterloh, M. Lewenstein, Fractional quantum hall states in ultracold rapidly rotating dipolar Fermi gases, *Phys. Rev. Lett. 94* (2004) 070404.
[47] D. Jaksch, P. Zoller, Creation of effective magnetic fields in optical lattices: The Hofstadter butterfly for cold neutral atoms, *New J. Phys. 5* (2003) 56.
[48] E.J. Mueller, Artificial electromagnetism for neutral atoms: Escher staircase and laughlin liquids, *Phys. Rev. A 70* (2004) 041603.
[49] A.S. Sørensen, E. Demler, M.D. Lukin, Fractional quantum hall states of atoms in optical lattices, *Phys. Rev. Lett. 94* (2005) 086803.
[50] K. Osterloh, M. Baig, L. Santos, P. Zoller, M. Lewenstein, Cold atoms in non-Abelian gauge potentials: From the Hofstadter moth to lattice gauge theory, *Phys. Rev. Lett. 95* (2005) 010403.
[51] J. Ruostekoski, G.V. Dunne, J. Javanainen, Particle number fractionalization of an atomic Fermi–Dirac gas in an optical lattice, *Phys. Rev. Lett. 88* (2002) 180401.
[52] J. Ruseckas, G. Juzeliūnas, P. Öhberg, M. Fleischhauer, Non-Abelian gauge potentials for ultracold atoms with degenerate dark states, *Phys. Rev. Lett. 95* (2005) 010404.
[53] G. Juzeliūnas, J. Ruseckas, P. Öhberg, M. Fleischhauer, Light-induced effective magnetic fields for ultracold atoms in planar geometries, *Phys. Rev. A 73* (2006) 025602.
[54] G. Juzeliunas, J. Ruseckas, M. Lindberg, L. Santos, P. Öhberg, Quasi-relativistic behavior of cold atoms in light fields, *Phys. Rev. A 77* (2008) 011802(R).
[55] L.V. Hau, Z. Dutton, C. Behrooz, Light speed reduction to 17 metres per second in an ultracold atomic gas, *Nature 397* (1999) 594.
[56] L. Allen, M. Padgett, M. Babiker, The orbital angular momentum of light, *Prog. Opt. 39* (1999) 291.
[57] L. Allen, S.M. Narnett, M. Padgett, Optical Angular Momentum, Institute of Physics, Bristol, 2003.
[58] R.G. Unanyan, M. Fleischhauer, B.W. Shore, K. Bergmann, Robust creation and phase-sensitive probing of superposition states via stimulated Raman adiabatic passage (stirap) with degenerate dark states, *Opt. Commun. 155* (1998) 144.
[59] S.L. Zhu, B. Wang, L.-M. Duan, Simulation and detection of Dirac fermions with cold atoms in an optical lattice, *Phys. Rev. Lett. 98* (2007) 260402.
[60] K.S. Novoselov, A.K. Geim, S.V. Morozov, D. Jiang, M.I. Katsnelson, I.V. Grigorieva, S.V. Dubonos, A.A. Firsov, Two-dimensional gas of massless Dirac fermions in graphene, *Nature 438* (2005) 197.
[61] E. McCann, V.I. Fal'ko, Landau-level degeneracy and quantum hall effect in a graphite bilayer, *Phys. Rev. Lett. 96* (2006) 086805.
[62] M.I. Katsnelson, K.S. Novoselov, A.K. Geim, Chiral tunneling and the Klein paradox in graphene, *Nature Phys. 2* (2006) 620.
[63] A.K. Geim, K.S. Novoselov, The rise of graphene, *Nature Mater. 6* (2007) 183.
[64] K.S. Novoselov, Room-temperature quantum hall effect in graphene, *Science 315* (2007) 1379.
[65] A. Matulis, F.M. Peeters, Appearance of enhanced Weiss oscillations in graphene: Theory, *Phys. Rev. B 75* (2007) 125429.
[66] R. Jackiw, S.-Y. Pi, Chiral gauge theory for graphene, *Phys. Rev. Lett. 98* (2007) 266402.
[67] J.B. Pendry, Negative refraction for electrons?, *Science 315* (2007) 1226.
[68] V.V. Cheianov, V. Fal'ko, B.L. Altshuler, The focusing of electron flow and a veselago lens in graphene $p-n$ junctions, *Science 315* (2007) 1252.

[69] V.G. Veselago, The electrodynamics of substances with simultaneously negative values of ε and μ, *Sov. Phys. Usp. 10* (1968) 509.

[70] J.B. Pendry, Negative refraction makes a perfect lens, *Phys. Rev. Lett. 85* (2000) 3966.

[71] G. Juzeliūnas, J. Ruseckas, P. Öhberg, M. Fleischhauer, Formation of solitons in atomic Bose–Einstein condensates by dark-state adiabatic passage, *Lithuanian J. Phys. 47* (2007) 357.

Index

A

Absorbing particles, 227, 231, 240
Absorption, 82, 84, 112, 122, 126, 172, 179, 218, 219, 241, 258, 303, 312
Acousto-optic deflectors, 114, 158–160
Adiabatic condition, 313, 314, 316, 318
Amplitudes, 1, 7, 63, 64, 68–72, 146, 152, 153, 173, 226, 230, 253, 254, 276, 280, 296, 297, 314–316
  complex, 143–145, 149, 153
Angle of convergence, 206–208, 212, 214, 216, 227
Angular momentum, 4–9, 16, 19, 20, 24, 59–63, 162, 163, 198–205, 210, 216–222, 226–228, 237, 238, 243, 249–252, 278
  content, 221, 224, 252, 253, 256
  density, 5, 6, 19, 200, 201, 229
  states, 241, 244, 286
  transfer, 218, 238, 240, 241, 245, 246, 249, 252
  transport of, 199, 202, 203, 206
Angular range, 188–190, 279
Angular velocity, 45–47, 56, 252
Aperture, 33, 37, 108, 110, 124, 143, 144, 239, 240, 255
Atomic motion, 298, 316, 324, 326, 328
Atomic states, 298, 324, 328, 329
Atoms
  dark-state, 312
  excited, 313, 314
  trapped, 170, 171, 296, 297, 317
Axial trapping, 211, 212, 214
Axis, 5, 6, 14, 20–22, 31, 32, 38, 39, 54, 59, 88, 90, 100, 178, 214, 215, 242, 287
Azimuthal
  angle, 45, 46, 51, 315, 323
  index, 30, 183, 204, 251, 254, 261

B

Beam
  axis, 2, 5, 7, 14, 37–39, 65, 74, 204, 210, 211, 214–216, 228–231, 240–242, 246, 279
  diameter, 143, 144, 241
  profile, 37, 71, 72, 171, 204, 206, 228, 276, 279
  waist, 3, 34, 122, 180, 186, 187, 283, 318
Beams, 1, 2, 4–7, 35–38, 46–48, 64–72, 142–146, 180–183, 206–219, 221, 226–230, 237–242, 252–256, 261–263, 315–319
  asymmetric, 239, 240
  azimuthally phased, 2
  bottle, 142
  classical, 37, 276
  collinear, 65, 69, 71, 72
  diffracted, 9, 11, 73, 143
  helically phased, 2, 8, 10
  hollow, 195, 211
  monochromatic, 34, 36, 38, 48, 196
  multiple, 145, 181, 182, 191, 192
  nonintegral, 73, 74
  nonparaxial, 198, 199, 204
  polychromatic, 42, 43
  rotating, 46, 47
Bessel beams, 7, 8, 31, 33, 68, 146, 208, 210, 241
Birefringent particles, 245, 252
Bose–Einstein condensate (BEC), 81, 170, 171, 285, 295, 299–306, 308, 316, 330

C

Cells, 100, 122, 125, 140, 169, 239, 264, 265, 268
Chiral particles, 220, 222
Chiral rotor, 222

Circular polarization, 4, 21, 26, 27, 36, 38–40, 44, 51, 209, 215, 216, 226, 252, 255, 256, 265
Cold atoms, 170, 295, 308, 321, 323–325, 327–329
Colloidal
  dispersions, 120, 131
  particles, 96, 115, 120, 122, 129, 159
Colloids, 117–119, 132
Complex electric field, 43, 45, 149
Complex fields, 33, 34, 41, 43
Conservation laws, 199, 201
Counterpropagating
  beam traps, 142, 161
  beams, 81, 172, 182, 183, 185, 191, 317
Critical radius, 215, 216
Crystal, photonic, 131, 132
Cube, polarizing beam splitter, 255, 256

D

Dark solitons, 305–307
Dark states, 310, 312–314, 318, 320–322, 325–328
  degenerate, 311, 329
Density matrix, 174, 289, 298
Dielectric resonator, 116, 119, 120, 130–132
Diffracted orders, 66–68, 73, 261, 262
Diffraction orders, 11, 148, 275
Diffractive optical element (DOE), 142, 144–147
Dipole, 83, 295–299, 311
  approximation, 83, 86, 296
  force, 142, 174, 175, 179, 299
  method, coupled, 113, 132
Direct search algorithm (DS), 156–158, 167
Doppler forces, 172, 173, 175, 177, 179
Doppler shift, 180, 181

E

Effective magnetic field, 296, 308, 314, 316, 317, 319, 322, 329
Eigenstates, 21, 51, 52, 54, 56, 305, 312, 326
  circular, 51, 52
Eigenvalues, 9, 20, 49, 52, 55, 56, 154
Electric field, 5, 8, 22–24, 28, 30–35, 40–42, 46, 80, 121, 124, 149, 150, 186, 226, 251
Electromagnetic
  angular momentum, 199
  field, 5, 21, 22, 82, 86, 94, 98, 100, 101, 103, 112, 135
Electronic states, 310, 312, 313
  excited, 312
Energy density, 21, 27, 29, 31, 35, 37, 38, 43, 45–47
Energy shift, 82, 84, 87, 92, 94
  induced, 85–87, 89–91, 93
Energy–momentum tensor, 200, 201
Ensemble, 99, 141, 149, 185, 295, 311
Entangled states, 272, 283, 284, 286, 288
Entanglement, 9, 272, 275, 276, 284, 285, 287
Entrance pupil, 143, 144
Equator, 53, 54
Evanescent
  fields, 107, 109, 110, 112, 113, 115, 117, 121, 127, 161, 162
  waves, 107–109, 122–126, 158
Eye fluid, 265, 266

F

Fermions, 299, 300
Field enhancement, 116, 119–122
Field operators, 27, 41, 302, 303
Fluid, 63, 117, 128, 245, 246, 249, 252, 253, 257–260, 264, 266–268
Forks, 66, 70, 73
Fourier
  lens, 143
  plane, 143, 144, 148, 150, 161
  transform, 143, 144, 148, 276, 279
Frequencies, corner, 159
Fresnel holographic optical trapping, 144, 145
Fringes, 2, 99, 115, 123

G

Gauge potential, 308–310, 314, 321
  light-induced, 308, 311, 316, 321, 323
Gaussian beam trap, 211–214, 227

Gaussian beams, 9, 143, 144, 198, 206, 212, 213, 222, 226, 228, 239, 262, 299, 318
polarized, 222, 230
Generalized adaptive additive algorithm (GAA), 155, 157, 158
Generalized Lorenz–Mie theory (GLMT), 112, 197, 198
Generalized phase contrast (GPC), 158, 160, 161
Gerchberg–Saxton algorithms, 153–158
weighted (GSW), 155–158
Gold nanoparticles, 113, 126–128, 142
Gouy phase, 9, 49–51, 59, 65, 74
Gratings, 11, 68, 144, 146, 148, 218
Gross–Pitaevskii equation, 303–305

H

Hamiltonian, 20, 27, 41, 49, 51, 296, 301, 310–312, 320, 326
Helmholtz wave equation, 27, 30
Hermite–Gaussian
beams, 9, 45, 98
modes, 2, 3, 9, 10, 12–14, 49, 50, 52–54, 239
Hermite–Laguerre sphere, 53–55
Holograms, 9, 11, 143–147, 149, 152, 153, 155–157, 222, 225, 261–263, 275–277, 279–281, 286, 289
Holographic
optical traps, 98, 144, 149, 153, 155, 157, 159
optical tweezers (HOT), 139, 147–149, 157, 158, 162, 245

I

Incident
beam, 113, 116, 121, 125, 222, 226, 243, 252
field, 121, 130, 196, 197, 202
laser beams, 121, 315
modes, 220, 221
Intensity
distribution, 54, 145, 146, 153
modulations, 158, 160
Interference, 65, 67, 99, 123, 158, 277
patterns, 9, 10, 65, 66, 70, 73, 129, 145, 222, 239, 240, 285
Interparticle forces, 79, 81

K

Kretschmann geometry, 109, 114, 115, 123

L

Λ scheme, 311, 313–315, 317, 319, 320
Ladder operators, 51–53, 57, 173
Lagrangian density, 40, 199, 200
Laguerre–Gaussian
beams, 6, 7, 9, 64, 65, 68, 69, 71, 96, 97, 130, 170, 171, 182, 183, 191, 205, 225–227, 231, 239–242
counterpropagating, 183, 185
superposition of, 69, 74
modes, 2–4, 9–14, 37, 40, 44, 45, 51–57, 66, 69, 175, 222, 242, 246, 251, 253, 254
Laser
beams, 2, 11, 110, 172, 237, 239, 242, 249–251, 255, 257, 296, 308, 310, 312
cooling, 172, 191, 295, 300
fields, 8, 173, 296, 299, 308, 312, 313, 322, 323, 327, 329
Lasers, 1, 33, 84, 109, 125, 126, 142, 143, 147, 153, 159, 295, 296, 299, 300, 310, 324, 327, 328
Lenses, 114, 115, 143, 144, 160, 198, 205, 242
LG, *see* Laguerre–Gaussian
Light beams, 2, 3, 5, 6, 19, 21, 22, 33, 60, 148, 149, 171, 172, 181, 277, 278, 280, 295, 304, 315
twisted, 186, 187
Light fields, 5, 19, 21, 33, 43, 64, 108, 112, 114, 127, 281, 282, 296, 309, 323, 324
Linear momentum, 5, 6, 19, 237, 238, 285
Linear polarization, 12, 13, 32, 36, 38, 47, 48, 287
direction of, 32, 39, 40, 221
position-dependent, 39, 40, 46
Liquid crystals, 38, 143, 147, 149, 170, 171, 186, 187, 191

twisted nematic, 188–190
Liquid-crystal (LC) layer, 147, 148
Lobes, 1, 70, 71
Loops, 74, 175, 177, 178
Lorenz–Mie solution, 196, 197

M

Magnetic fields, 5, 22–24, 28, 31, 34, 42, 43, 139, 226, 310, 315, 317, 319, 323
Mathieu beams, 7
Maxwell's equations, 19, 21, 24, 27, 28, 33, 34, 60
Methanol, 252, 256, 257
Micelle, 264, 265
Micromachines, optically driven, 140, 141, 219, 221–223, 243, 246
Micrometer-sized particles, 117, 122
Microparticles, 79, 96, 109, 110, 114, 126
Microscope slide, 262, 263, 265, 266
Microscopic particles, 131, 169, 195, 237, 249
Microviscometer, 250, 254, 260, 264, 266
Mie particles, 113, 130
Mie theory, 112
Modal composition of beam, 261–263
Mode
converter, cylindrical lens, 10, 13
functions, 6, 37, 38, 42, 46, 48, 50, 51, 53
indices, 57, 203, 204, 240
pattern, 38, 42, 46, 47, 51, 55, 56
pictures, 262, 263
transformations, 52, 68
Modes
high-order, 66, 68, 74, 240
plasmon, 117, 119
Momentum density, 6, 20–22, 27, 31, 34, 35
Monochromatic paraxial beams, 33, 50, 60
Multipole fields, 21, 27–29, 60
cylindrical, 19, 22, 30, 33

N

Nanoparticles, 81, 83, 96, 97, 101, 110, 130, 132
Near-field
optical micromanipulation, 107, 108, 110, 111, 116, 119, 124, 125, 127, 129, 136
optics, 107, 108, 114
Neutral atoms, 295, 296, 308, 329
Nonmonochromatic paraxial beam, 42
Nonparaxial optical vortices, 202, 205

O

OAM, *see* Orbital angular momentum
Objects, trapped, 108, 111, 117, 239, 241–243, 253, 254
Optical
beams, 72, 74, 170
binding, 80, 81, 98–102, 129, 130, 133, 134, 158, 169
forces, 98, 99, 101
crystals, 99, 100
element, diffractive, 142, 146, 224
fields, 9, 63, 64, 112, 125, 131, 139, 142, 145, 158, 161
forces, 79, 99, 109, 111–113, 116, 117, 119, 122, 126, 130, 139–141, 191, 198, 224, 250
lattices, 141, 144, 158, 308, 324, 325
manipulation, 169, 170, 295, 308, 329
molasses, 170, 172, 179, 181, 182, 191
spatial solitons, 130–132
torque, 140, 175, 218–221, 225, 251, 253, 257, 258, 268
trapping, 81, 98, 110, 111, 114–116, 119, 124, 129, 144, 145, 149, 195, 196, 198, 210, 237
traps, 97–99, 101, 110, 111, 114, 123, 124, 130, 140, 141, 146, 149, 160–162, 195, 249, 250, 261, 262, 304, 305
tweezers, 11, 79, 99, 112, 114, 116, 141–145, 161, 162, 211, 219, 237–246, 249–252, 254, 262, 263
vortex, 2, 4, 12, 19, 21, 55, 63, 64, 66–68, 72, 96, 97, 203, 204, 207, 209–212, 226–231
Optics, singular, 63, 64, 69
Orbital angular momentum, 20–22, 25, 36–38, 45, 46, 48, 51, 52, 55, 60, 170–172, 181, 182, 271–273, 275–281, 283, 285–287
conservation of, 275

content, 174, 181, 231, 254
density of, 36–38, 45
effects of, 171, 172, 191
exchange, 171, 172
in quantum communication, 271
modes, 271, 272, 277, 278, 284
of photons, 278
total, 5, 204
Orbital torque, 215, 225

P

Pair, entangled, 287, 289
Paraxial
light beams, 22, 57
optical vortex, 202, 203
beam, 203
wave equation, 33, 37, 42–44, 48–51, 64, 74, 81
Particle symmetry, effect of, 218
Particles
cylindrical, 81, 88, 94
elongated, 221
entangled, 275
interacting, 119, 123
nonspherical, 196
spinning, 266, 268
symmetric, 90, 92, 93, 219
Phase
factor, 39, 56, 58
fronts, 4, 5
geometric, 66, 68
hologram, 70, 146, 148, 154, 261, 262
imprinting, 303–306, 308, 330
levels, 147, 156
modulation, 146, 149, 153, 160
relative, 123, 221, 296, 308, 314, 320, 322, 323, 329
shift, 12–14, 74, 147, 149, 225, 277
singularities, 2, 19, 21, 22, 30, 37, 38, 54, 64, 68, 210, 242, 280
structure, 2–4, 7, 242, 280
vortex, 32, 226
Photon spin, 170, 203
Photonic crystal structures, 131
Photonics, 107, 108, 112, 116, 121
Photons, 26, 28, 29, 40, 44, 46, 47, 73, 82–84, 131, 228, 238, 252, 274–278, 283–287, 290
down-converted, 275, 276, 287
emitted, 79, 275, 276
entangled, 275, 276, 283, 286
laser, 84, 85
lower-frequency, 46
virtual, 82–84
Poincaré sphere, 13, 53, 55
Polarizability, 92, 94, 97, 101
Polarization, 12–14, 21, 22, 25, 26, 29, 30, 32, 38, 39, 42–44, 46–48, 51, 88, 112, 113, 129, 130, 203, 215, 216
degree of, 5, 6
direction, 38
plane of, 221, 230
singularities, 22, 60, 64
state, 12, 14, 129, 241, 242, 246, 250, 263, 272, 287
circular, 12, 13
vector, 26, 34, 36, 38, 39, 43, 53, 86, 90, 95, 205
vortex, 22, 32
Polarized beams, 5, 9, 221, 225–227, 230, 231, 241, 246, 251
Polarized light, 6, 9, 11, 221, 240, 251, 271
Polystyrene, 212, 214–217, 227, 250
Poynting vector, 21, 35, 66, 206, 226, 238, 240
Probability operators, 274, 275, 277, 278
Probe particle, 245, 250–253, 257, 258, 260, 264, 266–268
Propulsion, 114, 126–128
Protocols, 273, 286, 288
Pump, 244, 246, 275
beam, 275, 283

Q

Quantum
communication, 14, 271, 273, 275, 280, 286
cryptography, 272
gas, 299, 303, 304
information, 262, 272–275, 277, 280, 283, 286, 290

systems, 274, 290, 291
key distribution, 271, 286, 288
protocol, 272, 273, 286
states, 72, 273, 274, 283, 290, 300, 309
Qubits, 271, 272, 290, 291
Qunits, 272, 274, 286, 290, 291

R

Rabi frequency, 179, 181, 182, 297, 314
Radial
trapping efficiencies, 212, 213, 227
wave equation, 27, 30
Radiation
field, 23, 24, 26, 27, 33, 35, 36, 81, 171
forces, 116, 128, 191
pressure, 109, 114, 123, 126, 142, 299
Random superposition (SR) holograms, 153, 156, 157
Rayleigh particles, 215
Red blood cells, 115, 125, 239
Refraction, negative, 133, 324, 328
Refractive index, relative, 212–216, 227
Resonances, two-photon, 312, 313, 320, 325
Rotating wave, 173, 297, 298, 311
Rotational symmetry, 218, 220, 222, 229, 240, 277
discrete, 218, 219

S

Schrödinger equation, 3, 20, 48, 50, 297, 309, 321
Security in quantum cryptography, 272, 286, 289, 290
Single photon level, 9, 277, 278, 280
Single photons, 67, 72, 273, 277, 279, 285
Single-trap holograms, 152–154
Singularities, optical, 63, 65
SLM, *see* Spatial light modulator
Solitons, 305, 330
Solutions, polymer, 249, 250, 257
Spatial light modulator, 9, 11, 68, 70, 72, 74, 114, 142–149, 153, 160, 161, 234, 235, 242, 243, 262, 279, 280
liquid crystal (LC SLM), 147–149
plane, 143–145, 150, 153
reflective, 147, 148
Spherical particles, 94, 96, 97, 103, 192, 198, 218, 250
Spin
angular momentum, 4–6, 9, 21, 35, 36, 41, 60, 199–201, 203–205, 228, 229, 237, 238, 240, 241, 245, 246, 251–254, 263, 264
measurement, 254, 263
density, 36, 37, 44, 200, 201, 229
flux, 201, 205
tensor, 199, 200
torque, 225
States, 17–20, 42, 51–53, 55–60, 67, 68, 81, 272–275, 279–284, 286–288, 290, 297, 309–312, 321, 322, 327, 328
excited, 172, 297, 298, 311, 320, 321
linear, 12, 13
physical, 26, 328
steady, 175, 178, 182, 327
superposition of, 55, 56
Superposition
algorithms, 152, 156
of OAM states, 276, 277, 279, 280, 284, 285, 290
states, 273, 274, 276, 280–283, 288
Surface
plasmon polariton (SPP), 116, 117
modes, 116, 117
plasmons, 110, 116, 117, 122
Surface Enhanced Raman Scattering (SERS), 116

T

Torque, 95, 101, 102, 112, 172, 177–179, 191, 196–200, 202, 215, 218–222, 238, 240–243, 251–254, 262–264
density, 229
light-induced, 170, 172, 175, 178, 179, 181, 191
Total angular momentum, 201, 204–208, 210, 216, 218, 229
Total internal reflection, 108, 109, 111, 114, 122, 162
Transverse
electric (TE) fields, 23, 25, 28–32, 40–42, 129, 203

magnetic (TM) fields, 28–32
plane, 22, 31, 32, 34, 36, 39, 42–44, 47, 50, 55, 60, 74, 126, 153, 276
Trap arrays, 141, 145–147, 152, 156, 160, 161
optical, 141, 142, 161
Trapped particle, 99, 111, 112, 129, 159, 161, 195, 205, 213, 218, 241, 251, 254, 263
Trapping, 112–116, 125, 131, 132, 171, 172, 179, 180, 191, 192, 211, 212, 215, 216, 245, 246, 249, 299, 300, 303, 304, 312, 325, 326
beam, 111, 150, 195, 218, 219, 240, 243, 244, 262, 265
effective, 313, 318, 319
efficiency, 125, 231
external, 309, 313, 318, 322, 324
in vortex beams, 211
laser, 160, 250, 251, 255, 257, 258
nanoparticles, 113, 226
plane, 124, 143, 144, 146, 148, 153, 160, 161
potentials, 308, 310, 312–314
volume, 125, 145
Traps, 99, 124, 125, 129, 130, 140, 141, 145, 146, 148–151, 156–162, 196, 197, 211, 212, 231, 237, 239, 242, 266
evanescent wave, 114, 125
optical vortex, 195, 211–213, 231
Twisted beams, 184, 185
Twisted light, 186–188
Two-photon
photopolymerization, 223, 224
polymerization, 244, 246

U

Ultracold atoms, 295, 328, 329

V

Vaterite, 254, 264–266
particle, 256–258, 264–268
Vector Bessel beams, 210
Viscosity, 245, 246, 249–260, 263–268
Vortex, 4, 5, 21, 32, 54, 57, 58, 63, 66, 70, 71, 315
beams, 146, 211, 231, 323
optical, 195, 203, 204, 209
composite, 69, 71, 72, 74
pair, 58, 59
pancakes, 281, 282
structure, 27, 230, 231
VSWFs (vector spherical wave functions), 197–199, 202–204, 207, 208, 210, 218, 221, 226, 230
scattered, 218, 219, 222

W

Wave
equation, 2, 3, 8, 28, 33, 64, 75
fields, 63, 74, 75
functions, 19, 20, 50, 51, 56–58, 203, 305
vector, 4, 25, 26, 30, 33, 40, 81, 83, 100, 108, 173, 218, 299, 311, 326–329
Wave-packets, 327, 328
Wave-plates, 9, 12, 13, 240
Waveguide mode, 120–122, 126
Waveguides, 113, 122, 125–128, 133, 210
optical, 125–127, 134

Printed and bound by CPI Group (UK) Ltd, Croydon, CR0 4YY
08/05/2025
01864878-0001